Sustainable Practices in the Tannery Industry

Moulay Abdelazize Aboulhassan · Salah Souabi
Editors

Sustainable Practices in the Tannery Industry

Strategies for Eco-Friendly Leather Manufacturing

Springer

Editors
Moulay Abdelazize Aboulhassan
National School of Applied Sciences
Ibn Zohr University
Agidir, Morocco

Salah Souabi
Faculty of Science &
Techniques-Mohammadia
Hassan II Casablanca University
Mohammedia, Morocco

ISBN 978-3-031-92039-4 ISBN 978-3-031-92037-0 (eBook)
https://doi.org/10.1007/978-3-031-92037-0

This Springer imprint is published by the registered company Springer Nature Switzerland AG
The registered company address is: Gewerbestrasse 11, 6330 Cham, Switzerland

Preface

The leather industry has long played a vital role in the global economy, significantly contributing to employment, trade, and value-added production. From high-end luxury goods to everyday essentials, leather has remained a preferred material due to its durability, versatility, and timeless appeal. However, this enduring demand comes with a considerable environmental cost. Traditional leather manufacturing processes often involve excessive consumption of natural resources, toxic chemicals, and the generation of hazardous waste—particularly from tannery operations. In the face of growing environmental awareness and increasingly stringent regulations, the leather industry is at a crossroads, requiring an urgent shift toward cleaner, greener, and more sustainable practices.

The environmental challenges posed by tannery operations extend far beyond the industry itself, with far-reaching consequences for public health, ecosystems, and the conservation of natural resources. Tannery effluents, often laden with heavy metals such as chromium and other toxic pollutants, frequently contaminate water bodies and soil. This pollution poses severe risks to aquatic life, agriculture, and human health, particularly in communities near tannery facilities. Compounding this issue is the absence of robust waste management systems, which amplifies leather production's environmental footprint and jeopardizes sustainability efforts.

Sustainable Practices in the Tannery Industry: Strategies for Eco-Friendly Leather Manufacturing offers a timely and comprehensive response to the complex challenges faced by the sector. It provides a detailed roadmap for achieving sustainability within the leather industry by addressing critical environmental issues and presenting actionable solutions. The book explores state-of-the-art advancements in wastewater treatment, pollution control, resource recovery, and the integration of circular economy principles. By delving into these key areas, the book is an invaluable resource for industry professionals, researchers, policymakers, and stakeholders seeking to transition the leather industry into an environmentally responsible and economically viable sector.

This book underscores the urgency of adopting innovative technologies and sustainable practices to mitigate environmental harm while maintaining the industry's economic competitiveness. Through a series of well-curated chapters, it offers practical strategies for tackling pressing issues, including effluent management, advanced wastewater treatment, and the sustainable utilization of tannery solid wastes. Furthermore, it highlights the potential of circular economy principles to establish closed-loop systems that minimize waste, maximize resource efficiency, and generate long-term environmental and economic value.

This book seeks to inspire collaboration and innovation across the leather industry by providing actionable insights and a forward-looking perspective. It encourages stakeholders to align their practices with global sustainability goals, enabling the industry to reinvent itself as a leader in environmental stewardship. With a collective commitment to eco-friendly manufacturing, the leather industry can balance its rich legacy with the pressing need for sustainable solutions, meeting the evolving demands of consumers and markets worldwide.

Agidir, Morocco — Moulay Abdelazize Aboulhassan
Mohammedia, Morocco — Salah Souabi

Introduction

The leather industry has long been an integral part of human civilization, with leather products ranging from clothing to accessories and upholstery forming an essential part of daily life. As the global demand for leather continues to grow, so too does the environmental toll associated with its production. The tanning process, which transforms raw hides into durable leather, involves the use of various chemicals and energy-intensive procedures that can lead to significant environmental pollution. Water contamination, air pollution, hazardous waste, and the depletion of natural resources are just a few of the pressing issues linked to leather manufacturing. In light of these challenges, it is more crucial than ever for the leather industry to embrace sustainable practices that not only meet the needs of consumers but also preserve and protect the planet's resources for future generations.

This book, *Sustainable Practices in the Tannery Industry: Strategies for Eco-Friendly Leather Manufacturing*, aims to provide a comprehensive exploration of the environmental challenges faced by the leather industry, while highlighting cutting-edge solutions and innovative strategies for achieving sustainability in tannery operations. It is designed to serve as a valuable resource for industry professionals, environmentalists, researchers, and policymakers interested in reducing the negative environmental footprint of leather manufacturing. The book offers an in-depth look at sustainable practices, focusing on waste management, effluent treatment, pollution reduction, and the adoption of circular economy principles to create a more sustainable future for the industry.

Organized into four distinct parts, the book covers a wide array of topics and approaches, from addressing the immediate environmental impact of tannery effluents to promoting long-term systemic changes in manufacturing practices. Each section builds on the previous one, offering a structured approach to understanding and tackling the complex challenges of sustainable leather production.

Contents

About the Editors

Prof. Moulay Abdelazize Aboulhassan holds a Ph.D. in Science and Technology from Hassan II University, Faculty of Science and Technology Mohammedia, with a specialization in water treatment. An accomplished researcher, he has published over 45 peer-reviewed articles in indexed journals. He is currently a professor at the National School of Applied Sciences (ENSA), Ibn Zohr University, Morocco.

In addition to his teaching role, Prof. Aboulhassan serves as the Director of the Laboratory of Mechanic Process Energy and Environment. His research interests include water and waste treatment, wastewater plant operations, leachate treatment, solid waste valorization, and landfill management.

From 2016 to 2021, he led the Process and Environmental Engineering program and currently chairs the Process Engineering Department at ENSA Agadir. Widely recognized as a national expert, Prof. Aboulhassan also serves as a reviewer for numerous scientific journals.

Dr. Salah Souabi is a Professor at Hassan II University of Casablanca, Morocco, Faculty of Science and Technology Mohammedia, Laboratory of Process and Environmental Engineering. He holds a Ph.D. in water chemistry and microbiology from the University of Poitiers, France, and a Ph.D. in water treatment from Moulay Ismail Maknes University, Morocco. Prof. Salah Souabi treats liquid and solid waste in a team. He is an expert in several scientific journals in environmental sciences. Mr. Souabi has participated in several waste management projects and recovery projects. He has adequate experience in water, wastewater, waste management, environmental pollution, and the realization of various R&D projects. Prof. Souabi has participated in several industrial projects to manage and recover liquid and solid discharges. He is responsible for the engineering and Mastar Water and Environmental Engineering courses. He is an expert at the level of several international newspapers. Prof. Souabi has supervised 28 doctoral students for Ph.D. degrees in water, wastewater treatment, environmental pollution, solid waste management, and wastewater treatment plant operations and sustainability. He has published 165 peer-reviewed articles, 30 book chapters, 11 edited books, and 310 communications.

Abbreviations

AI	Artificial Intelligence
BOD	Biochemical oxygen demand
CE	Circular economy
COD	Chemical oxygen demand
CPCB India	Central Pollution Control Board India
GOI	Government of India
SDG	Sustainable development goals
SS	Suspended solids
TDS	Total dissolved solids
USEPA	United States Environmental Protection Agency

Part I
Tannery Effluents and Environmental Impact

The first part of this book focuses on one of the most critical environmental challenges faced by the leather industry: the management of tannery effluents. Leather tanning, especially when using traditional methods such as chromium-based tanning, produces effluents that can be highly toxic, containing a range of harmful chemicals that pose severe risks to both human health and the environment. The pollution of water bodies and soil, as well as the contamination of groundwater, has become a serious concern for communities living near leather production facilities.

Chapter 1 explores the health risks associated with chromium contamination, providing a thorough health risk assessment for communities living near tanning facilities. This chapter underscores the urgency of addressing these risks through better monitoring, stricter regulations, and more sustainable practices. Chapter 2 further investigates the ecological footprint of leather production, highlighting the various stages of the tanning process that contribute to environmental degradation. This chapter also outlines potential mitigation strategies to reduce the industry's overall impact. In Chap. 3, the environmental and marine ecosystem impacts of tannery effluents are analyzed, with a focus on how pollution loads, effluent variability, and the use of bioindicators can help monitor the health of aquatic ecosystems. Chapter 4 addresses the speciation of heavy metals in industrial discharges from tanneries and explores methods to reduce metallic pollution, laying the groundwork for more sustainable effluent management.

Chapter 1
Health Risk Assessment for Contamination of Water by Chromium Near Leather Tanning Facilities

Arti and Rohit Mehra

Abstract In the recent years, the world has faced major environmental concerns due to rapid urbanization and industry, which has increased groundwater contamination. The industrial areas around Tannery Industries and its surroundings pose a significant risk of groundwater contamination, endangering both human health and the delicate balance of the ecology due to the presence of Leather and Sports goods industries. This industry employs a substantial amount of chromium (Cr) and compounds derived from chromium. Chromium concentration can be estimated using a Triple Quad Inductively Coupled Plasma Mass Spectrometer, an Atomic Absorption Spectrophotometer, and many other sophisticated Techniques that have been discussed in this chapter. The USEPA-proposed approach uses the Water Quality Index (WQI) and Heavy Metal Pollution Index (HPI) to determine groundwater quality and the associated human health risks for adults. It has been reported in many studies that the tanning industries have high levels of chromium contamination around them. Health risks associated with chromium contamination of drinking water pose challenges to the sustainability of the tanning industry. Many techniques are being used nowadays to remediate chromium from drinking water. The current study supports the Global goal no. 3 of the Sustainable Development Goals (SDGs) of the 17 interlinked Global goals that are designed to be a "blueprint for achieving a better and more sustainable future for all."

Keywords Water contamination · Chromium · Health risk assessment · Leather processing · Remediation of tannery effluent · Tannery industries

Arti · R. Mehra (✉)
Department of Physics, Dr B R Ambedkar National Institute of Technology, Jalandhar, Punjab, India
e-mail: mehrar@nitj.ac.in

Arti
e-mail: artisharma20212021@gmail.com

M. A. Aboulhassan and S. Souabi (eds.), *Sustainable Practices in the Tannery Industry*,
https://doi.org/10.1007/978-3-031-92037-0_1

1.1 Introduction

The importance of water for life cannot be neglected as it forms the most important component for any life form. Life in any place is substantially based on freshwater availability for consumption and daily household chores. Many civilizations in Earth's history were based on this important resource's availability. But, in the recent days, these freshwater sources have started to face problems due to the excessive use and exploitation by human activities, be it surface water or groundwater. Many studies have reported this over-exploitation in different places (Chidambaram et al. 2012; Hashem et al. 2015). All the forms of sources of water, be it rivers, groundwater, or ponds, have started to suffer from this problem worldwide (Chidambaram et al. 2012; Krishna et al. 2009; Coyte et al. 2018). Thus, there is an urgent need for proper assessment and monitoring of these water sources to evaluate the health risks to the people consuming this contaminated water. Along with catering to the need to improve the quality using proper control and treatment measures, assessing the water quality is of utmost importance. A need for research on the major sources of pollution to these water resources reveals that industrialization is one of the major contributors to this issue. Industrialization has undoubtedly improved the lifestyle of people, but it has been brought about at the cost of over-exploitation of the natural resources available, which are limited (Fig. 1.1). A need for thus, sustainable management of these resources is needed, which can be brought about only by the proper management and remediation techniques (Tan and Wang 2022). Many industries, namely the textile industry, tanning industry, hand-tools industry, etc., have been found to be the major contributors to the exploitation and contamination of soil and water in the vicinity (Kabir et al. 2012; Rao et al. 2011).

Tanning is one of the important steps in leather production to increase the durability and quality of leather goods. Many types of tanning processes are being used in the modern world, chrome tanning being one of the most widely used methods. About 80–90% of the leather in today's world is produced by Chrome tanning. Chrome tanning is not only one of the quickest methods, but it also allows the leather products to be colored into different beautiful colors, makes the leather much more durable, and is cheaper than vegetable-tanned leather. But, to a surprise, this process involves a large amount of water; about 90% of this water is discharged chiefly out of the tanning facilities into the drains nearby or directly into the soil. The effluent from these facilities, thus, can be a potential cause of pollution of the environment (Arti and Mehra 2023; Chowdhury et al. 2015). The effluent from the tanning industry mainly carries chromium and other heavy metals into the water sources and the soil. Researchers in many different studies in countries like Bangladesh, India, China, etc., have characterized effluent characteristics from the tanning industries (Hasan et al. 2021; Hasnat et al. 2013). The effluent has been found to contain heavy metals like Copper, Mercury, Cadmium, etc., above the safety limits by World Health Organization (WHO) and United States Environmental Protection Act (USEPA) (USEPA 1992; WHO 2017).

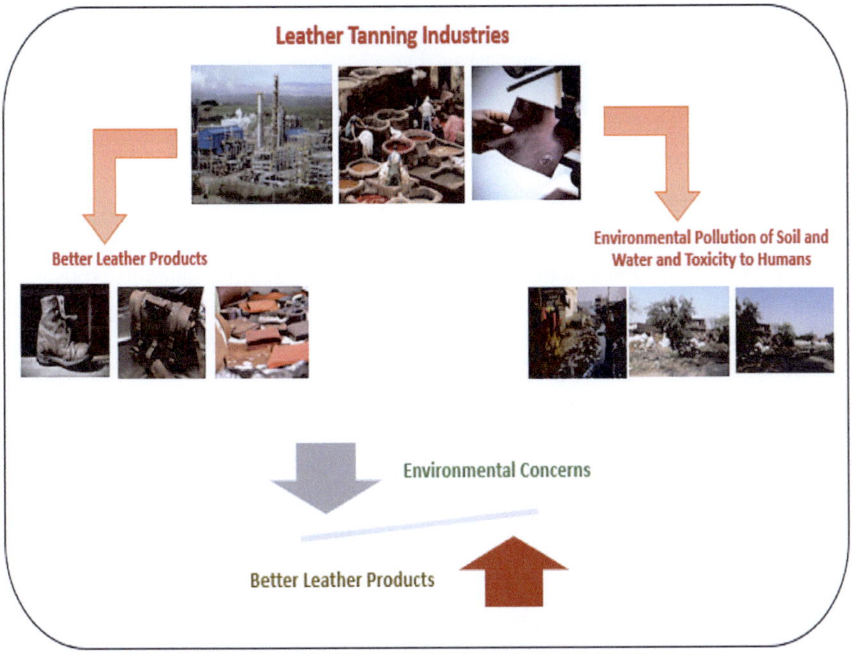

Fig. 1.1 Two sides in the figure represent the need for a balance between good products and environmental sustainability

Not only the water but soil has also been found to be contaminated by the discharge of effluent from tanning facilities in their vicinity (Hasan et al. 2021; Tariq et al. 2006). The heavy metal concentration of soils can be determined by many easily available methods these days, such as Inductively coupled Plasma Mass Spectrometer (ICP-MS) and Atomic Absorption Spectrometer (AAS). Thus making it easy to assess the environments for heavy metals and other contaminants (Bettinelli et al. 2000; Jan et al. 2010). Soil forms an important component of ecosystem for an organism. All the biotic components are dependent on this vital source for many micronutrients and macronutrients. All the plants directly and animals indirectly or directly depend on the soil for their nutrients and minerals requirements and an increased level of heavy metals can pose dangerous health implications for them all (Fu and Xi 2020; Juel et al. 2016). Therefore, all the components of the biota are being affected by the heavy metals. Sustainability cannot be brought about without proper assessment. So, analysis of the effluents from these tanning facilities and the environs is required continuously (Gosh and Hossain 2019; Iyer et al. 2006). Many studies have already reported increased levels of heavy metals Cr, Cu, Cd, Ni, Hg, etc., above the guidelines values in the vicinity of these tanning facilities (Ali et al. 2013; Chowdhury et al. 2015).

Chromium is mostly stable in its trivalent and hexavalent states, with the hexavalent states having better adsorption inside the cell (Fig. 1.2). Bronchial asthma, lung

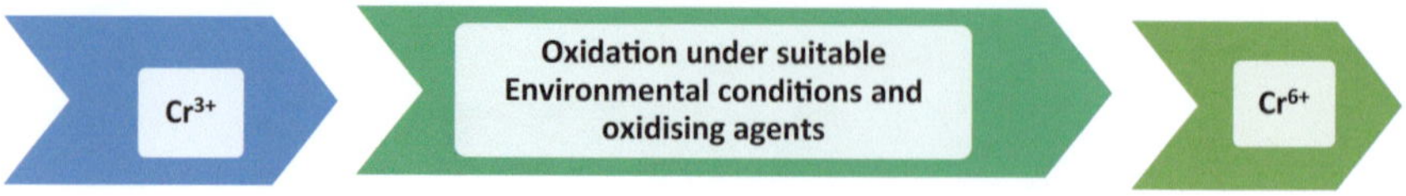

Fig. 1.2 Oxidation of Cr^{3+} to Cr^{6+} under suitable conditions

and nasal ulcers and cancers, reproductive problems, skin problems, and developmental problems are some of the most common health hazards related to exposure to chromium. Along with the carcinogenic nature of chromium, the potential threat of death in case of excess exposure is a cause of concern for all (Georgaki et al. 2023; Shekhawat et al. 2015; Xie 2024). Health risks associated with chromium exposure from drinking water, particularly higher risks of cancer occurrence in a case study in Orissa, India, have been estimated (Naz et al. 2016). Chronic toxicity associated with exposure to Chromium has also been reported in the Ranipet area of Tamil Nadu, India. The carcinogenic risks in the area were found to be higher than the acceptable limit of 1×10^{-6}. Higher chromium intake via ingestion, being the leading cause (Srinivasa Gowd and Govil 2008). Health risks associated with chromium exposure have also been reported in Awash River sub-basin, Ethiopia, sub-Saharan Africa. The Hazard Quotient for chromium for Women and Children has been found to be greater than unity (HQ > 1), indicating the potential for detrimental health in them (Aklilu et al. 2024). Along with the popular pathways of ingestion and inhalation exposure to chromium, today, even food grains have been found to contain chromium contamination. A study in Northern China has reported that the hazard index due to Chromium pollution is greater than unity in wheat grains, and this opens up a new challenge to human health (Liao et al. 2024).

1.1.1 Pathway of Chromium from Tanning Facilities to Groundwater

Chromium, especially trivalent chromium (Cr^{3+}), is an important chemical in tanning, where chromium sulfate is applied to leather to give it more strength and value. However, not all the chromium attaches to the leather, so much of it goes into waste either in wastewater or solid waste. The chromium-laden waste thus becomes the entry point for its pathway into the environment. Untreated or poorly treated wastes from tanning industries are released into the local waterbodies or lands. Also, solid wastes from the tanning processes, i.e., leather shavings and sludge are usually left on the landfill grounds or open areas to degrade in the environment. Sometimes, it can find its way into the air and then settle onto water or land. When released into the environment, it undergoes many transformations (Fig. 1.3). Under aerobic conditions and catalyzed by suitable catalysts (Manganese oxide and other soil components) that facilitate oxidation, Cr^{3+} ions are catalyzed to Cr^{6+} ions. A part of this chromium may be adsorbed by the soil particles or is tightly bound to the soil particles. The hexavalent chromium forms are also rarely mobile due to their high charge density. However, leaching processes, driven by rainwater or irrigation, can increase mobility by dissolving and transporting chromium through the soil.

Once in the groundwater, chromium disperses according to the hydraulic properties of the aquifer. Hexavalent chromium is especially hazardous because it presents serious health risks if the groundwater is used for drinking or irrigation. The degree of contamination depends on the efficiency of the waste treatment systems, the chemical and physical properties of the soil and aquifer, and the environmental conditions at the site. Proper treatment of tannery effluents, including reducing Cr^{6+} to Cr^{3+} and ensuring safe containment of wastes, is crucial in preventing contamination. Regulatory compliance and monitoring can also go a long way in ensuring that chromium does not penetrate groundwater systems.

1.1.2 Health Risks Associated with Chromium Tanning

Chromium tanning is the most widely used tanning process for leather. It is known to have severe health effects caused mainly by the emission of chromium (Cr^{6+}), which is highly toxic and carcinogenic. It is produced through the oxidation of trivalent chromium (Cr^{3+}), which is mostly used in tanning (Fig. 1.4).

Inhalation of Cr^{6+} leads to respiratory problems like nasal irritation, ulcers, and even perforation of the nasal septum. Prolonged exposure increases the risk of lung cancer substantially. In case of skin contact, chromium (Cr^{6+}) exposure can cause allergic dermatitis, resulting in rashes, redness, and eczema. In extreme cases, it may also lead to ulcerations. Ingestion of Cr^{6+} through contaminated water is associated with kidney and liver damage, stomach ulcers, and other gastrointestinal problems.

Although trivalent chromium (Cr^{3+}) is less toxic, excessive exposure can still cause adverse effects. For instance, prolonged skin contact with Cr^{3+} present in tanning waste can cause irritation and allergic reactions in some people. Moreover, under specific environmental conditions, Cr^{3+} may be oxidized to more toxic Cr^{6+}, thus increasing toxicity. Workers handling chromium compounds and in close contact with tanning effluents in tanneries are prone to chromium exposure. Occupational exposure might lead to chronic respiratory problems associated with inhaling chromium-contaminated dust or fumes and occupational dermatitis following repeated skin contact. The surrounding communities are also at risk in case the Chromium-containing wastes are not disposed off properly and end up contaminating both soil and water sources.

There is a great health danger to the people who draw their drinking water from the source contaminated with Cr^{6+}, including the risk of cancer and kidney damage. Irrigation using water that contains chromium could lead to its bioaccumulation in crops. This would pose a further dietary risk to the community. Long-term exposure to Chromium, particularly Cr^{6+}, has significant implications, such as carcinogenic effects, as it is classified by the International Agency for Research on Cancer (IARC)

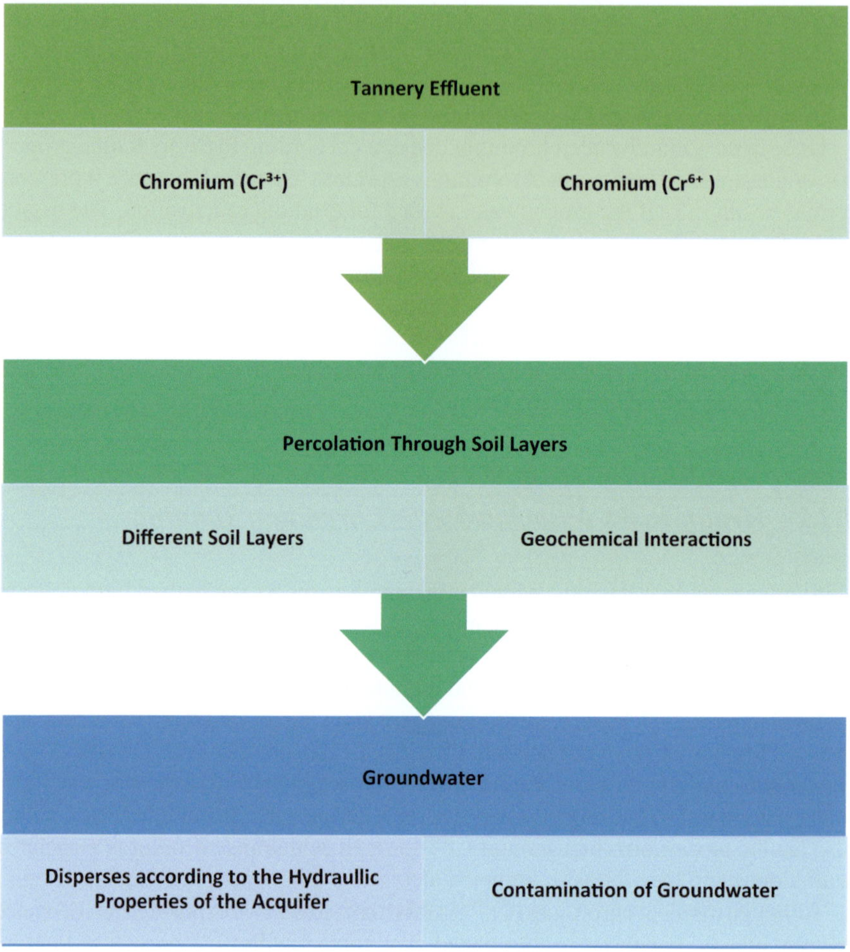

Fig. 1.3 Transport of chromium through the soil to groundwater

as a Group 1 carcinogen. There is also evidence that chromium exposure could affect reproductive health and cause developmental issues. Mitigation strategies are crucial to addressing these risks. Proper waste treatment in tanneries is essential to prevent Cr^{6+} formation and release. Workers should be equipped with personal protective equipment and subjected to periodical health monitoring. Regulatory compliance and public awareness of risks related to chromium contamination can ensure the protection of human health and the environment.

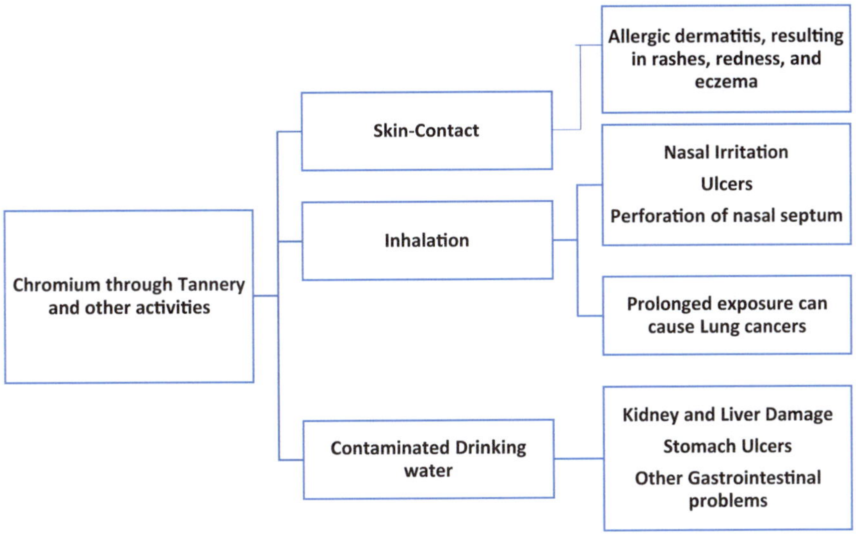

Fig. 1.4 Health effects of chromium (VI) exposure through different pathways

1.1.3 Tanning Industries in India and the Process of Tanning

India is among the top tanners and leather product producers, continuing to make up an impressive chunk of the global market. The tanning industry holds a significant share of the Indian economy, with a population of over 1.4 billion people, the majority of them being from weaker economic backgrounds. Kanpur, Chennai, Kolkata, Jalandhar, and Agra are the significant tanning locations, and each place specializes in certain kinds of leather processing. India is one of the world's largest leather producers, accounting for about 13% of global production. Most of its production is exported to Europe, the USA, and Southeast Asia. The industry has a huge population of livestock, thus ensuring a steady supply of raw material for tanning.

However, most tanning plants have been criticized for their environmental degradation, especially for polluting water bodies with untreated chromium-rich waste. Tanning is the chemical and physical process of changing animal hides and skins into a tough, flexible, and durable leather. The beginning processes of tanning involve pre-tanning such as soaking, liming, and fleshing. The main purpose of these pre-tanning processes is to clean and prepare the raw hides. In liming, chemicals like lime and sodium sulfide loosen hair and proteins; fleshing then removes excess fat and flesh. These are then delimed and bated with enzymes to soften them and make them pliable for further treatment. The actual tanning process stabilizes the collagen Fibers in the hides, thus resisting decomposition. Chrome tanning is the most widely practiced in India and involves the use of chromium salts, such as chromium sulfate, to yield heat and moisture-resistant, flexible leather. The second one, vegetable tanning, employs natural tannins from tree bark and fruits, yielding an eco-friendly product but is less commonly used since it is time-consuming.

Other methods used include aldehyde tanning and combination tanning for special products, such as softer or eco-friendly leather. Post-tanning processes, make the leather's quality and appearance better. Neutralization will remove excess tanning agents, and dyeing involves adding color by synthetic or natural dyes. Fat liquoring introduces oils and fats to increase the softness and flexibility of the leather. Finally, the leather is dried, stretched, and finished to achieve the desired texture, sheen, and durability. While the tanning industry contributes significantly to India's economy, it also raises environmental concerns. Chromium-based tanning generates toxic effluents that can contaminate water and soil if untreated. Many tanneries face challenges in adopting advanced wastewater treatment systems due to high costs and limited regulatory enforcement. Addressing these issues is crucial for balancing economic benefits with environmental sustainability.

1.1.4 Techniques Commonly Used for the Estimation of Chromium in Water

With the advancements in technology, several techniques are available today, for the precise determination of Chromium in water. Figure 1.5 includes some of those techniques, they have also been discussed below.

1. **UV–Vis Spectrophotometry**

It has been a widely used method for estimating hexavalent chromium (Cr^{6+}) in water because of its sensitivity and simplicity. The reaction with 1,5-diphenylcarbazide (DPC) in acidic conditions of Cr^{6+} resulted in the formation of purple-red complex, which showed the absorbance maximum at 540 nm. The intensity of this color is directly proportional to Cr^{6+}, thereby making it possible to measure its concentration. The process includes making up a filtered water sample that is acidified down to a pH of approximately 2 with sulfuric or phosphoric acid. Subsequently, the DPC reagent is added and allowed to react for about 5–10 min. Then, its absorbance at 540 nm is measured with a UV–Vis Spectrophotometer.

The concentration in the sample can be obtained by using calibration plots made with standard Cr^{6+}solutions. This method is very sensitive and can be used to detect as low as 0.01 mg/L Cr^{6+} concentration. If the samples are prepared properly, it is specific for Cr^{6+}, but it does not detect trivalent chromium (Cr^{3+}) directly. Additional steps, like oxidizing Cr^{3+} to Cr^{6+}, are needed for a total chromium analysis. The field applications are usually portable spectrophotometers for in situ testing. Although the method has advantages such as cost-effectiveness and ease of use, the method is susceptible to interference from other colored compounds or metals in the sample, and handling acids and reagents requires care for safety. UV–Vis spectrophotometry is highly applied in monitoring drinking water, wastewater, industrial effluents, and environmental samples for chromium contamination.

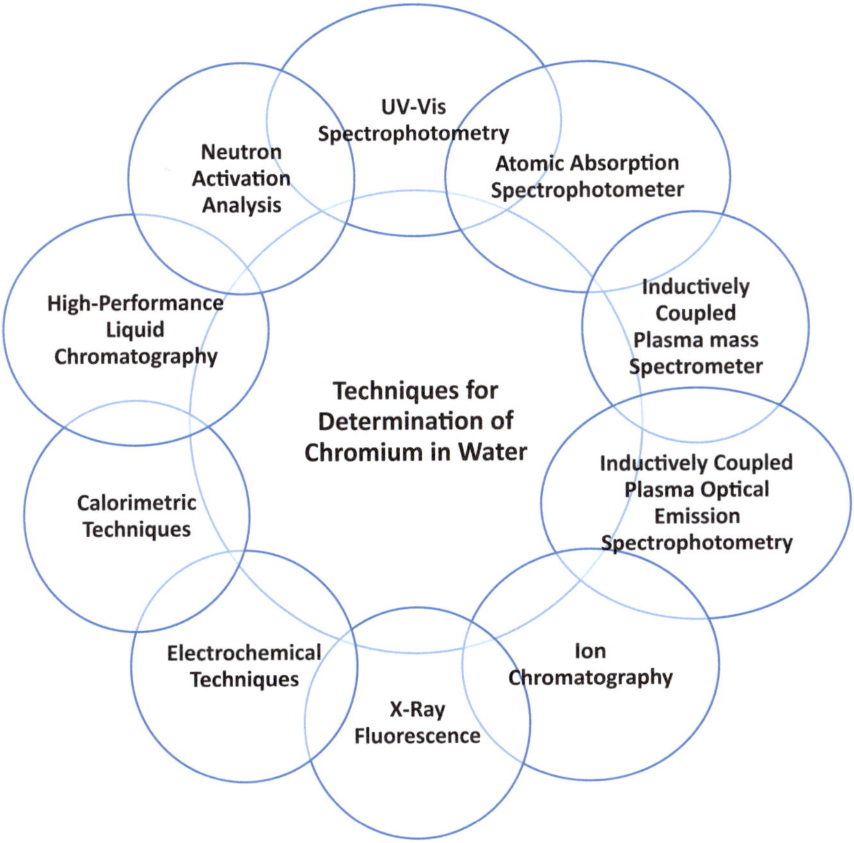

Fig. 1.5 Techniques for the determination of chromium in water

2. **Atomic Absorption Spectroscopy (AAS)**

Atomic Absorption Spectroscopy (AAS) is a reliable and widely used method for estimating chromium in water. It is particularly effective for detecting total chromium, which includes both trivalent chromium (Cr^{3+}) and hexavalent chromium (Cr^{6+}). The principle of AAS involves measuring the absorption of light by free atoms in the gaseous state. A chromium-specific hollow cathode lamp emits light at a characteristic wavelength (357.9 nm for chromium), which the chromium atoms absorb in the sample. Thus, the extent of the absorbed light will be a direct reflection of the amount of Chromium in the sample (Fig. 1.6).

3. **Inductively Coupled Plasma Mass Spectrometry (ICP-MS)**

Inductively Coupled Plasma Mass Spectrometry (ICP-MS) is a powerful analytical technique for the estimation of chromium in water. It is valued particularly for its ability to detect ultra-trace levels of chromium, often in parts-per-trillion (ppt) range. The process works on the principle by ionizing the sample using the ICP source, where a high-temperature plasma is generated about 10,000 K. Then, these ions are injected into the mass spectrometer, where separation and quantification take place based on their mass-to-charge ratio. For chromium analysis, ICP-MS measures total chromium in the sample, comprising both trivalent chromium (Cr^{3+}) and hexavalent chromium (Cr^{6+}). In case speciation is desired, pre-treatment steps would be necessary to separate or convert the species before analysis. For instance, Cr^{6+} can be selectively extracted or reduced to Cr^{3+} for determination of each species separately (Fig. 1.7).

4. **Inductively Coupled Plasma Optical Emission Spectroscopy (ICP-OES)**

Inductively Coupled Plasma Optical Emission Spectroscopy (ICP-OES) is the technique for estimating chromium levels in water. It offers robustness and versatility in estimating the trace to moderate concentrations of chromium. The method is based on the principle of optical emission spectroscopy, where the sample is introduced into a high-temperature inductively coupled plasma (about 10,000 K). The plasma excites atoms and ions of chromium, which start to emit light at characteristic wavelengths. For chromium, those common emission lines are around 267.716 and 357.869 nm.

5. **Ion Chromatography (IC)**

Ion chromatography (IC) is a powerful and highly effective technique for the analysis of chromium in water, especially for the separation and quantification of hexavalent chromium (Cr^{6+}) and trivalent chromium (Cr^{3+}). IC is capable of detecting chromium species at trace levels and is widely used in environmental monitoring and water quality assessments.

6. **X-Ray Fluorescence (XRF)**

XRF is a non-destructive analytical technique, usually found in elemental composition determination for different materials. However, this technique is specialized within water analysis, such as estimation for chromium. XRF involves bombarding the sample with high-energy X-rays that, consequently, excite the atoms, which emit secondary or fluorescent X-rays at characteristic energies distinct to each element, from which these emitted X-rays are detected and calculated according to the presence and quantification of elements, in particular, chromium.

7. **Electrochemical Techniques**

Electrochemical techniques are valuable in detecting and quantifying chromium in water, being relatively simple, sensitive, and having the potential for site application. These techniques have found particular utility in separating between the two main forms of oxidation of chromium. Among the electrochemical techniques, most widely used are voltametry, amperometry, and potentiometry.

8. **Colorimetric Techniques**

Due to the simplicity, cost efficiency, and applicability of such techniques both in field conditions and laboratory, the chromic concentration in water was more often estimated using colorimetric techniques. This includes assessing color intensity, which

Fig. 1.6 Atomic absorption spectrophotometer

can be correlated directly to concentration, hence its detection, especially in the case of hexavalent ion.

9. **High-Performance Liquid Chromatography (HPLC)**

The advanced analytical technique used to detect and quantify chromium in water is high-performance liquid chromatography. Particularly, it is useful in chromium speciation for separately determining hexavalent chromium and trivalent chromium (Cr^{3+}) in a single analysis. Basically, the method separates species in a liquid sample by use of a chromatographic column coupled to a suitable detector.

10. **Neutron Activation Analysis (NAA)**

Neutron Activation Analysis (NAA) is a sensitive and accurate estimation of chromium in water; this is a non-destructive method where the elements within the sample, including the estimation of chromium, are subjected to irradiation with thermal neutrons, are subjected to irradiation with thermal neutrons, thus making those elements radioactive. The radioactive isotopes subsequently decay, emitting gamma rays at characteristic energies.

1.1.5 USEPA's Approach to Assessing Water Quality Index and Heavy Metal Pollution Index

The USEPA uses several tools for assessing water quality, such as the Water Quality Index (WQI) and the Heavy Metal Pollution Index (HPI). These indices are simplified and comprehensive ways of evaluating the overall health of water bodies by aggregating data from multiple physical, chemical, and biological parameters (Fig. 1.8).

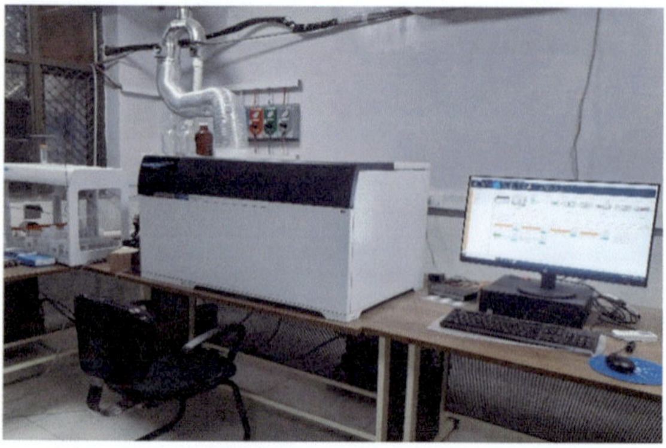

Fig. 1.7 Inductively coupled plasma mass spectrometer

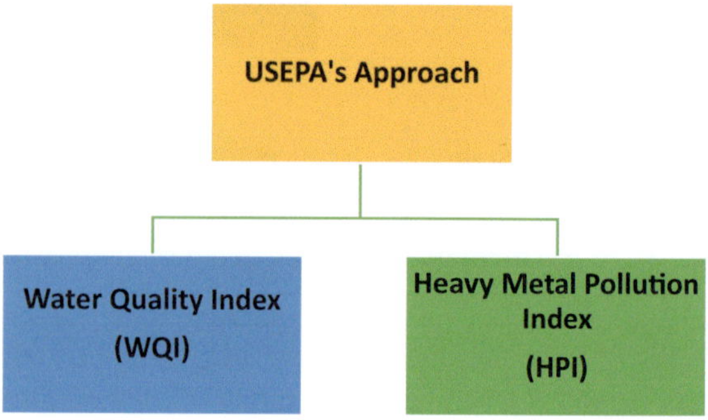

Fig. 1.8 USEPA's approach for assessing water quality

The Water Quality Index (WQI) is designed to represent the overall quality of water by combining scores from key indicators such as dissolved oxygen (DO), pH, turbidity, total dissolved solids (TDS), nitrates, phosphates, and biological oxygen demand (BOD). The USEPA's approach to WQI involves selecting relevant parameters, assigning weights to each based on their importance, and normalizing data to a common scale. This means that cross location and time comparisons make sense. The ultimate WQI value provides the water quality with classes descriptive of "Excellent", "Good", "Fair", or "Poor" categories, so that trends are more accessible and areas to be intervened are identified better.

The Heavy Metal Pollution Index (HPI) aims at the pollution of waters by heavy metals, which has a great impact on the health of humans and on the aquatic ecosystem. The USEPA approach on HPI includes the selection of toxic metals such as lead (Pb), cadmium (Cd), chromium (Cr), mercury (Hg), arsenic (As), zinc (Zn), copper (Cu), and nickel (Ni). Each of these metals is given a weight inversely proportional to the permissible limit and reflects their potential risk to health and the environment. The sub-index values for each metal are then summed into a single HPI score, which categorizes water quality based on heavy metal pollution levels from safe to highly contaminated. The WQI and HPI are important tools used by the USEPA in monitoring and managing water quality. These indices track pollution trends, ensure compliance with regulatory standards, and guide policies in water resource management. They play an essential role in safeguarding the health of the public and ensuring ecological soundness of water bodies. Another reason is the increase in technological advancements, particularly GIS, which have been of immense help in studying and visualizing water quality across varied regions.

1.1.6 Comparison of Previous Studies in Different Tannery Industrial Areas

A large number of studies of drain water and, ultimately, groundwater contamination have been reported in the recent years. Many studies in India itself have revealed the potential of chromium toxicity and health risks to the people living in the nearby tannery industrial areas. Hence, some of these studies that have been reported in the literature have been summarized in Table 1.1.

These studies suggest that chromium from the tannery effluents is a source of pollution for both soil and groundwater in the present world. So, there is an immediate need for proper treatment of effluents from these industries before disposal into the environment in order to make the industry sustainable.

1.1.7 Recent Remediation Studies in the Last Decade

A large number of remediation studies have been conducted worldwide to treat tannery effluents right before their disposal into the drains. Some of these recent studies in the last decade have been summarized in Table 1.2.

Apart from these recent innovative solutions present in the recent times, many remedial solutions have already been established and implemented in many countries. The importance of these treatment plants cannot be neglected. The proper treatment of effluents from the tannery industry, before their release into the environment is necessary to make the process sustainable. Not only the chromium content, but other heavy metals have also been found to be above the safety limits in tannery effluent that needs to be treated for safe disposal into the environment (Arti and Mehra 2023) (Fig. 1.9).

1.1.8 Reaching Sustainability Challenges and Goals in Tanning Facilities

Sustainability in tanning facilities requires overcoming several environmental, social, and regulatory challenges toward specific sustainability goals. One of the biggest

Table 1.1 Different studies conducted worldwide on soil and water pollution due to tanning industries

S. No.	Country	Location	Year	Major issue raised	References
1.	India	Ganga River water	2015	River pollution and toxicity to biotic lifeforms	Nagpure et al. (2015)
2.	Morocco	Mnasra region	2024	Groundwater pollution and toxicity to people	Sanad et al. (2024)
3.	India	Dindigul, Tamil Nadu	2005	Exploitation of groundwater and health risk assessment	Mondal et al. (2005)
4.	Saudi Arabia	Riyadh	2015	Assessment of tannery sludge and changes in the soil profile	Sallam et al. (2015)
5.	India	Sukinda Chromite Mine, Odissa	2016	Health risk assessment for exposure to chromium through drinking water	Naz et al. (2016)
6.	Bangladesh	Hazaribagh	2015	Characterization of effluent from leather processing facilities possesses heavy metals beyond acceptable limits	Chowdhury et al. (2015)
7.	Nigeria	Kano	2007	Effluent from tannery industry carries huge amount of heavy metals	Akan et al. (2007)
8.	Pakistan	Sialkot	2013	Effluents from tannery industry contain heavy metals above the safety limits	Ali et al. (2013)
9.	China	28 provinces of China	2024	Chromium in the food grains (wheat)	Liao et al. (2024)
10.	India	Kanpur	2005	Oxidation of Cr(III) to Cr(IV) in tannery sludge dumped sites	Apte et al. (2005)
11.	Bangladesh	Dhaleshwari river	2022	High level of Cr contamination in the sediment samples	Rahman et al. (2022)
12.	Bangladesh	Bangladesh	2023	Chromium in leather shavings possesses great potential for environmental degradation	Shaibur (2023)
13.	Turkey	Kula	2010	Pollution of soil and groundwater near tannery facility	Tarcan et al. (2010)
14.	Bangladesh	Hemayetpur	2022	Environmental pollution in the vicinity of the shifted tannery industrial area	Wahiduzzaman et al. (2022)
15.	Italy	NE Italy	2008	Chromium contamination of the environment due to tannery effluents	Bini et al. (2008)
16.	India	Ranipet, Tamil Nadu	2011	Groundwater contamination at a tannery dumpsite	Rao et al. (2011)

(continued)

Table 1.1 (continued)

S. No.	Country	Location	Year	Major issue raised	References
17.	Bangladesh	Dhaka	2020	Soil and groundwater contamination near tannery industries	Alam et al. (2020)
18.	India	Kanpur	2017	Soil and groundwater contamination after irrigation with tannery effluents	Dotaniya et al. (2017)
19.	India	Chennai	2012	Groundwater contamination near tanning industries	Brindha and Elango (2012)
20.	Turkey	Dil Creek	2008	Declining water quality due to chromium disposal	Oktor et al. (2008)

challenges is water usage; the tanning process requires much water, which leads to scarcity and pollution issues. Another major concern is chemical waste: the use of hazardous chemicals such as chromium, solvents, and dyes has led to environmental damage. High energy usage for heating, drying, and chemical processing also adds to carbon emissions. Another environmental risk is air and water pollution from the emissions of VOCs and the discharge of untreated wastewater. To overcome these challenges, sustainability goals in tanning facilities focus on water conservation and effective wastewater management through systems like zero-liquid discharge (ZLD) to minimize freshwater use and reduce pollution. Another key objective is to reduce the use of harmful chemicals, especially chromium, and replace them with safer alternatives. Energy efficiency is promoted through the adoption of renewable energy sources and energy-saving technologies. Ethical sourcing and sustainable use of raw materials are also prioritized to ensure animal welfare and reduce dependency on non-renewable resources. Another critical approach to reducing landfill use is by implementing circular economy principles in the management and recycling of leather by-products. Strategies toward sustainability are collaboration with stakeholders, transparency in sustainability reporting, and continuous process improvement through technological advancement such as artificial intelligence and automation. Overall, the benefits of sustainable practices in tanning facilities include cost reduction, enhancement of market competitiveness, compliance with regulations, and improvement in reputation with consumers and stakeholders.

1.2 Conclusions and Future Perspectives

This chapter demands a very urgent and critical issue with far-reaching implications for public health and environmental sustainability regarding the contamination of water sources by the tanning industry. The tanning industry processes animal hides into leather, using hazardous chemicals to produce good leather products. Chromium is one of those important chemicals. Chromium-rich effluents that are discharged into

Table 1.2 Recent remedial studies for the treatment of tannery wastewater

S. No.	Year	The strategy suggested for remediation	References
1.	2015	Phytoremediation, plants, and its related microorganisms to sterilize the low and modesty debased locales	Chauhan et al. (2015)
2.	2016	Combined chemical–biological treatment system	Ahmed et al. (2016)
3.	2017	Physical, chemical, biological, or a combination of these methods. Biological treatment of wastewater is more favorable	Kumar et al. (2017)
4.	2018	Bacteria, fungi, biofilms, algae, genetically engineered microbes, and immobilized microbial cells for the removal of heavy metals	Igiri et al. (2018)
5.	2019	Preservation of original skins by freezing, fresh skin processing, low salt preservation, hair saving, high-absorption chrome tanning and low chrome tanning, and reduction of salt dosage in tanning	Zhao and Chen (2019)
6.	2020	Ozonation, ozonation coupled with phytoremediation	Saranya and Shanthakumar (2020)
7.	2021	Advanced oxidation processes and the use of microalgae, coupling of advanced oxidation processes with biological processes, mainly microalgae, is seen as a viable biotechnological strategy	Urbina-Suarez et al. (2021)
8.	2022	Coagulation and flocculation, adsorption, biological treatment, membrane filtration, advanced oxidation process	Zhao et al. (2022)
9.	2023	KOH-modified biochar, *Bacillus cereus*, and *Aspergillus flavus* biomass	Gajendiran et al. (2023)
10.	2024	Activated sludge, chemical precipitation, coagulation, coagulation-aided sedimentation, membrane filtration, and advanced oxidation process (AOP)	Husain et al. (2024)

waterways pollute freshwater resources while at the same time providing health risks to the local communities and ecosystems. Chromium, especially in its hexavalent form, has been established as a carcinogen, and exposure to the element can lead to health problems such as respiratory problems, skin irritations, and even cancer. The presence of this toxic substance in water supplies will be devastating to human populations, especially children and the elderly. Furthermore, the pollution goes beyond human health. It disrupts aquatic ecosystems and harms fish and other wildlife, which can further degrade biodiversity in affected areas.

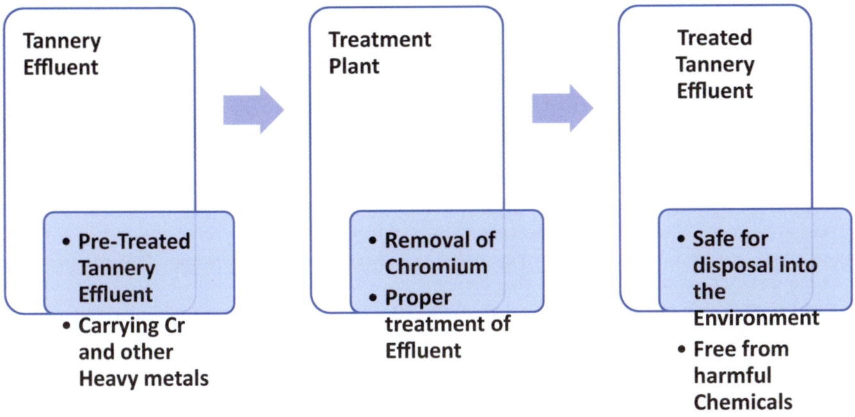

Fig. 1.9 Treatment of tannery effluent before disposal into the environment

Given the seriousness of the problem, the article also underscores the urgent necessity of introducing sustainable management practices in the tanning industry. These involve adopting cleaner production techniques that do not consume many harmful chemicals and also minimize waste generation. There is also the need to develop and implement proper waste treatment solutions. These solutions can include advanced filtration systems, the bioremediation of processes, and recycling waters used in tanning so that harmful substances are suitably treated before they escape into the environment.

Summing up, the issue of water contamination requires an immediate call for action. This becomes possible by giving precedence to sustainable practices and, therefore, investment in developing innovative waste treatment technologies as a way of reducing adverse industrial pollution effects on freshwater, protecting public health, and conserving our ecosystem's integrity over time. It is a big challenge, but it is one that industry, policymakers, and society collectively need to take.

References

Ahmed E, Abdulla HM, Mohamed AH, El-Bassuony AD (2016) Remediation and recycling of chromium from tannery wastewater using combined chemical-biological treatment system. Process Saf Environ Prot 104:1–10. https://doi.org/10.1016/j.psep.2016.08.004

Akan JC, Moses EA, Ogugbuaja VO, Abah J (2007) Assessment of tannery industrial effluents from Kano Metropolis, Kano State, Nigeria. J Appl Sci 7(19):2788–2793

Aklilu T, Sahilu G, Ambelu A (2024) Public health risks associated with drinking water consumption in the Upper Awash River sub-basin, Ethiopia, sub-Saharan Africa. Heliyon 10(3):e24790. https://doi.org/10.1016/j.heliyon.2024.e24790

Alam MS, Han B, Al-Mizan, Pichtel J (2020) Assessment of soil and groundwater contamination at a former Tannery district in Dhaka, Bangladesh. Environ Geochem Health 42(7):1905–1920. https://doi.org/10.1007/s10653-019-00457-6

Ali Z, Malik RN, Qadir A (2013) Heavy metals distribution and risk assessment in soils affected by tannery effluents. Chem Ecol 29(8):676–692

Apte AD, Verma S, Tare V, Bose P (2005) Oxidation of Cr(III) in tannery sludge to Cr(VI): field observations and theoretical assessment. J Hazard Mater 121(1–3):215–222

Arti, Mehra R (2023) Analysis of heavy metals and toxicity level in the tannery effluent and the environs. Environ Monit Assess 195(5):554. https://doi.org/10.1007/s10661-023-11154-4

Bettinelli M, Beone GM, Spezia S, Baffi C (2000) Determination of heavy metals in soils and sediments by microwave-assisted digestion and inductively coupled plasma optical emission spectrometry analysis. Anal Chim Acta 424(2):289–296

Bini C, Maleci L, Romanin A (2008) The chromium issue in soils of the leather Tannery district in Italy. J Geochem Explor 96(2–3):194–202

Brindha K, Elango L (2012) Impact of tanning industries on groundwater quality near a metropolitan city in India. Water Resour Manage 26(6):1747–1761

Chauhan S et al (2015) Implementation of phytoremediation to remediate heavy metals from tannery waste: a review. Adv Appl Sci Res 6(3):119–128. http://pelagiaresearchlibrary.com/advances-in-applied-science/vol6-iss3/AASR-2015-6-3-119-128.pdf

Chidambaram S, Karmegam U, Prasanna MV, Sasidhar P (2012) A study on evaluation of probable sources of heavy metal pollution in groundwater of Kalpakkam region, South India. Environmentalist 32(4):371–382

Chowdhury M et al (2015) Characterization of the effluents from leather processing industries. Environ Process 2(1):173–187

Coyte RM et al (2018) Large-scale uranium contamination of groundwater resources in India. Environ Sci Technol Lett 5(6):341–347

Dotaniya ML et al (2017) Geo-accumulation indices of heavy metals in soil and groundwater of Kanpur, India under long term irrigation of tannery effluent. Bull Environ Contam Toxicol 98(5):706–711

Fu Z, Xi S (2020) The effects of heavy metals on human metabolism. Toxicol Mech Methods 30(3):167–176. https://doi.org/10.1080/15376516.2019.1701594

Gajendiran K et al (2023) Tannery wastewater remediation competence of metal tolerant bacteria and fungi under the influence of chemically modified water hyacinth biochar: an in vitro evaluation. Biomass Convers Bioref 0123456789. https://doi.org/10.1007/s13399-023-05126-4

Georgaki MN et al (2023) Chromium in water and carcinogenic human health risk. Environments MDPI 10(2):1–26

Gosh PK, Hossain MD (2019) Assessment of tannery effluent: a case study of Dhaleshwari river in Bangladesh. Proc Int Conf Plan Archit Civ Eng 2(7):461–464

Hasan SMM et al (2021) Chromium contamination from tanning industries and phytoremediation potential of native plants: a study of savar tannery industrial estate in Dhaka, Bangladesh. Bull Environ Contam Toxicol 106(6):1024–1032. https://doi.org/10.1007/s00128-021-03262-z

Hashem MA, Islam A, Mohsin S, Nur-A-Tomal MS (2015) Green environment suffers by discharging of high-chromium-containing wastewater from the tanneries at Hazaribagh, Bangladesh. Sustain Water Resour Manag 1(4):343–347

Hasnat A, Rahman I, Pasha M (2013) Assessment of environmental impact for tannery industries in Bangladesh. Int J Environ Sci Dev 217–220

Husain A et al (2024) Overview of treatment methods of tannery wastewater. IOP Conf Ser Earth Environ Sci 1326(1)

Igiri BE et al (2018) Toxicity and bioremediation of heavy metals contaminated ecosystem from tannery wastewater: a review. J Toxicol 2018

Iyer GV, Mastorakis NE, Theologou AI (2006) Assessment of pollution load from unsafe chromium leather tanneries in India. WSEAS Trans Environ Dev 2(3):207–215. http://www.wseas.us/e-library/conferences/2006evia/papers/516-512.pdf

Jan F, Akbar M, Ishaq II, Asim SM (2010) Multivariate statistical analysis of heavy metals pollution in industrial area and its comparison with relatively less polluted area: a case study from the city of Peshawar and district Dir Lower. J Hazard Mater 176(1–3):609–616

Juel MAI, Chowdhury ZUM, Ahmed T (2016) Heavy metal speciation and toxicity characteristics of tannery sludge. AIP Conf Proc 1754

Kabir E et al (2012) Current status of trace metal pollution in soils affected by industrial activities. Sci World J 2012

Krishna AK, Satyanarayanan M, Govil PK (2009) Assessment of heavy metal pollution in water using multivariate statistical techniques in an industrial area: a case study from Patancheru, Medak district, Andhra Pradesh, India. J Hazard Mater 167(1–3):366–373

Kumar R, Sharma AK, Ahluwalia SS (2017) Advances in environmental biotechnology

Liao X et al (2024) Distribution and health risk of chromium in wheat grains at the national scale in China. J Hazard Mater 474:134846. https://doi.org/10.1016/j.jhazmat.2024.134846

Mondal NC, Saxena VK, Singh VS (2005) Assessment of groundwater pollution due to tannery industries in and around Dindigul, Tamilnadu, India. Environ Geol 48(2):149–157

Nagpure NS et al (2015) Assessment of pollution of river Ganges by tannery effluents using genotoxicity biomarkers in murrel fish, *Channa punctatus* (Bloch). Indian J Exp Biol 53(7):476–483

Naz A, Mishra BK, Gupta SK (2016) Human health risk assessment of chromium in drinking water: a case study of Sukinda chromite mine, Odisha, India. Expo Health 8(2):253–264

Oktor K, Yılmaz S, Türker G, Erkuş E (2008) Speciative determination of Cr(III) and Cr(VI) in dyeing waste water of Dil Creek discharge to Izmit Gulf (Izmit-Kocaeli, Turkey) by ICP-AES. Environ Monit Assess 141(1–3):97–103

Rahman MS et al (2022) Assessment of heavy metal contamination in sediment at the newly established tannery industrial estate in Bangladesh: a case study. Environ Chem Ecotoxicol 4:1–12

Rao GT et al (2011) Assessment of groundwater contamination from a hazardous dump site in Ranipet, Tamil Nadu, India. Hydrogeol J 19(8):1587–1598

Sallam AS et al (2015) Environmental assessment of tannery wastes in relation to dumpsite soil: a case study from Riyadh, Saudi Arabia. Arab J Geosci 8(12):11019–11029

Sanad H et al (2024) Ecological and health risk assessment of heavy metals in groundwater within an agricultural ecosystem using GIS and multivariate statistical analysis (MSA): a case study of the Mnasra region, Gharb Plain, Morocco. Water (Switzerland) 16(17)

Saranya D, Shanthakumar S (2020) An integrated approach for tannery effluent treatment with ozonation and phycoremediation: a feasibility study. Environ Res 183:109163. https://doi.org/10.1016/j.envres.2020.109163

Shaibur MR (2023) Heavy metals in chrome-tanned shaving of the tannery industry are a potential hazard to the environment of Bangladesh. Case Stud Chem Environ Eng 7:100281. https://doi.org/10.1016/j.cscee.2022.100281

Shekhawat K, Chatterjee S, Joshi B (2015) Chromium toxicity and its health hazards manuscript info abstract. Int J Adv Res 3:167

Srinivasa Gowd S, Govil PK (2008) Distribution of heavy metals in surface water of Ranipet industrial area in Tamil Nadu, India. Environ Monit Assess 136(1–3):197–207

Tan X, Wang X (2022) Water treatment and environmental remediation applications of carbon-based nanomaterials. In: Emerging nanomaterials for recovery of toxic and radioactive metal ions from environmental media, pp 229–311

Tarcan G, Akıncı G, Ali Danışman M (2010) Assessment of the pollution from tannery effluents upon waters and soils in and around Kula vicinity, Turkey. Water Air Soil Pollut 213(1–4):199–210

Tariq SR et al (2006) Multivariate analysis of trace metal levels in tannery effluents in relation to soil and water: a case study from Peshawar, Pakistan. J Environ Manage 79(1):20–29

Urbina-Suarez NA, Machuca-Martínez F, Barajas-Solano AF (2021) Procesos de Oxidación Avanzada y Alternativas Biotecnológicas Para El Tratamiento de Aguas Residuales de Tenerías. Molecules 26(11):1–25

USEPA (1992) Guidelines for exposure assessment. US Environmental Protection Agency, Risk Assessment Forum, Washington, DC (EPA/600/Z-92/001)

Wahiduzzaman M, Hossain MF, Parveen Z (2022) Assessment of environmental quality of an area adjacent to the relocated tannery industries at Hemayetpur, Bangladesh. Dhaka Univ J Biol Sci 31(1):29–43

WHO (2017) 55 proceedings of the royal society of medicine biochemical aspects

Xie S (2024) Water contamination due to hexavalent chromium and its health impacts: exploring green technology for Cr(VI) remediation. Green Chem Lett Rev 17(1):1–19. https://doi.org/10.1080/17518253.2024.2356614

Zhao C, Chen W (2019) A review for tannery wastewater treatment: some thoughts under stricter discharge requirements. Environ Sci Pollut Res 26(25):26102–26111

Zhao J et al (2022) Tannery wastewater treatment: conventional and promising processes, an updated 20-year review. J Leather Sci Eng 4(1). https://doi.org/10.1186/s42825-022-00082-7

Chapter 2
Unveiling Tanning Troubles and Mitigating the Ecological Footprint of Leather Production

Amna Tehreem, Nazia Jamil, and Rida Batool

Abstract Leather industry is one of the key economic drivers, creating million of jobs and generating significant foreign exchange but this industry also faces environmental challenges. Many industries are known for the release of hazardous waste but tanneries are notorious for releasing slow poisonous substances to the biosphere and undermining the quality life of humans, plants and animals. Tanneries use significant amount of chromium salts during tanning operation, which leads to unwanted release of large quantities of effluents and solid waste, i.e., sludge, containing chromium derivatives. These contaminants ruin the ecological balance and survival of various life forms. This chapter delves into the physical characteristics of tannery effluents, focusing on liquid and solid waste streams. It identifies main pollutants such as sulfides, ammonium salts, chlorides, biological oxygen demand (BOD), and heavy metals. Impacts of the assorted tanning processes on effluent composition is also described. Solid waste, primarily in the form of sludge, unveils additional environmental challenges due to heavy metal contamination, particularly in developing countries such as Pakistan, where inadequate waste management practices are prevalent. This chapter explores the detailed environmental impact of tanning practices including water contamination, soil degradation and air emissions. Considering the long-term consequences of inadequate waste disposal in tanneries highlights need for eco-friendly tanning approach and opportunities for sludge valorisations. Conventional tanning practices include chrome-based tanning which are effective but exhibits irreversible results. This review addresses the setbacks in current practices and explores emerging and innovative eco-friendly alternatives such as chrome-free

A. Tehreem · N. Jamil · R. Batool (✉)
Institute of Microbiology and Molecular Genetics, University of the Punjab, Quaid-e-Azam Campus, Lahore, Pakistan
e-mail: rida.mmg@pu.edu.pk

A. Tehreem
e-mail: amnarana951@gmail.com

N. Jamil
e-mail: nazia.mmg@pu.edu.pk

M. A. Aboulhassan and S. Souabi (eds.), *Sustainable Practices in the Tannery Industry*,
https://doi.org/10.1007/978-3-031-92037-0_2

tanning, waterless tanning, enzymatic tanning and nanotechnology that use non-hazardous steps. Moreover, this chapter discusses the treatment of tannery wastewater and the potential integration of treated wastewater reuse to reduce freshwater consumption in water-scarce regions of world and environmental impacts. Tannery waste water treatment includes reverse osmosis (RO), nanofiltration (NF) and ultrafiltration (UF) also hybrid treatment technique that comply many physical, chemical and biological techniques found more effective and producible. Conclusively, this study emphasizes the urgent call for the authorities and leather industry to adopt sustainable and retained regulatory measures for improved waste management, technological innovation and stricter environmental regulations to ensure the life safety and protection of environment.

Keywords Leather industry · Chromium salts · Environmental degradation · Sludge valorisation · Sustainable tanning · Technological innovation · Resource conservation

2.1 Introduction

The leather industry is one of the leading industries alongside agriculture, transportation, mining, energy, etc. It has significant role in global economy as millions of people rely on leather production and processing for employment. Numerous countries worldwide produce various leather products that occupy significant space in the sector like footwear, fashion, upholstery, leather-made protective and defensive products, music and entertainment items and automobile products (Belay 2010).

There are many major leather producing countries like China, India, Brazil, Russia, Italy, Argentina, Vietnam and Pakistan. Pakistan holds a noteworthy position in leather production with 2.5% export per annum globally, while China being the largest exporter of leather with approx. 25% (United Nations 2023). Pakistan is rich in agricultural and livestock population which plays an important role in its economy by producing large quantities of hides and skins that empower leather industry (Bhatti et al. 2016).

Leather industry is economically important due its foreign exchange earnings and employment opportunities. While engaging millions of people to this sector and economic significance, it also has major environmental footprints. Leather manufacturing involves transforming raw hides into leather using various techniques. Due to poor disposal of the waste generated as a result of tanning lead serious environmental impacts which ultimately affect human beings, wildlife, water bodies and agricultural yield. This includes many salts, lime sludge and heavy metals primarily chromium.

Chromium is a key component in leather tanning industry that exists notably in oxidation states of $+2$, $+3$, and $+6$. The chromium $+3$ (trivalent) is the most stable state and a trace metal in mammals. While chromium $+6$ (hexavalent) is highly toxic state of chromium which is carcinogenic, mutagenic and teratogenic.

In tanning, only a fraction of chromium salt used absorbed by raw hides, while the rest of the salts enter waterbodies which ends up in wastewater sludge, containing indispensable Cr^{+6} salts (Belay 2010). This environmental contamination due to overlooked principles for the disposal of tannery effluents. World Health Organization (WHO) along with other organization including environmental protection authority (EPA) and national bodies have provided standards to be followed for tannery effluents disposal but disappointingly improper disposal practices continue due to poor legislative enforcement. The current situation clearly highlights the unethical implementation and poor monitoring by authorities and workforce in leather industry for tannery effluent disposal.

Given significant input in economics globally and demand of leather, leather industries cannot be shut down simply due to its serious and irreversible health consequences. Therefore, it is imperative for legislative authorities, workforce in industry and consumers to adopt collective measures for transparent and sustainable tanning practices that save both people and planet.

2.2 Characterization of Tannery Effluents

Leather processing and manufacturing involves series of steps, typically followed by pre-tanning, tanning, post-tanning and refining (Heidemann 1993; Madhan et al. 2001). Each step generates variable and large quantity of waste, which needs proper disposal. The waste generated can be broadly characterized into many categories followed by (a) tanned and untanned waste, (b) solid and liquid effluents and (c) tannery waste with physical, chemical, biological and toxicological parameters.

Untanned waste generated either solid or liquid in the initial steps of leather processing, referred as pre-tanning operation, involves removal of unwanted parts of raw hides mainly neck, tail, body fat/flesh conforming to raw hides and hairs also rotted or spoiled hides (not processed in time) (Tahiri and DeLaGuardia 2009). Pre-tanning generate bulk amount of waste followed by trimming, soaking, liming, de-liming, pickling, scudding. The non-removal of unwanted body parts reduces the efficiency of leather operations. These preparative phases incorporate unwanted effluents, microorganic waste, dirt or contaminants associated with hides.

Tanned waste include waste generated following tanning operations which mainly includes hides-cuts, chrome-trimmings involves cutting of pleats and edges (estimated at 5.5–6% of wet weight of hides) that may hinder tanning operation; chrome-splits: a very thin splits that are unusable accounts for 4–6% of total waste, chrome shaving accounts for 1–2% on the weight of the raw hide (Rao et al. 2002), leather crust shavings, buffing dust, finished leather trimmings, waste water sludge. The tanning agents, i.e., salts like chromium sulfate, synthetic tannins, residue of vegetable tanning operation, other chemicals used meanwhile like sodium chloride, acid and bases to maintain pH and dehairing, heavy metals like chromium, zinc, cadmium and arsenic used during tanning processing and organic load in waste. General flowchart of tannery waste generation described as in Fig. 2.1.

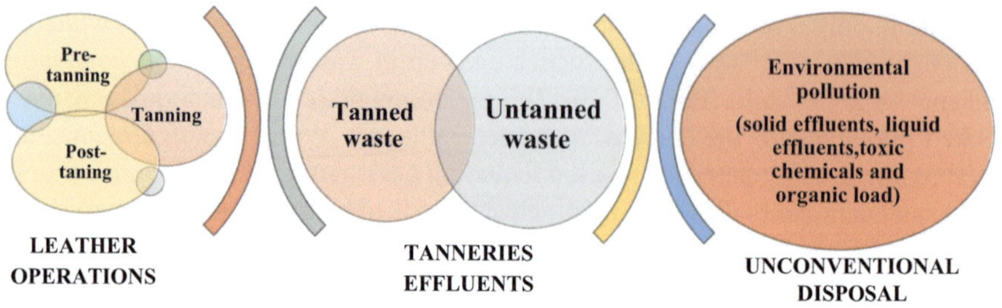

Fig. 2.1 Waste-generating hotspots in leather processing operations

Tanneries incorporate great number of solids, liquid and gaseous waste in the environment, thus violating standard guidelines. Each waste substance exhibits specific physical, chemical, biological and toxicological nature. The need to understand the nature of toxic substance is crucial as it comply with environmental regulations. Based on the flowchart in Fig. 2.2, provide a detailed description of the inputs and outputs involved in the leather industry process.

2.3 Chromium Speciation in Tannery Effluents

The tannery sector uses great amount of chromium (III) specific salts for tanning. The trivalent chromium (III) salts primarily include chromium sulfate ($Cr_2(SO_4)_3$) and basic chromium sulfate (BCS) $Cr(OH)(SO_4)$, while these chromium acetate and chromium formate usually used for re-tanning, leather dying or post-tanning procedures. Only about 20% of the large number of chemicals used in the tanning process is absorbed by leather and the rest is released as wastes (Verma et al. 2008; Dhungana and Yadav 2009; Tabesh et al. 2011).

The chromium tanning in tanneries is primarily revolves around the process called leather tanning, which transforms raw hides of animals into durable and flexible leather also prevent from microbial attack. In chrome-tanning, the Cr^{+3} ions cross-links the collagen fibers which ultimately strengthen leather and make it suitable for long-term usage. Chromium exists primarily in oxidation states of + 2, + 3 and + 6 (Sharma et al. 2021). Despite of calculated usage of trivalent chromium (Cr^{+3}) in large- and small-scale tanneries for tanning; various researches showed the presence and abundance of hexavalent chromium Cr^{+6} in tanneries wastewater and wastewater sludge (Younas et al. 2023). This is due to the inadequate or no access of treatment plant for tannery waste water and sludge (Wang et al. 2021a).

Chromium (III) is less toxic than hexavalent chromium (VI). The abundance of Cr^{+6} in the soil and wastewater is due to the oxidation of Cr^{+3} to Cr^{+6}. The presence of oxidizing and reducing agents in the surrounding environment like soil and water determines the bioaccumulation and bioavailability of chromium species (Dereje Homa et al. 2016). In water, the reduction and oxidation of chromium depends upon

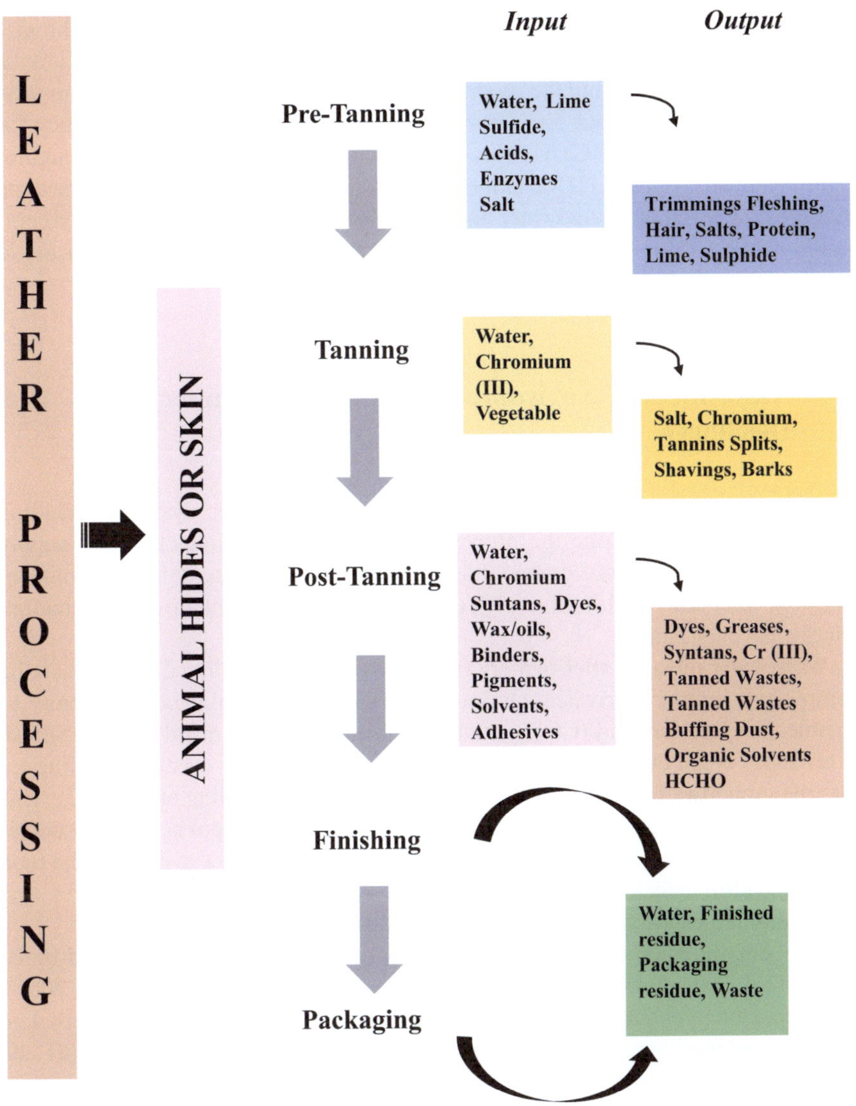

Fig. 2.2 A systemic analysis of the raw materials, chemicals and energy inputs, alongside the products, emissions and waste outputs in the leather industry

the water chemistry parameters like pH, redox potential, dissolved oxygen (DO), biological oxygen demand (BOD), total dissolved solids (TDS) and redox coupling agents like ferrous iron (Fe(II)), sulfide (S^{2-}) and organic carbon. The Fe(II) and S^{2-} may reduce Cr(VI) into Cr(III), while Cr(III) oxidizes into Cr(VI) by oxidizing agents, such as manganese oxide (MnO_2), hydrogen per oxide (H_2O_2) and dissolved oxygen (DO) (Apollaro et al. 2019; Vasileiou et al. 2019). The unregulated chrome-contamination (such as Cr^{+3} and Cr^{+6}) of soil and waterbodies ultimately leads to the flow of toxic chromium into the food chain; damaging living life including humans, wildlife, plants and aquatic life (Bozlur Rahman 2004). International Agency for Research on Cancer (IARC) has classified chromium (VI) compounds as "carcinogenic to humans" (Group 1) (Kazakis et al. 2017). Cr^{+6} is soluble in water that mediate its easy flow through groundwater and soil, which leads to long distance contamination, while Cr^{+3} is less toxic than Cr^{+6}, but persistent and excessive exposure to trivalent chromium also pose serious hazards to life.

2.4 Assessment of Environmental Impact of Tannery Discharge

Tannery discharges impart significant footprints on natural ecosystems, including water, soil and air. Despite the process employed for the production and supply chain of leather, hazardous compounds continue to accumulate in the environment and pose long-term health affects by causing respiratory problems, skin irritation/allergy and even cancer.

Wastewater released after tannery operations is heavily polluted by the metallic components, primarily trivalent chromium (III), sulfides and high concentrations of chemical oxygen demand (COD) and biochemical oxygen demand (BOD) (Suresh et al. 2001; Islam et al. 2014). This untreated, polluted water eventually infiltrates into the groundwater and enters the aquatic ecosystem (Schilling et al. 2012). This causes nutrient imbalance in waterbodies, thus leading to hypoxia (oxygen depletion), affecting aquatic organisms, animals (Siqueira et al. 2011), disrupting the food chain, and ultimately a serious threat to human well-being. Moreover, it reduces seed germination in extensively cultivated pulse crops (Altaf et al. 2008).

Soil contamination of chromium is a major concern due to improper land-based disposal of tanned solid waste and wastewater sludge (Dhayalan et al. 2007). The deposition of chromium salts and residues of other salts or acids used in tanning imparts a great change in soil characteristics. The change in soil pH and chemical composition renders land unfit for agriculture (Mwinyihija 2010). As a result, it reduces soil fertility, loss of soil biodiversity, and crop contamination by toxic substances. The long-term exposure of these toxins poses serious health consequences both in humans and animals.

Incineration of tannery sludge emits hazardous gases, including dioxins, sulfur dioxide and nitrogen oxides, and the release of VOCs, sulfur compounds and ammonia during tanning operations deteriorates the air quality. This causes serious health problems in the local population, such as respiratory problems and severe skin conditions due to prolonged exposure to contaminated air.

2.5 Technologies Used in Leather Production and Their Environmental Impact

Leather-making operations have evolved over time, integrating both conventional and innovative methods. However, these technologies have their own ecological footprints. Traditional or chemical processing begins with soaking of raw hides/skins in salt solution aims to restore cured hides/skin into correct dimensions, using water and preservatives (Thanikaivelan et al. 2005). This incorporates sodium chloride into the waste water, difficult to remove. Soak liquor separation would greatly reduce salinity in environment (Thanikaivelan et al. 2005). This followed by liming, cause swelling of hides to remove excess hair, fat or flesh using alkaline solution primarily calcium hydroxide and sodium sulfide, release high chemical load to environment. Following liming, there comes bating which involves wholehearted removal of unattractive interfibrillar materials and short hairs using enzymatic treatments. This step is pH specific as enzymes are functional in a pH range of 8.2–8.8 or 3.5–4.5 (Thanikaivelan et al. 2005). This adds proteineous or small hair in tannery effluent thus enhance COD and BOD (Guo et al. 2006). Use of less toxic and eco-friendly enzyme might reduce environment biological and chemical load. Pre-tanning is followed by tanning, process of stabilizing leather-making protein against any stress (Krishnamoorthy et al. 2012). It involves many methods such as vegetable tanning and chrome-tanning. Despite ecological stress, chrome-tanning is a common practice commercially for its ability to produces warmer, invincible, bright-shade leathers with unusual wet heat resistance in a shorter period of time (Thanikaivelan et al. 2005; Kiliç et al. 2011; Black et al. 2013). Chrome-tanning involves use of chromium III salts (typically in the form of chromium sulfate) which are used to treat hides. The chromium ions penetrate the collagen threads, generating cross-links that stabilize the hides and transform them into leather. Because of its bluish tinge, the leather during this stage is known as "wet blue". This emerges with heavy chromium pollution in ecosystem, that needs to be addressed and manage appropriately (Black et al. 2013; Tasca and Puccini 2019). The final step of leather-making operation is post-tanning that adds great intrinsic values and pleasing elegance in leather products (Wang et al. 2021b). Post-tanning operation involves re-chroming, pH, neutralization, re-tanning, dyeing (particularly azo-dyes) (Clarke and Motschi 2009) and wax/oils to give great finish to leather (Black et al. 2013; Tasca and Puccini 2019).

Conventional chrome-tanning is energy intensive process (Navarro et al. 2020). Requirement of excess use of water and high chemical demand in each step is leaving intergenerational ecological consequence. High chemical load in the environment is increasing water pollution, toxic waste generation and a large carbon footprint (Joseph and Nithya 2009). Reuse, proper standards and management would reduce ecological stress (Black et al. 2013; de Aquim et al. 2019).

2.6 Innovations in Eco-Friendly Leather Processing

High time to employ innovative and eco-friendly operations in leather industry to reduce ecological stress. Exploring sustainable and applicable strategies in leather-making operation would definitely reduce economic burden and environmental impact. Some of the innovations or alternatives to chrome-tanning are described below.

Vegetable tanning, less efficient and time taking in contrast to chrome-tanning, produces less effluents. It involves tannins extracted from plants. Most common tannins include bark extracts, leaves, fruit and roots, all include the fragrant, organic component phenol necessary for tanning leather (Madhan et al. 2001). Vegetable tanning is a labor-intensive method that takes longer than chrome-tanning, which can be completed in a single day but is most likely much more beneficial to the environment (Kanagaraj et al. 2020).

Aldehyde tanning is used to achieve chrome-free outcome. The process employs aldehydes instead of chrome yields a white leather base (wet-white leather) while eliminating the need for toxic heavy metals (Serra et al. 1991; Wojdasiewicz et al. 1992). The other chrome-free approach is glutaraldehyde tanning. A non-toxic alternative that minimizes chromium's environmental hazards while also creating biodegradable leather.

Bioleather technologies use only biodegradable tanning agents, which results in leather that decomposes organically at the end of its existence. When discarded, these alternatives degrade faster than regular leather. Enzymes are used to expedite tanning processes, reduce water usage, and eliminate harsh chemicals. Enzymatic systems have revolutionized liming and bating procedures by displacing old chemical-intensive approaches. New dry tanning technologies need only minimal water, considerably dropping the high-water consumption associated with leather processing. These techniques additionally mitigate wastewater and pollutants.

Use of plant-based natural dyes instead of synthetic non-degradable dyes (Clarke and Motschi 2009) and water finishing in the refining stages of leather makes tanning an environment friendly approach (Kanagaraj et al. 2005). With the advancement in every stage of tanning operation, nano tannage is the promising approach from the last decade. Nanomaterials offer advantages in leather production processes. Metal oxide nanoparticles (Fe_3O_4) contain magnetic properties have been conjugated with enzymes and used for removing hair and opening skin fibers without utilizing brutal osmotic forces (Murugappan et al. 2016). Moreover, Ag, ZnO and CuO have broad-spectrum antimicrobial activity (Ren et al. 2017; Muthukrishnan et al. 2019) and can eliminate microbes that cause skin putrefaction. Nanomaterials and their role have a significant impact on the finished products, which have self-cleaning properties, among other things. The application of the same at each stage may improve the durability, quality and stability of the desired leather product. This method may reduce the chemical load exposed to the finished leather product (Dixit et al. 2015).

2.6.1 Treatment of Tannery Waste Water

Waste water containing toxic components generated as an undesired by product of leather processing. Tannery waste water employs several of treatment procedures, with each step crucial in detoxification of tannery effluent. Treatment involves primary, secondary and hybrid/innovative strategies.

In primary treatment, the waste water is usually subjected to sedimentation and filtration that mediate the removal of suspended solids like organic debris, sludge and sand in the effluent (Zereen et al. 2013). Followed by chemical treatment, coagulation/precipitation aids the removal of minute residue by adding coagulants like alum ($Al_2(SO_4)_3$) or ferric chloride ($FeCl_3$) and precipitation of fine particle particularly heavy metals like chromium from waste (Yang et al. 2015). Furthermore, pretreatment of wastewater can be done by different techniques like adsorption method, micro-electrolysis technology, photo-catalyze method, electrodialysis method, each specific for the removal of particular component in defined operation.

Secondary treatment is a biological process, which primarily includes aerobic treatment and anaerobic treatment. Aerobic processes degrade organic matter present in tannery wastewater using oxygen dependent microbes. The most common methods include sludge method, biofilm method (Zupančič and Jemec 2010; Zhao and Chen 2019). While anaerobic treatment relies on oxygen-independent microorganisms which is more advantageous because of low sludge production, low consumption of energy, smaller requirement of floor space, minimum investigation, strong resistance to shock loading and stable technology. The secondary treatment is followed by number of variable processes each with its specificity and results, mainly include oxidation ditch technology, constructed wetlands, sequencing batch reactor activated sludge method, moving bed biofilm reactor method, micro-bio-remediation method, etc. (Bai et al. 2013).

Many advanced and stabilized techniques for effluent treatment have been employed such as membrane filtration technologies like reverse osmosis (RO), nanofiltration (NF) and ultrafiltration (UF) to separate dissolved pollutants from waste water, use of adsorbing agents like activated carbon or zeolites, electrochemical treatment for the removal of chromium form waste water.

Emerging and hybrid treatment technologies involve combining multiple physical, chemical and biological processes to enhance wastewater treatment. These hybrid and emerging technologies offer cost-effective and high-efficient results for reducing both pollutant loads and the environmental footprints of tannery wastewater. This helps in water reuse and ensure compliance with increasingly stringent environmental regulations.

2.7 Innovations in Eco-Friendly Leather Processing

High time to employ innovative and eco-friendly operations in leather industry to reduce ecological stress. Exploring sustainable and applicable strategies in leather-making operation would definitely reduce economic burden and environmental impact. Some of the innovations or alternatives to chrome-tanning are described below.

Vegetable tanning, less efficient and time taking in contrast to chrome-tanning, produces less effluents. It involves tannins extracted from plants. Most common tannins include bark extracts, leaves, fruit and roots, all include the fragrant, organic component phenol necessary for tanning leather (Madhan et al. 2001). Vegetable tanning is a labor-intensive method that takes longer than chrome-tanning, which can be completed in a single day but is most likely much more beneficial to the environment (Kanagaraj et al. 2020). Aldehyde tanning is used to achieve chrome-free outcome. The process employs aldehydes instead of chrome yields a white leather base (wet-white leather) while eliminating the need for toxic heavy metals (Serra et al. 1991; Wojdasiewicz et al. 1992). The other chrome-free approach is glutaraldehyde tanning. A non-toxic alternative that minimizes chromium's environmental hazards while also creating biodegradable leather. Bioleather technologies use only biodegradable tanning agents, which results in leather that decomposes organically at the end of its existence. When discarded, these alternatives degrade faster than regular leather.

Enzymes are used to expedite tanning processes, reduce water usage and eliminate harsh chemicals. Enzymatic systems have revolutionized liming and bating procedures by displacing old chemical-intensive approaches. New dry tanning technologies need only minimal water, considerably dropping the high-water consumption associated with leather processing. These techniques additionally mitigate wastewater and pollutants. Use of plant-based natural dyes instead of synthetic non-degradable dyes (Clarke and Motschi 2009) and water finishing in the refining stages of leather makes tanning an environment friendly approach (Kanagaraj et al. 2005).

With the advancement in every stage of tanning operation, nano tannage is the promising approach from the last decade. Nanomaterials offer advantages in leather production processes. Metal oxide nanoparticles (Fe_3O_4) contain magnetic properties have been conjugated with enzymes and used for removing hair and opening skin fibers without utilizing brutal osmotic forces (Murugappan et al. 2016). Moreover, Ag, ZnO, and CuO have broad-spectrum antimicrobial activity (Ren et al. 2017; Muthukrishnan et al. 2019) and can eliminate microbes that cause skin putrefaction. Nanomaterials and their role have a significant impact on the finished products, which have self-cleaning properties, among other things. The application of the same at each stage may improve the durability, quality and stability of the desired leather product. This method may reduce the chemical load exposed to the finished leather product (Dixit et al. 2015).

2.7.1 Treatment of Tannery Waste Water

Waste water containing toxic components generated as an undesired by product of leather processing. Tannery waste water employ several of treatment procedures, with each step crucial in detoxification of tannery effluent. Treatment involves primary, secondary and hybrid/innovative strategies. In primary treatment, the waste water is usually subjected to sedimentation and filtration that mediate the removal of suspended solids like organic debris, sludge and sand in the effluent (Zereen et al. 2013). Followed by chemical treatment, coagulation/precipitation aid the removal of minute residue by adding coagulants like alum ($Al_2(SO_4)_3$) or ferric chloride ($FeCl_3$) and precipitation of fine particle particularly heavy metals like chromium from waste (Yang et al. 2015). Furthermore, pre-treatment of wastewater can be done by different techniques like adsorption method, Micro-electrolysis technology, Photo-catalyze method, Electrodialysis method, each specific for the removal of particular component in defined operation.

Secondary treatment is a biological process, which primarily includes aerobic treatment and anaerobic treatment. Aerobic processes degrade organic matter present in tannery wastewater using oxygen dependent microbes. The most common methods include sludge method, biofilm method (Zupančič and Jemec 2010; Zhao and Chen 2019). While anaerobic treatment rely on oxygen-independent microorganisms is more advantageous because of low sludge production, low consumption of energy, smaller requirement of floor space, minimum investigation, strong resistance to shock loading, and stable technology. The secondary treatment is followed by number of variable processes each with its specificity and results, mainly include oxidation ditch technology, constructed Wetlands, sequencing batch reactor activated sludge method, Moving bed biofilm reactor method, Micro bio-remediation method etc. (Bai et al. 2013).

Many advanced and stabilized techniques for effluent treatment have been employed such as Membrane filtration technologies like reverse osmosis (RO), nanofiltration (NF), and ultrafiltration (UF) to separate dissolved pollutants from waste water, use of adsorbing agents like activated carbon or zeolites, electrochemical treatment for the removal of chromium form waste water.

Emerging and hybrid treatment technologies involve combining multiple physical, chemical, and biological processes to enhance wastewater treatment. These hybrid and emerging technologies offer cost-effective and high-efficient results for reducing both pollutant loads and the environmental footprints of tannery wastewater. This helps in water reuse and ensure compliance with increasingly stringent environmental regulations.

2.8 Reuse of Treated Tannery Wastewater

The recycling of treated tannery wastewater represents a viable strategy for promoting water conservation, effectively decreasing the consumption of freshwater and mitigating environmental repercussions. Treated wastewater can be integrated into variable tannery processes such as soaking, washing, re-tanning and dyeing, reducing the usage of fresh water and lowering operational costs. Additionally, it can be used in agricultural irrigation, especially in water-scarce regions like Pakistan, India, Turkey and parts of Africa, though strict water quality standards must be met to prevent disastrous effects on crop and soil (UN 2024). Moreover, treated water can also serve urban uses like park irrigation and street cleaning, reducing pressure on potable water sources. However, obstacles such as high treatment costs, compliance with water quality standards, public perception and infrastructure limitations can hinder implementation. In regions like India, Turkey and South Africa, successful examples show that up to 70–80% of wastewater can be reused in tanneries, significantly lowering freshwater demand and effluent discharges (Alemu et al. 2019; Bhardwaj et al. 2023). These practices offer cost savings, environmental benefits and compliance with stricter regulations on water use. Innovative technologies, including membrane bioreactors and anaerobic digestion, offer promising avenues for effective treatment and reuse. With appropriate management strategies, the reuse of tannery wastewater can significantly enhance sustainable water practices and lessen the ecological impact of tannery operations.

2.9 Conclusion

The leather industry, while a cornerstone of the global economy, has significant environmental repercussions, primarily due to chromium-laden waste and poor waste management practices. The current state of operations, especially in developing countries, underscores the need for immediate intervention to mitigate the ecological and health hazards associated with tannery effluents. Innovative technologies such as enzymatic tanning, nanotechnology and waterless tanning offer promising solutions for reducing ecological footprints and improving process efficiency. Similarly, contemporary wastewater treatment methods provide pathways for mitigating contamination and reusing resources sustainably. However, successful implementation of these strategies requires an integrative effort involving regulatory authorities, industry stakeholders, and consumers. It is imperative to enforce stricter environmental regulations, invest in research and development for environmentally conscious technologies and adopt circular economy principles within the leather industry. By prioritizing sustainable practices, the leather sector can balance economic growth with environmental stewardship, ensuring a safer and healthier future for both people and the planet.

References

Alemu T et al (2019) Integrated tannery wastewater treatment for effluent reuse for irrigation: encouraging water efficiency and sustainable development in developing countries. J Water Process Eng 30:100514

Altaf MM et al (2008) Impact of long-term application of treated tannery effluents on the emergence of resistance traits in Rhizobium sp. isolated from *Trifolium alexandrinum.* Turk J Biol 32(1):1–8

Apollaro C et al (2019) Release and fate of Cr(VI) in the ophiolitic aquifers of Italy: the role of Fe(III) as a potential oxidant of Cr(III) supported by reaction path modelling. Sci Total Environ 660:1459–1471

Bai J et al (2013) Combined industrial and domestic wastewater treatment by periodic allocating water hybrid hydrolysis acidification reactor followed by SBR. Biochem Eng J 70:115–119

Bhatti MS et al (2016) Current trends and identification of critical factors that are hindrance in the progress of leather industry in Pakistan. Int J Multidiscip Curr Res 4:1176–1180

Belay AA (2010) Impacts of chromium from tannery effluent and evaluation of alternative treatment options. J Environ Protect 1.1:53–58

Black M et al (2013) Best available techniques (BAT) reference document for the tanning of hides and skins. Eur Comm Database 46:2013

Bhardwaj A, Sanjeev K, Davinder S (2023) Tannery effluent treatment and its environmental impact: a review of current practices and emerging technologies. Water Qual Res J 58(2):128–152

Bozlur Rahman M (2004) Design and fabrication of column filtration reactor for the removal of dissolved dyes from tannery wastewater

Clarke E, Motschi H (2009) Regulatory and legislative aspects of relevance to high performance pigments: Europe. High Perform Pigm 431–465

de Aquim PM et al (2019) Water reuse: an alternative to minimize the environmental impact on the leather industry. J Environ Manage 230:456–463

Homa D, Haile E, Washe AP (2016) Determination of spatial chromium contamination of the environment around industrial zones. Int J Anal Chem 1:7214932

Dhayalan K et al (2007) Biodegradability of leathers through anaerobic pathway. Waste Manage 27(6):760–767

Dhungana T, Yadav P (2009) Determination of chromium in tannery effluent and study of adsorption of Cr(VI) on saw dust and charcoal from sugarcane bagasses. J Nepal Chem Soc 23:93–101

Dixit S et al (2015) Toxic hazards of leather industry and technologies to combat threat: a review. J Clean Prod 87:39–49

Guo Z-R et al (2006) Enhanced chromium recovery from tanning wastewater. J Clean Prod 14(1):75–79

Heidemann E (1993) Fundamentals of leather manufacture. Eduard Roether KG. ISBN 3-7929-0206-0

Islam B et al (2014) Evaluation and characterization of tannery wastewater. J For Prod Ind 3(3):141–150

Joseph K, Nithya N (2009) Material flows in the life cycle of leather. J Clean Prod 17(7):676–682

Kanagaraj J et al (2005) Alternatives to sodium chloride in prevention of skin protein degradation—a case study. J Clean Prod 13(8):825–831

Kanagaraj J et al (2020) Trends and advancements in sustainable leather processing: future directions and challenges—a review. J Environ Chem Eng 8(5):104379

Kazakis N et al (2017) Origin of hexavalent chromium in groundwater: the example of Sarigkiol Basin, Northern Greece. Sci Total Environ 593:552–566

Kiliç E et al (2011) Environmental optimization of chromium recovery from tannery sludge using a life cycle assessment approach. J Hazard Mater 192(1):393–401

Krishnamoorthy G et al (2012) Green chemistry approaches to leather tanning process for making chrome-free leather by unnatural amino acids. J Hazard Mater 215:173–182

Madhan B et al (2001) Improvements in vegetable tanning-can acrylics be co-tanning agents? J Am Leather Chem Assoc 96(4):120–126

Murugappan G et al (2016) A novel approach to enzymatic unhairing and fiber opening of skin using enzymes immobilized on magnetite nanoparticles. ACS Sustain Chem Eng 4(3):828–834
Muthukrishnan L et al (2019) Bio-engineering and cellular imaging of silver nanoparticles as weaponry against multidrug resistant human pathogens. J Photochem Photobiol B 194:119–127
Mwinyihija M (2010) Ecotoxicological diagnosis in the tanning industry. Springer Science & Business Media
Navarro D et al (2020) Life cycle assessment and leather production. J Leather Sci Eng 2:1–13
Rao JR et al (2002) Green route for the utilization of chrome shavings (chromium-containing solid waste) in tanning industry. Environ Sci Technol 36(6):1372–1376
Ren G et al (2017) Chromium (VI) adsorption from wastewater using porous magnetite nanoparticles prepared from titanium residue by a novel solid-phase reduction method. Sci Total Environ 607:900–910
Schilling K et al (2012) Adapting the Austrian Edict on wastewater emissions for tanneries as consequence of foam formation on surface waters. Environ Sci Policy 23:68–73
Serra A et al (1991) Carbonyl compounds in tanning process. AQEIC Bol Tec 42(24):3
Sharma N et al (2021) Heavy metal pollution: insights into chromium eco-toxicity and recent advancement in its remediation. Environ Nanotechnol Monit Manage 15:100388
Siqueira IR et al (2011) Anxiety-like behaviour in mice exposed to tannery wastewater: the effect of photoelectrooxidation treatment. Neurotoxicol Teratol 33(4):481–484
Suresh V et al (2001) An improved product-process for cleaner chrome tanning in leather processing. J Clean Prod 9(6):483–491
Tabesh M et al (2011) Quality management of water distribution networks by optimizing dosage and location of chlorine injection. Int J Environ Res 5(2):321–332
Tahiri S, DeLaGuardia M (2009) Treatment and valorization of leather industry solid wastes: a review. J Am Leather Chem Assoc 104(02):52–67
Tasca AL, Puccini M (2019) Leather tanning: life cycle assessment of retanning, fatliquoring and dyeing. J Clean Prod 226:720–729
Thanikaivelan P et al (2005) Recent trends in leather making: processes, problems, and pathways. Crit Rev Environ Sci Technol 35(1):37–79
UN (2024) Wastewater: new black gold. United Nations. Retrieved from https://www.un.org/africarenewal/news/wastewater-new-black-gold
United Nations (2023) UN COMTRADE database. https://comtrade.un.org
Vasileiou E et al (2019) Expounding the origin of chromium in groundwater of the Sarigkiol basin, Western Macedonia, Greece: a cohesive statistical approach and hydrochemical study. Environ Monit Assess 191:1–34
Verma T et al (2008) Quality assessment of treated tannery wastewater with special emphasis on pathogenic *E. coli* detection through serotyping. Environ Monit Assess 145:243–249
Wang Q et al (2021a) Novel recycling of phosphorus-recovered incinerated sewage sludge ash residues by co-pyrolysis with lignin for reductive/sorptive removal of hexavalent chromium from aqueous solutions. Chemosphere 285:131434
Wang Y et al (2021b) Advanced collagen nanofibers-based functional bio-composites for high-value utilization of leather: a review. J Sci Adv Mater Devices 6(2):153–166
Wojdasiewicz W et al (1992) Tanning with hides to the wet white stages. J Am Leather Chem Assoc 87(1992):121
Yang Q et al (2015) Screening of bioflocculant and preliminary application to treatment of tannery wastewater. J Resid Sci Technol 12(3)
Younas F et al (2023) Unveiling distribution, hydrogeochemical behavior and environmental risk of chromium in tannery wastewater. Water 15(3):391
Zereen A et al (2013) Effect of tannery wastewater on the growth and yield of sunflower (*Helianthus annuus L.*). Bangladesh J Bot 42(2):279–285

Zhao C, Wuyong C (2019) A review for tannery wastewater treatment: some thoughts under stricter discharge requirements. Environ Sci Pollut Res 26: 26102–26111

Zupančič GD, Anita J (2010) Anaerobic digestion of tannery waste: semi-continuous and anaerobic sequencing batch reactor processes. Bioresour Technol 101(1) 26–33

Chapter 3
Environmental and Marine Ecosystem Impacts of Tannery Effluent: Pollution Load, Variability, and Bioindicator Analysis

Soukaina Harif, Rachid Faouzi, and Moulay Abdelazize Aboulhassan

Abstract Water pollution is a significant global challenge, with severe implications for both the environmental and human health. This study evaluated the pollution load and environmental effects of tannery effluent on the marine ecosystem. The most polluting procedures were identified to be pre-tanning stage, with chemical oxygen demand (COD) concentration reaching as high as 30,000 mg L^{-1}. Wastewater composition varied considerably through the production steps. The analysis of tannery effluents revealed substantial temporal variability, with COD concentration ranging from 700 to 3400 mg L^{-1} and suspended solids from 500 to 8000 mg L^{-1}, corresponding to average pollution load of 114 and 358 kg/d, respectively. Chromium concentrations reached from 40 to 115 mg L^{-1}, giving a daily load of 6.5 kg, while sulfide concentrations ranged from 65 to 160 mg L^{-1}. The low biodegradability index (BOD_5/COD: 0.1–0.2) indicates limited biodegradability of the effluent. To evaluate the impact on marine organisms, bioindicators including the mussel *Mytilus galloprovincialis* and algae *Ulva lactuca* and *Corallina officinalis* were examined. Significant chromium accumulation was observed in areas near wastewater discharge points, highlighting the effluent's detrimental effects on marine ecosystems.

Keywords Tannery wastewater · Characterization · Impact · Marine environment

---

S. Harif (✉) · R. Faouzi · M. A. Aboulhassan
Laboratory of Mechanic Process Energy and Environment, National School of Applied Sciences, Ibn Zohr University, Agadir, Morocco
e-mail: s.harif@uiz.ac.ma

R. Faouzi
e-mail: rachid.faouzi@edu.uiz.ac.ma

M. A. Aboulhassan
e-mail: a.aboulhassan@uiz.ac.ma

M. A. Aboulhassan
Laboratory of Engineering Sciences, Engineering Sciences Team, Management and Educational Innovation—ESTEM, Casablanca, Morocco

M. A. Aboulhassan and S. Souabi (eds.), *Sustainable Practices in the Tannery Industry*,
https://doi.org/10.1007/978-3-031-92037-0_3

3.1 Introduction

The global leather industry has experienced significant growth over recent decades, driven by rising demand for leather goods. As one of the world's oldest industries, leather production involves tanning, the process of treating proteinaceous skin/hide flayed from dead animals to create leather, an economically important commodity with a wide variety of uses (Lofrano et al. 2013; Garg and Chandra 2022). Tanning processes are categorized into vegetable tanning and chrome tanning (Giaccherini et al. 2017).

Vegetable tanning, the oldest and most complex method, uses natural tannins extracted from plant materials such as oak, chestnut, and mimosa. However, chrome tanning accounting for 70–80% of global leather production is simpler and yields leather with superior properties, including enhanced softness, durability, fullness, and heat resistance (Appiah-Brempong et al. 2020). Despite its efficiency and cost-effectiveness, chrome tanning generates significant environmental impacts, consuming approximately 35–40 L of water per kilogram of skin. Up to 90% of the water used in chrome tanning is discharged as effluent, often containing pollutants such as chromium, sulfides, and organic matter. According to Dave et al. (2022), global leather production processes approximately 1.5×10^{10} kg of hides and skins annually. This results in a daily average discharge of over 1.5×10^{10} kg of wastewater and an annual generation of 6×10^{9} kg of solid waste (Hu et al. 2011).

The tanning process involves pre-tanning, tanning, and post-tanning operations (Fig. 3.1). These steps involve the use of nearly 100 different chemicals including sodium hydroxide, sodium hypochlorite, potassium dichromate, lime, chlorides, sulfuric acid, formic acid, surfactants, sodium sulfide, and sodium and ammonium salts (Kanagaraj et al. 2015). Additionally, 50–70% of chromium used during tanning stage, leaving concentrations as high as 246 mg/L in the effluent, exceeding permissible environmental thresholds (Dave et al. 2022). As a result, tanning process wastewaters are a significant source of chromium pollution in the environment (Husain et al. 2024). Furthermore, these wastewaters often have high COD (> 10,000 mg/L), total dissolved solids (TDS) (> 14,000 mg/L), and low biodegradability, making conventional treatment challenging (Tasca and Puccini 2019).

The tanning industry can be viewed as a process that repurposes potential slaughterhouse waste to produce leather. However, its environmental consequences are significant. Despite the socioeconomic benefits of the tanning industry, including job creation and revenue generation, it is often perceived negatively by society due to the pollution it generates. The tanning process can produce a lot of waste, including toxic chemicals and wastewater. If these pollutants enter rivers or oceans, they can harm aquatic life and disrupt ecosystems. For instance, fish and other marine animals may suffer from diseases or die due to contaminated water.

This study aims to characterize tannery wastewater, focusing on its physicochemical properties and pollutant load, and assess its impact on marine ecosystems using bioindicators.

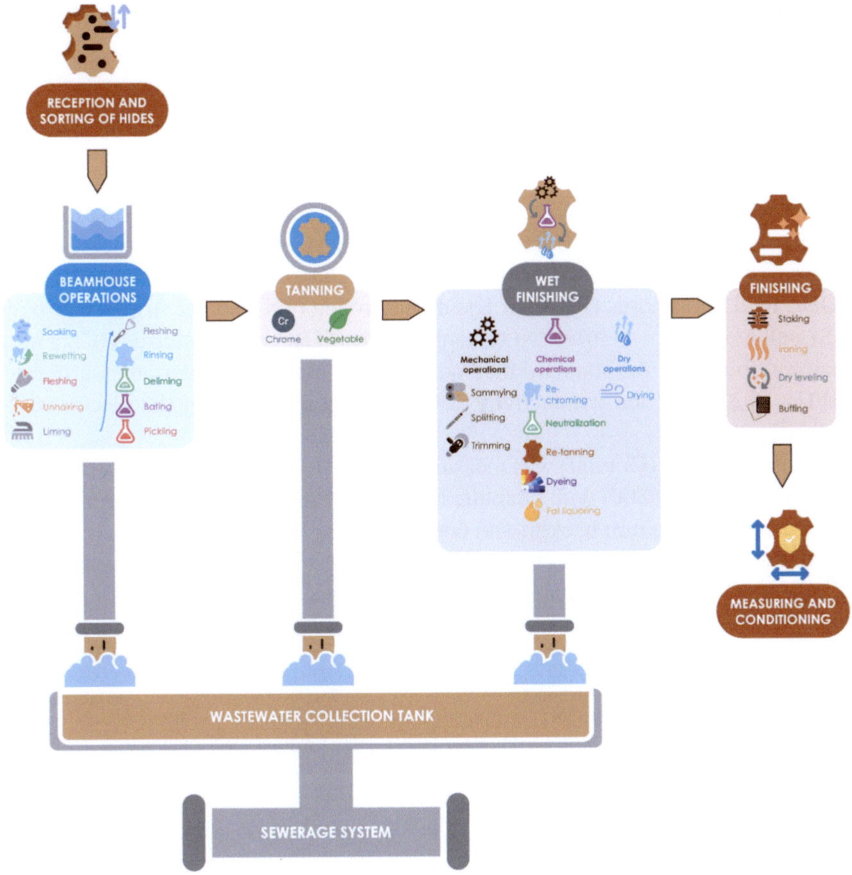

Fig. 3.1 Leather production processes and wastewater generation

3.2 Material and Methods

3.2.1 Characteristics of Tannery Wastewater

3.2.1.1 Tannery Wastewater Sampling

The leather tanning process includes several stages: preservation with common salt (i.e., sodium chloride) and storage, beam house or post-tanning operations, tanning, wet finishing and finishing (Fig. 3.1).

Tannery effluents samples were collected from the output of the production stage and the collection wastewater tank, which received all industrial wastewater (Fig. 3.1). Two sampling techniques were employed: grab sampling, capturing instant effluent characteristics, and composite sampling, integrating hourly samples over a

24h period to yield a representative temporal average. Tannery effluents samples were collected and stored in a refrigerator at 4 °C before the experiments.

3.2.1.2 Tannery Wastewater Characterization

The average characterization of samples is reported in Table 3.1, showing the limit value of each parameter according to the Moroccan standards. The physicochemical analyses in terms of chemical oxygen demand (COD), biochemical oxygen demand BOD_5, total phosphorus (Pt), total Kjeldahl nitrogen (TKN), total suspended solids (SS), sulfates (SO_4^{2-}), nitrate (NO_3^-), phosphate (PO_4^{3-}), and chlorides (Cl^-) were analyzed following the standard methods for examination of water and wastewater (APHA 2022). pH and conductivity (EC) were measured using the electrometric method.

Total chromium (Cr total) was measured using Atomic Absorption Spectrometer (SHIMADZU AA-6800) with Graphite Furnace (GFA-EX7). The measurement was made using a Deuterium background correction.

Hydrogen sulfide (H_2S) concentrations were measured using a selective electrode method employing the Oldham MX 2000 field device.

Table 3.1 Characterization of tannery effluent

	Instantaneous samples	Average sample	Flow (kg/day)	Limit value (Moroccan standards)	International standards	
					(USEPA)	WHO
pH	6.2 ± 0.5	5.6	–	6.5–8.5	6.5–8.5	6–9
T (°C)	23.4 ± 1.1	–	–	35		
Conductivity (mS cm^{-1})	17.4 ± 10	41.2	–	–		
Turbidity (NTU)	295 ± 86	240	–	–		50
TSS (mg L^{-1})	4476 ± 3944	550	358.1	600		
SO_4^{2-} (mg L^{-1})	2353 ± 1352	6801	188.314	400	250	500
Cl^- (mg L^{-1})	1834 ± 672	2485	146.7	–	250	250
PO_4^{3-} (mg L^{-1})	4.5 ± 1	1.9	0.4	–	5	
Pt (mg L^{-1})	10 ± 2	10.3	0.8	10		
NO^{3-} (mg L^{-1})	81.1 ± 31	92.5	6.5	–	50	50
TNK (mg L^{-1})	221.9 ± 54	283.7	17.8	–		
BOD_5 (mg L^{-1})	253 ± 209	369	20.2	500	30	
COD (mg L^{-1})	1429 ± 317	2723	114.4	1000	500	250
BOD_5/COD	0.2 ± 0.1	0.13		–		
Total chromium (mg L^{-1})	81.3 ± 34	47.2	6.5	2	0.1	0.05
H_2S (mg L^{-1})	116.7 ± 48	90.2	9.3	–	0.05	0.05

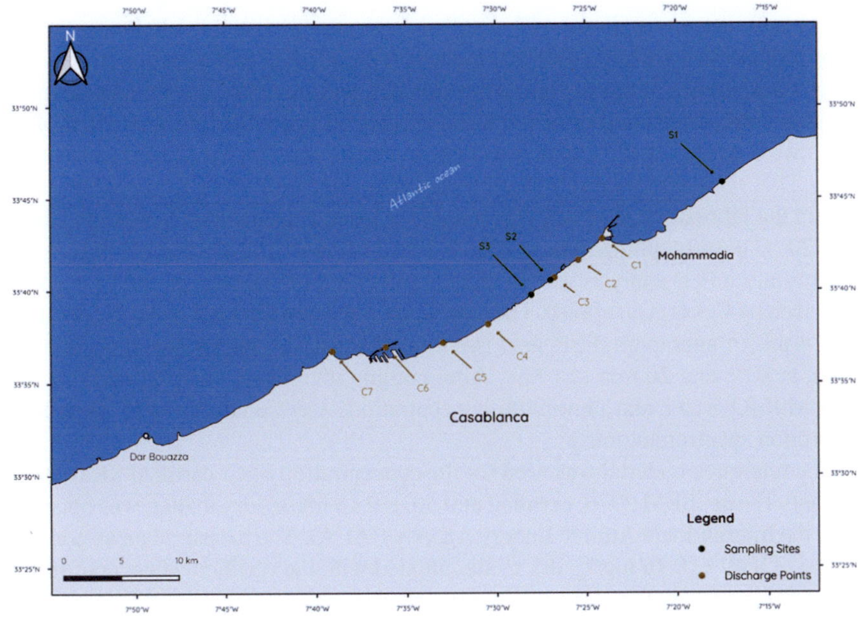

Fig. 3.2 Location of the sampling points

3.2.2 Marine Organism Analysis

3.2.2.1 Sampling Sites

Three studied sites in Mohammedia site were selected to collect mussels and algae based on their proximity to primary industrial effluent discharge zones (Fig. 3.2).

Seasonal sampling campaigns were conducted at three sites:

- **S1**: Located 8 km north of Mohammedia, in proximity to recreational and agricultural areas.
- **S2**: Positioned near the C3 discharge point, receiving effluents from petrochemical, textile, and tanning industries.
- **S3**: Situated 5 km south of the C3 discharge point, within a recreational area.

3.2.2.2 Sampling Procedures

Mussels (*Mytilus galloprovincialis*) were hand-collected from the lower mid-littoral zone. Approximately 40 adult individuals (average size: 3–5 cm) were gathered, sorted on-site, and stored in polyethylene containers filled with water from the sampling location. Following a depuration period of up to 36 h, the soft tissues were separated from the shells for further analysis.

The soft tissue was rinsed with distilled water, drained, and finely homogenized using a stainless-steel knife mill. It was then lyophilized for 3–4 days until a constant weight was achieved, ground into a fine powder, and stored in airtight containers.

After collecting, samples of the algae species *Ulva lactuca* and *Corallina officinalis* were washed with seawater to remove debris and epiphytes attached to their thalli. They were then placed in plastic containers filled with water from the sampling site. In the laboratory, they were thoroughly washed with distilled water and stored at − 20 °C until lyophilization.

Chromium (Cr) analysis was conducted following the mineralization protocol as described by Cossa et al. (1980). The sample was digested for 1 h at room temperature with 4 mL of suprapure nitric acid (Merck), followed by microwave digestion for 5 min at 80% and 20 min at 100%. After cooling, the digest was diluted to 50 mL with Milli-Q water and chromium concentrations were analyzed with an atomic absorption spectrophotometer.

The reference material was used for the concentration of Cr certified SRM 2976, Mussels Tissue. SRM 2976, certified at 0.50 ± 0.16 mg/kg dry matter, was obtained from the International Atomic Energy Agency (IAEA). The measured mean concentration of 0.49 ± 0.16 mg/kg dry matter aligned with the certified value, confirming analytical accuracy (Fig. 3.3).

3.3 Results and Discussion

3.3.1 Physicochemical Characterization

The results of physicochemical analysis of tannery effluents are presented in Table 3.1 and Figs. 3.4 and 3.5.

3.3.1.1 Effluents from Manufacturing Process Stages

Figure 3.4 shows the variations in pH, conductivity, suspended solids, COD, total phosphorus, and sulfate ions across effluents from different stages of the leather manufacturing process.

The results indicate significant variation in the analyzed parameters between production stages, with relatively minor fluctuations within the same stage.

- *Assessment of Pre-tanning Effluents*

The pre-tanning process, or beamhouse operations stage (Fig. 3.1), removed excess muscles, tissue or fat adhering to the hide so that uniform thickness may be acquired. Effluents from pre-soaking (Pr), soaking, and rewetting (R1, R2) are slightly alkaline, exhibit high conductivity, and contain elevated sulfate concentrations. These effluents showed high concentrations of suspended solids, phosphorus, and organic matter,

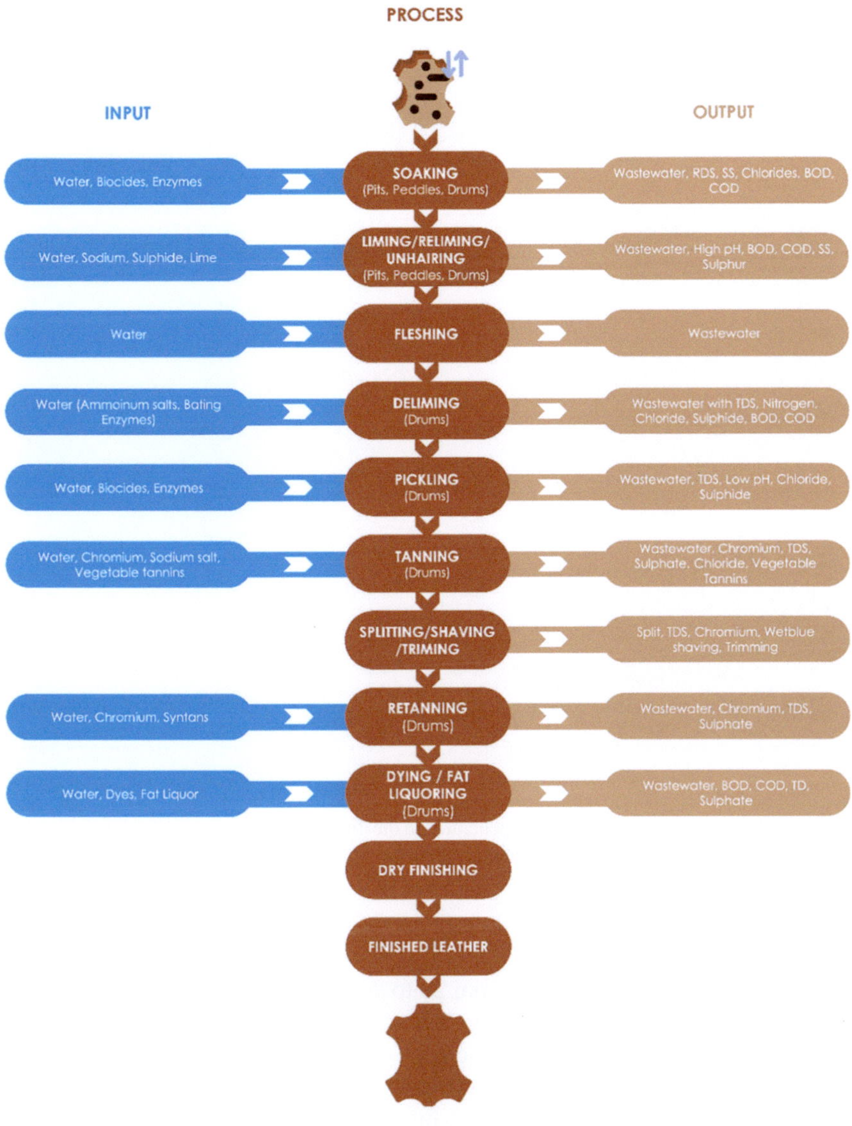

Fig. 3.3 Leather processing steps and their respective inputs and outputs

with COD levels exceeding 4000 mg L^{-1}. In a study conducted by Dixit et al. (2015), the COD was ranging from 22 to 33 kg/ton of skin/hide. These high loads of salts, organics compounds, and solids would be due to the removal of preservation agents (e.g., sodium chloride), impurities such as dirt, manure, and blood, proteinaceous debris, and dissociated hair in the wastewater.

The pH from the unhairing-liming (E/P) stage ranges between 11 and 12. In this step, the effluent presents high concentration of suspended solids, sulfates, and

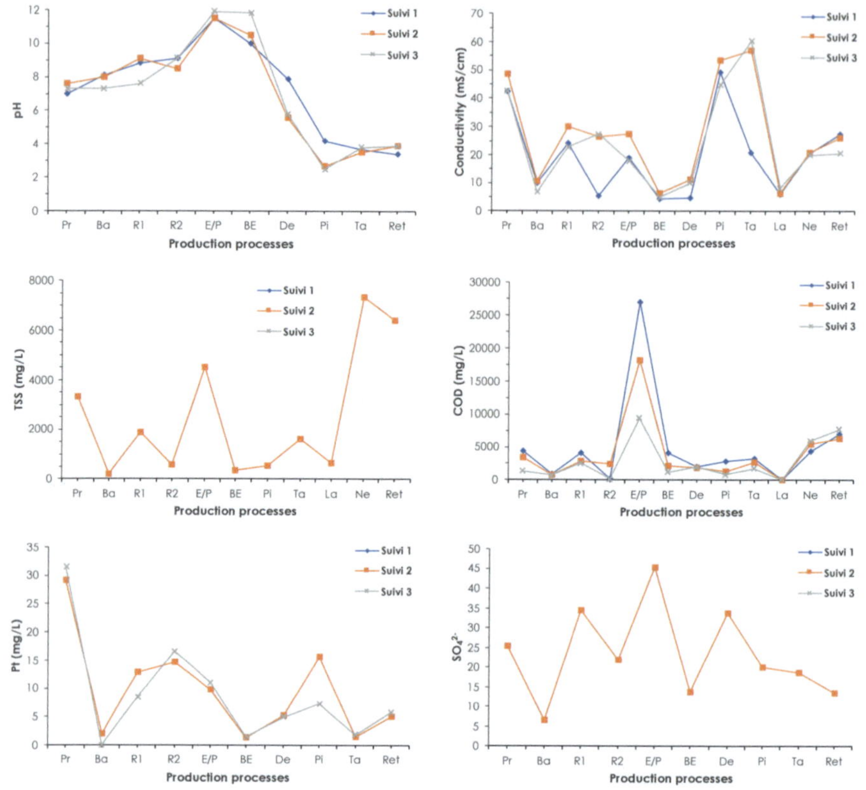

Fig. 3.4 Characteristics of the effluents generated in the different operations involved in the fabrication of leather (Pr: pre-soaking; Ba = rinsing bath; R1 = soaking; R2 = rewetting; E/P = unhairing/liming; BE = fleshing; De = deliming; Pi = pickling; Ta = tanning; La = washing before re-tanning; Ne = neutralization; Ret = re-tanning)

organic loads, with COD ranging from 20 to 30 g L^{-1}. Additionally, Munim et al. (2025) revealed that liming effluent holds higher concentration of sulfide (6533 ± 12.30 mg/L), BOD_5 (3270 ± 11.23 mg/L), COD (13,312 ± 18.22 mg/L), TDS (9.5 ± 1.2 g/L), EC (17.55 ± 0.06 mS/cm), and salinity (9.60 ± 0.08 ppt). Liu et al. (2021) demonstrated that the unhairing-liming process generates highly toxic and alkaline wastewater, very high proteins, sulfides, suspended solids, and salts. During this stage, the rehydrated hide is treated with an alkaline solution that hydrolyzes keratins, the proteins forming the epidermis and hair. Approximately 60–70% of the total pollution in leather manufacturing originates from conventional unhairing (liming) and re-liming processes in the beamhouse (Hashem et al. 2016).

Deliming effluents (De) are less polluted, with near-neutral pH and mineral salts. As shown in Fig. 3.4, they exhibit lower COD and phosphorus concentrations but high sulfate concentrations. Aluminum sulfate is used in this stage to neutralize the hide and remove salts by converting lime or soda into water-soluble compounds.

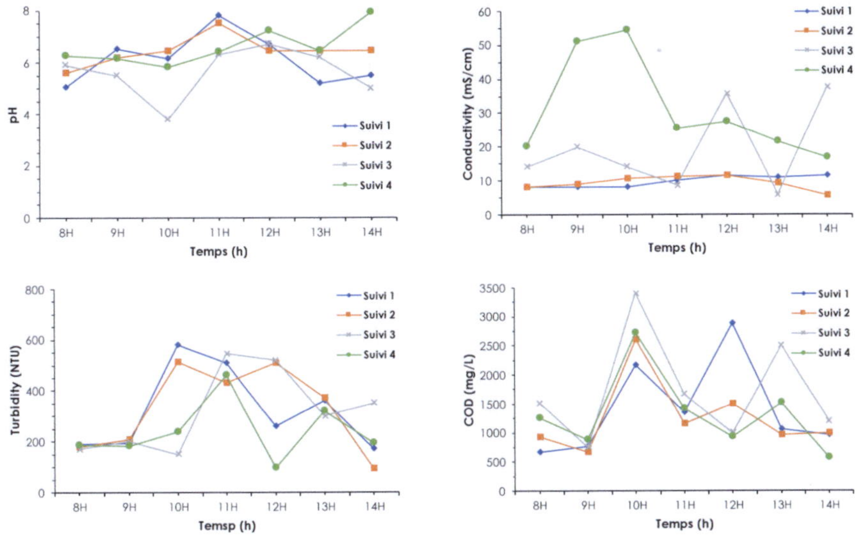

Fig. 3.5 Diurnal variations of pH, conductivity, turbidity, and chemical oxygen demand (COD) in the combined tannery effluent

$$Al_2(SO_4)_3 + 6NaCl \rightarrow 2AlCl_3 + 3Na_2SO_4$$

$$H_2SO_4 + 2NaOH \rightarrow Na_2SO_4 + 2H_2O$$

The pickling process (Pi) generates highly acidic, saline wastewater with pH values between 2 and 4 and conductivity exceeding 50 mS/cm (Fig. 3.4). This step occurs in an acidic medium with neutral salt (NaCl) to prevent hide swelling.

- *Assessment of Pollution from the Tanning Process*

The tanning process involves treating hides with agents that displace intercellular water and bond with collagen fibers, enhancing their resistance to heat, hydrolysis, and microbial degradation. This process stabilizes collagen as the tanning agents combine with skin proteins to produce leather. While chromium salts are the most used tanning agents, vegetable and synthetic tannins are also employed (Nazer et al. 2006). The factory employs two tanning methods: chrome tanning and vegetable tanning. Wastewater generated in this stage is considerably lower in volume than pre-tanning effluents. As shown in Fig. 3.4, tanning effluents are highly acidic ($pH < 4$), with conductivity up to 60 mS/cm. These effluents showed higher sulfates concentration and COD value more than 3200 mg/L. The high pollutant load arises from the acidic conditions and elevated chrome sulfate concentrations, particularly in chrome tanning, where a significant portion is discharged in wastewater. Chrome tanning effluent is a saline solution containing depleted chrome, fungicide residues, and a pH between 3 and 5 (Zhang et al. 2016).

- *Evaluation of Finishing Stage Effluents*

Effluents from hides placed on racks or in piles during resting periods, as well as drainage from wringing, are produced in small volumes but exhibit high pollutant loads. As shown in Fig. 3.4, wastewater from neutralization and re-tanning stages contains elevated COD and TSS levels, reaching 7500 mg/L and 7000 mg/L, respectively. In chrome tanning, these effluents may also contain significant amounts of chromium oxide.

3.3.1.2 Physicochemical Characterization of the Combined Tannery Effluent

Table 3.1 presents the analysis of wastewater collection tank, which collects the total wastewater from all the processes. Results include an average of four samples and a composite sample. Variations in wastewater quality are attributed to the differing nature and composition of effluents from each production unit. Tannery wastewater is characterized by substantial organic matter, high salinity, sulfate high, and important suspended solids. In the main, the effluents do not meet the prescribed international and Moroccan standards (Table 3.1).

pH is a significant parameter in the tannery wastewater due to its effect on physicochemical parameters and metal ion availability. Table 3.1 gives that the average pH of the wastewater is slightly acidic (Table 3.1) and agrees with that reported by Ali et al. (2013). Daily variations of pH in the collection wastewater tank are shown in Fig. 3.5. It is observed that pH ranges between strongly acidic (3) and highly basic (8), with predominantly acidic pH values, due to different quantities of salts used in the tanning processes. This pH value of this study is below Moroccan and international standards (Table 3.1). Vilardi et al. (2018) and Ake et al. (2022) showed acidic pH values of their studies (Table 3.2). Several authors Módenes et al. (2012), Saxena et al. (2018a, b), Moradi and Moussavi (2019), and Zhao et al. (2022) revealed basic pH values of their analyzed samples. However, neutral pH was recorded by Saxena et al. (2018b) and Pasciucco et al. (2024) (Table 3.2).

The extent of pollution in tannery effluent was assessed through chemical oxygen demand (COD) measurements, which serve as a key indicator of the organic load. High COD values reflect the substantial presence of degradable and non-degradable organic matter, underscoring the environmental impact of tannery discharges. In this study, the COD of effluents exhibits significant levels, with an average daily flux of 114 kg/day. Figure 3.5 shows that the COD concentrations range from 700 to 3400 mg/L, with the highest COD values observed at 10h. These higher levels are attributed to wastewater from beamhouse operations, which contain high concentrations of organic matter. This is also confirmed by many other studies (Mandal et al. 2010; Ali et al. 2013; Roy et al. 2015; Vilardi et al. 2018; Zhao and Chen 2019; Ake et al. 2022; Yang et al. 2023) (Table 3.2). Meanwhile, Pasciucco et al. (2024) revealed much lower COD concentration (380 ± 8.5 mg/L) of their analyzed samples (Table 3.2). In a study conducted by (Módenes et al. 2012; Saxena et al.

Table 3.2 Comparison of the chemical oxygen demand (COD) and total chromium content of various tannery effluents

COD (mg L^{-1})	BOD (mg L^{-1})	pH	Chrome total (mg L^{-1})	References
939.33 ± 26.46–7739.55 ± 297.68	24–42	2.69–3.65	200–500	Ake et al. (2022)
2480	992	6.69	592.20	Ali et al. (2013)
2533	977	7.9–9.2		Mandal et al. (2010)
11,878		8.3	72.1 ± 0.5	Módenes et al. (2012)
20,600		11.41	30.11	Moradi and Moussavi (2019)
380 ± 8.5	–	7.7 ± 0.04	–	Pasciucco et al. (2024)
2986	526	7.6	9.21	Roy et al. (2015)
18,800–23,000	2540–2960	11.98–12.54		Saxena et al. (2018a)
11,000–12,800	1040–1100	12.2–12.5		Saxena et al. (2018a)
8800–10,080	2800–3200	7.2–7.5		Saxena et al. (2018b)
1823.23 ± 2.04		3.74 ± 0.08	69.12 ± 0.12	Vilardi et al. (2018)
3000–4000	1500–2000	8–11	60–100	Zhao and Chen (2019)
700–3400	253 ± 209	6.2 ± 0.5	10–128	This work

2018a, b; Moradi and Moussavi 2019) revealed high levels of COD (> 8800 mg/L). Furthermore, the wastewater produced has an average flow of about 18 kg/day of TKN (Table 3.1). These results could be explained by the toxic nature of the wastewater and the presence of biologically resistant organic substances. It has been shown that a substantial part of the chemicals used in the tanning process is discharged into the environment and is not absorbed by hides (Amanial 2016).

Electrical conductivity (EC) is an important indicator of ions dissolved concentration in wastewater and water. Figure 3.2 illustrates the EC of the effluents. It was ranging between 9 and 31 and reached 55 mS cm^{-1}. However, the high EC values can be attributed to various dissolved chemicals used in the tanning industry, including organic and inorganic compounds as well as salts, particularly sodium salt, aluminum sulfate and chromium sulfate employed in tanning stage. In acidic or oxygen-deficient environments, sulfate ions can lead to the formation of hydrogen sulfide. An average flux of 9 kg/day of H_2S was observed. Acidic pH is the primary cause of the production of highly toxic hydrogen sulfide (H_2S) gas (Ali et al. 2013). The results showed high chloride concentration (1–2.5 g/L) in tanneries wastewater (Table 3.2). These concentrations are primarily due to the use of sodium chloride as an additive for preservation hides. Roy et al. (2015) showed that the tannery effluent had load chloride of around 187 mg/L and much lower than this study. However, Ali et al. (2013) found 2176.40 mg/L and Pasciucco et al. (2025) demonstrated that Chloride concentration of tannery wastewater can reached 3363.7 ± 42 mg/L.

Chromium (Cr) is among the most significant environmental contaminants generated by tanneries (Husain et al. 2024). In the environment, chromium exists in two stable forms: trivalent (Cr^{3+}) and hexavalent (Cr^{6+}). Due to its high solubility, mobility in aquatic and soil environments, Cr^{6+} is significantly more hazardous than Cr^{3+} (Marsh and McInerney 2001; Shukla and Rai 2006). The results of this study showed that the levels of chromium range from 40 to 115 mg/L (Table 3.1) with an average pollutant load reaching 6.5 kg/day. According to other authors (Ali et al. 2013; Ake et al. 2022), total Cr concentrations in tannery effluents can reach significantly higher levels (Table 3.2). The presence of Cr is consistent with the leather manufacturing process involving Cr. Cr is used by tanners to obtain good quality leather. But the release of Cr into the environment can cause damage to fauna and flora. Roy et al. (2015) revealed a low concentration of total Cr (9.21 mg/L). Nevertheless, Módenes et al. (2012), Vilardi et al. (2018), Moradi and Moussavi (2019), Zhao and Chen (2019) revealed different concentrations of total Cr 72.1 ± 0.5; 69.12 ± 0.12, 30.11 and 60–100 mg/L, respectively; these values are comparable to that of this study (Table 3.2). Chromium present in effluents can spread into the environment, posing substantial risks to human health and ecosystems. However, in the presence of MnO_2 and organic molecules, Cr(III) can be re-oxidized to Cr(VI), complicating efforts to achieve permanent toxicity reduction (Pettine et al. 1991; Shukla and Rai 2006; Yen et al. 2025).

In addition, the tanneries wastewater shows high TSS concentration between 550 and 8000 mg/L (Table 3.2), and an average pollutant load of 358 kg/day. These high TSS concentrations are significantly due to chemical compounds such as sodium, chlorides, and sulfates used in leather production. According to Amanial (2016), TSS values below 100 mg/L indicate weak wastewater, values between 100 and 220 mg/L indicate medium-strength wastewater, and value more than 220 mg/L represent high-strength wastewater. As a result, these effluents could be classified as strong wastewater. Moreover, diurnal turbidity in tannery effluents varied widely, ranging from 100 to 600 NTU (Fig. 3.5) because the unstable quality of effluents.

The BOD_5/COD ratio ranges from 0.1 to 0.2 (Table 3.1). The results are comparable to those of Saxena et al. (2018a, b). The biodegradability index (BOD_5/COD) below 0.3 indicates the organic matter of a wastewater is slightly or non-biodegradable and exhibiting high treatment costs (Saxena et al. 2018a). In this study, the effluents from these industrial units are not biodegradable. Saxena et al. (2018a, b) achieved similar results with the BOD_5/COD ratio of 0.14 and 0.33, respectively. This result is attributed to the bio-refractory organics substances and microbial activity inhibitors in tannery wastewater.

Tannery wastewater effluent (TWE) contains significant amounts of recalcitrant compounds, ranging from simple halogenated hydrocarbons to complex polymers (Korpe and Rao 2021). These synthetic compounds, often slowly biodegradable or non-biodegradable, primarily originate from wet processing activities in tanneries. Furthermore, their presence contributes to environmental toxicity, posing challenges in complying with effluent standards for discharge into sewer systems or receiving water bodies (Korpe and Rao 2021). Tanning processes, particularly those conducted in drums, as in the industrial unit studied, involve high water and chemical consumption (Fig. 3.3). To facilitate the penetration and reaction of these chemicals with collagen, reagents are added in excess, with only partial uptake by the hides.

3.3.2 Environmental Impact of Tannery Effluents on Marine Ecosystems

Analysis of wastewater from the collection basin revealed a high pollutant load, particularly in organic matter, suspended solids, and salts, exceeding the prescribed Moroccan standards. Our results share a high similarity with several authors' findings reduced dissolved oxygen levels and elevated concentrations of nitrogen compounds and orthophosphates (Kouali et al. 2022). The high organic and inorganic content, including phosphates and nitrates, in tannery wastewater is a major contributor to eutrophication in many water bodies. Additionally, untreated pollution discharged into the marine environment has an adverse effect on the aquatic life present inside. Hexavalent chromium is the primary contributor to many health problems (Uc-Peraza et al. 2021).

Figure 3.2 illustrates the evaluation of chromium contamination in bioindicator species near the discharge points of these effluents, conducted to assess their impact on the marine environment. Table 3.3 presents the spatial variations in chromium concentrations in the mussel *M. galloprovincialis* and the two algal species *U. lactuca* and *C. officinalis*. The average chromium concentrations in mussels are relatively high compared to other regions, such as the Moroccan Mediterranean coast, the French coastline, and the Pacific Ocean (Table 3.4).

Table 3.3 The average organism chromium was contaminated at different locations along the Casablanca-Mohammadia ocean coastline (mg kg^{-1} dry wt.)

Sites	*M. galloprovincialis*	*U. lactuca*	*C. officinalis*
S1	6.4 ± 2.99 (1.88–11.14)	2.35 ± 1.56 (0.55–5.15)	5.25 ± 3.73 (2.65–16.60)
S2	9.77 ± 4.34 (4.41–17.89)	3.02 ± 1.93 (0.27–6.90)	6.64 ± 3.36 (3.66–15.20)
S3	8.38 ± 3.97 (1.63–16.53)	3.76 ± 2.72 (0.86–9.09)	7.56 ± 3.96 (3.09–17.30)

Values include the mean, standard deviation, minimum, and maximum

Table 3.4 Chromium contents (mg kg^{-1} dry wt.) in mussels from various world areas

Localization	Chromium (mg/kg P.S.)	References
Littoral Casablanca—Mohammedia, Morocco	6.40–18.71	Present study[a]
Strait of Messina—Messina	0.51 ± 0.14	Bruno et al. (2024)[a]
Vitoria Bay—Brazil	0.89 ± 0.39	Vieira et al. (2021)[b]
Rio de Janeiro, Brazil	1.0–7.5	Lino et al. (2016)[b]
Santos Bay, Brazil	0.5–2.5	Campolim et al. (2017)[b]
Europe	1.8–1.9	Orescanin et al. (2006)[a]
Homa lagoon—Turkish	0.57	Bilgin and Uluturhan-Suzer (2017)[a]
China	2.18–10.93	Fung et al. (2004)[d]
Venezuela	0.22 ± 0.01	Rojas de Astudillo et al. (2005)[d]
Eljadida—Morocco	2.05–54.30	Kouali et al. (2022)[a]
Baja California, México	1.58–0.50	Munä Oz-Barbosa et al. (2000)[c]

[a]*Mytilus galloprovincialis*
[b]*Perna*
[c]*M. californianus*
[d]*Perna viridis*

The results highlight site S2, located near the discharge point of textile and tannery industries, as a significant source of chromium contamination, although the other two stations also show significant levels. Chromium accumulation in bivalve mollusks does not exhibit significant variations based on the age or size of the individuals studied. This dissimilarities with other metals, such as cadmium and copper, which demonstrate concentrations inversely proportional to the size of mussels collected in the same study area (Campolim et al. 2017; Kouali et al. 2022). The chromium values found in our samples may be due to contamination of Cr in the marine environment.

For the two algal species, a comparison of mean concentrations indicates a highest accumulation of chromium in *C. officinalis*. This difference may be attributed to the ecology and morphology of each species, their immersion and exposure periods (Fung et al. 2004), and the dynamics of phytoplankton blooms. Additionally, an interaction between species and site factors was observed, aligning with organotrophic studies on mussels reported by various authors (Orescanin et al. 2006; Campolim et al. 2017; Uc-Peraza et al. 2021; Kouali et al. 2022).

Results showed that chromium accumulation followed the order: *Mytilus galloprovincialis* > *Corallina officinalis* > *Ulva lactuca* (Table 3.3). These differences in metal sensitivity and accumulation between mussels and algae can be attributed to factors such as variations in metal accumulation mechanisms (available forms), organism size, and their respective ecological and biological characteristics

(Table 3.4). Particularly, polysaccharide polymers in algae are a significant substrates for metal attachment (Kouali et al. 2022).

3.4 Conclusion

The study highlights that the effluents from tanneries significantly affect marine ecosystems. Specifically, it was found that the concentration of chromium (Cr) in marine organisms, such as mussels (*M. galloprovincialis*) and algae (*C. officinalis* and *U. lactuca*), is notably higher than previously reported in the literature. This indicates a substantial anthropogenic contribution to chromium levels in these organisms. The research indicates that the mussel species *M. galloprovincialis* accumulates more chromium compared to the two types of algae studied. This suggests that different species have varying capacities to absorb and retain pollutants, which is crucial for understanding the ecological impact of tannery effluents. The analysis of wastewater from various stages of leather production revealed that pollutant concentrations vary significantly depending on the specific stage of the tanning process. This variability complicates the selection of appropriate treatment methods for these effluents, as their characteristics can change over time. The study emphasizes the high pollution load associated with chromium in tannery effluents, which poses serious environmental risks. The presence of such pollutants can lead to detrimental effects on marine ecosystems, necessitating effective pre-treatment of these effluents before they are discharged into natural water bodies. Tannery effluents are described as complex mixtures of high organic and inorganic compounds, particularly chromium. Their non-biodegradable nature and fluctuating characteristics over time present significant challenges for environmental management and pollution control. These conclusions underscore the urgent need for improved treatment processes for tannery effluents to mitigate their harmful effects on marine environments and to protect aquatic life.

References

Ake AHJ, Hafidi M, Ouhdouch Y, Fels LEL (2022) Physico-chemical characterization and pollutant charge of industrial effluents from tanneries under semi-arid climate. Desalin Water Treat 260:148–160. https://doi.org/10.5004/dwt.2022.28431

Ali Z, Malik RN, Qadir A (2013) Heavy metals distribution and risk assessment in soils affected by tannery effluents. Chem Ecol 29:676–692. https://doi.org/10.1080/02757540.2013.810728

Amanial HR (2016) Physico chemical characterization of tannery effluent and its impact on the nearby river. OALib 03:1–8. https://doi.org/10.4236/oalib.1102427

American Public Health Association (2022) Standard methods for the examination of water and wastewater, 24th edn. American Water Works Association

Appiah-Brempong M, Essandoh HMK, Asiedu NY et al (2020) An insight into artisanal leather making in Ghana. J Leather Sci Eng 2:25. https://doi.org/10.1186/s42825-020-00039-8

Bilgin M, Uluturhan-Suzer E (2017) Assessment of trace metal concentrations and human health risk in clam (*Tapes decussatus*) and mussel (*Mytilus galloprovincialis*) from the Homa Lagoon

(Eastern Aegean Sea). Environ Sci Pollut Res 24:4174–4184. https://doi.org/10.1007/s11356-016-8163-2

Bruno F, Nava V, Zappalà S et al (2024) Mineral composition in mussel *Mytilus galloprovincialis* and clam *Tapes decussatus* from Faro Lake of Messina: risk assessment for human health. Front Toxicol 6. https://doi.org/10.3389/ftox.2024.1494977

Campolim MB, Henriques MB, Petesse ML et al (2017) Metal trace elements in mussels in Urubuqueçaba Island, Santos Bay, Brazil. Pesq Agropec Bras 52:1131–1139. https://doi.org/10.1590/S0100-204X2017001200001

Cossa D, Bourget E, Pouliot D et al (1980) Geographical and seasonal variations in the relationship between trace metal content and body weight in *Mytilus edulis*. Mar Biol

Dave H, Vajpayee M, Ledwani L (2022) Photocatalytic and other similar green technologies for reducing environmental impacts of leather industries, pp 477–511

Dixit S, Yadav A, Dwivedi PD, Das M (2015) Toxic hazards of leather industry and technologies to combat threat: a review. J Clean Prod 87:39–49. https://doi.org/10.1016/j.jclepro.2014.10.017

Fung CN, Lam JCW, Zheng GJ et al (2004) Mussel-based monitoring of trace metal and organic contaminants along the east coast of China using *Perna viridis* and *Mytilus edulis*. Environ Pollut 127:203–216. https://doi.org/10.1016/j.envpol.2003.08.007

Garg S, Chandra A (2022) Green photocatalytic semiconductors. Springer International Publishing, Cham

Giaccherini F, Munz G, Dockhorn T et al (2017) Carbon and energy footprint analysis of tannery wastewater treatment: a global overview. Water Resour Ind 17:43–52. https://doi.org/10.1016/j.wri.2017.03.001

Hashem MA, Nur-A-Tomal MS, Bushra SA (2016) Oxidation-coagulation-filtration processes for the reduction of sulfide from the hair burning liming wastewater in tannery. J Clean Prod 127:339–342. https://doi.org/10.1016/j.jclepro.2016.03.159

Hu J, Xiao Z, Zhou R et al (2011) Ecological utilization of leather tannery waste with circular economy model. J Clean Prod 19:221–228. https://doi.org/10.1016/j.jclepro.2010.09.018

Husain A, Ansari MA, Khan AH et al (2024) Overview of treatment methods of tannery wastewater. IOP Conf Ser Earth Environ Sci

Kanagaraj J, Senthilvelan T, Panda RC, Kavitha S (2015) Eco-friendly waste management strategies for greener environment towards sustainable development in leather industry: a comprehensive review. J Clean Prod 89:1–17

Korpe S, Rao PV (2021) Application of advanced oxidation processes and cavitation techniques for treatment of tannery wastewater—a review. J Environ Chem Eng 9:105234. https://doi.org/10.1016/j.jece.2021.105234

Kouali H, Chaouti A, Achtak H et al (2022) Trace metal contents in the mussel *Mytilus galloprovincialis* from Atlantic coastal areas in northwestern Morocco: levels of contamination and assessment of potential risks to human health. Mar Pollut Bull 179. https://doi.org/10.1016/j.marpolbul.2022.113680

Lino AS, Galvão PMA, Longo RTL et al (2016) Metal bioaccumulation in consumed marine bivalves in Southeast Brazilian coast. J Trace Elem Med Biol 34:50–55. https://doi.org/10.1016/j.jtemb.2015.12.004

Liu H, Li X, Li M et al (2021) A simple and sustainable beamhouse by the recycling of waste-water from KCl-dispase synergistic unhairing in leather making. J Clean Prod 282. https://doi.org/10.1016/j.jclepro.2020.124535

Lofrano G, Meriç S, Zengin GE, Orhon D (2013) Chemical and biological treatment technologies for leather tannery chemicals and wastewaters: a review. Sci Total Environ 461–462:265–281

Mandal T, Dasgupta D, Mandal S, Datta S (2010) Treatment of leather industry wastewater by aerobic biological and Fenton oxidation process. J Hazard Mater 180:204–211. https://doi.org/10.1016/j.jhazmat.2010.04.014

Marsh TL, McInerney MJ (2001) Relationship of hydrogen bioavailability to chromate reduction in aquifer sediments. Appl Environ Microbiol 67:1517–1521. https://doi.org/10.1128/AEM.67.4.1517-1521.2001

Módenes AN, Espinoza-Quiñones FR, Borba FH, Manenti DR (2012) Performance evaluation of an integrated photo-Fenton—electrocoagulation process applied to pollutant removal from tannery effluent in batch system. Chem Eng J 197:1–9. https://doi.org/10.1016/j.cej.2012.05.015

Moradi M, Moussavi G (2019) Enhanced treatment of tannery wastewater using the electrocoagulation process combined with UVC/VUV photoreactor: parametric and mechanistic evaluation. Chem Eng J 358:1038–1046. https://doi.org/10.1016/j.cej.2018.10.069

Munä Oz-Barbosa A, Gutieâ Rrez-Galindo EA, Flores-Munä Oz G (2000) *Mytilus californianus* as an indicator of heavy metals on the northwest coast of Baja California, Mexico

Munim F, Bin Faruk M, Hashem MA, Miem MM (2025) Treatment of pre-treated liming wastewater in tannery by mixed banana peel adsorbent in column filtration studies. Sustain Chem Environ 9:100203. https://doi.org/10.1016/j.scenv.2024.100203

Nazer DW, Al-Sa'ed RM, Siebel MA (2006) Reducing the environmental impact of the unhairing–liming process in the leather tanning industry. J Clean Prod 14:65–74. https://doi.org/10.1016/j.jclepro.2005.04.002

Orescanin V, Lovrencic I, Mikelic L et al (2006) Biomonitoring of heavy metals and arsenic on the east coast of the Middle Adriatic Sea using *Mytilus galloprovincialis*. Nucl Instrum Methods Phys Res B 245:495–500. https://doi.org/10.1016/j.nimb.2005.11.050

Pasciucco E, Pasciucco F, Iannelli R, Pecorini I (2024) A Fenton-based approach at neutral and un-conditioned pH for recalcitrant COD removal in tannery wastewater: experimental test and sludge characterization. Sci Total Environ 926:172070. https://doi.org/10.1016/j.scitotenv.2024.172070

Pasciucco E, Pasciucco F, Panico A et al (2025) Optimization of a Fenton-based process as a tertiary treatment of tannery wastewater through response surface methodology. J Water Process Eng 69. https://doi.org/10.1016/j.jwpe.2024.106588

Pettine M, Millero FJ, La Noce T (1991) Chromium (III) interactions in seawater through its oxidation kinetics

Rojas de Astudillo L, Chang Yen I, Bekele I (2005) Heavy metals in sediments, mussels and oysters from Trinidad and Venezuela. Rev Biol Trop 53:41–53

Roy S, Nagarchi L, Das I et al (2015) Cytotoxicity, genotoxicity, and phytotoxicity of tannery effluent discharged into Palar River basin, Tamil Nadu, India. J Toxicol 2015. https://doi.org/10.1155/2015/504360

Saxena S, Rajoriya S, Saharan VK, George S (2018a) An advanced pretreatment strategy involving hydrodynamic and acoustic cavitation along with alum coagulation for the mineralization and biodegradability enhancement of tannery waste effluent. Ultrason Sonochem 44:299–309. https://doi.org/10.1016/j.ultsonch.2018.02.035

Saxena S, Saharan VK, George S (2018b) Enhanced synergistic degradation efficiency using hybrid hydrodynamic cavitation for treatment of tannery waste effluent. J Clean Prod 198:1406–1421. https://doi.org/10.1016/j.jclepro.2018.07.135

Shukla OP, Rai UN (2006) Hexavalent chromium induced changes in growth and biochemical responses of chromate-resistant bacterial strains isolated from tannery effluent. Bull Environ Contam Toxicol 77:96–103. https://doi.org/10.1007/s00128-006-1037-4

Tasca AL, Puccini M (2019) Leather tanning: life cycle assessment of retanning, fatliquoring and dyeing. J Clean Prod 226:720–729. https://doi.org/10.1016/j.jclepro.2019.03.335

Uc-Peraza RG, Gutiérrez-Galindo EA, Delgado-Blas VH, Muñoz-Barbosa A (2021) Total mercury content in the California ribbed sea mussel *Mytilus californianus* from the west coast of Baja California, México: levels of contamination and human health risk. Mar Pollut Bull 170. https://doi.org/10.1016/j.marpolbul.2021.112585

USEPA (2021) Drinking water contaminants. National Primary Drinking Water Regulations, Washington

Vieira KS, Delgado JF, Lima LS et al (2021) Human health risk assessment associated with the consumption of mussels (*Perna perna*) and oysters (*Crassostrea rhizophorae*) contaminated with metals and arsenic in the estuarine channel of Vitória Bay (ES), Southeast Brazil. Mar Pollut Bull 172:112877. https://doi.org/10.1016/j.marpolbul.2021.112877

Vilardi G, Di Palma L, Verdone N (2018) On the critical use of zero valent iron nanoparticles and Fenton processes for the treatment of tannery wastewater. J Water Process Eng 22:109–122. https://doi.org/10.1016/j.jwpe.2018.01.011

WHO (1996) Guidelines for drinking-water quality: health criteria and other supporting information. World Health Organization

Yang F, Wang XB, Shan Y et al (2023) Research recap of membrane technology for tannery wastewater treatment: a review. Collagen Leather 5

Yen C-C, Chen K-Y, Ahmed MMM et al (2025) Photochemical oxidation of Cr(III) to Cr(VI) in the presence of Fe(III): influence of Fe(III) concentration and UV wavelength. J Hazard Mater 485:136852. https://doi.org/10.1016/j.jhazmat.2024.136852

Zhang C, Lin J, Jia X, Peng B (2016) A salt-free and chromium discharge minimizing tanning technology: the novel cleaner integrated chrome tanning process. J Clean Prod 1055–1063

Zhao C, Chen W (2019) A review for tannery wastewater treatment: some thoughts under stricter discharge requirements. Environ Sci Pollut Res 26:26102–26111. https://doi.org/10.1007/s11356-019-05699-6

Zhao J, Wu Q, Tang Y et al (2022) Tannery wastewater treatment: conventional and promising processes, an updated 20-year review. J Leather Sci Eng 4

Chapter 4
Industrial Discharges from Tanneries: Speciation of Heavy Metals and Reduction of Metallic Pollution

Roukaya Bouyakhsass, Imane Illou, Khalid Digua, Hajar Bakraoui, Moulay Abdelazize Aboulhassan, and Salah Souabi

Abstract Tannery effluents are complex discharges, highly loaded with organic and mineral matter, especially chromium. They have variable characteristics and biodegradability over time. In this study, the concentrations of total heavy metals, especially total Cr in tannery wastewater samples, were successfully determined by the ICP-AES technique. The excesses of total metal concentrations indicate that the studied waters were so polluted considering the total concentration. This pollution causes various serious negative environmental effects, which are carried out on the human body through the food chain. The speciation of metallic elements in tannery discharges remains very interesting to prevent the action plan to reduce the pollution of these discharges. The speciation results indicate that most analyzed metals are abundant in the particulate phase. The percentage of Cr, Cu, Pb, Hg, and Ni in the dissolved phase is of the order of: Cr: 0.05–8%; Cu: 7.1–47.2%; Pb: 4.4–50%; Hg: 2–11%, Ni: 0–14%. Meanwhile, cadmium presents a much more significant fraction: Cd: 60–100%. The presence of metals in particulate form follows the following decreasing order: Cr > Ni > Hg > Cu > Fe > Pb > Zn > Cd. The observed changes in the physical speciation of trace metals highlight the importance of distinguishing the permeable phase (truly dissolved) from all dissolved species to better assess environmental releases. However, traditional laboratory methods do not separate colloids from the dissolved phase. Such a lack of separation between the colloidal and permeable fractions can lead to inaccuracies in the results of studies on the fate and behavior of contaminants in the aquatic environment. This fraction should be the subject of particular interest to arrive at a better ecotoxicological assessment of releases into aquatic ecosystems. Determining the total concentration of a potentially toxic element provides little information on its actual toxicity. Indeed, the toxicity of an element is directly linked to the chemical form in which the element is present (and

R. Bouyakhsass · I. Illou · K. Digua · H. Bakraoui · S. Souabi (✉)
Faculty of Sciences and Techniques Mohammed, Hassan II University Casablanca, Casablanca, Morocco
e-mail: salah.souabi@gmail.com

M. A. Aboulhassan
ENSA, Ibn Zohr University, Agadir, Morocco

M. A. Aboulhassan and S. Souabi (eds.), *Sustainable Practices in the Tannery Industry*,
https://doi.org/10.1007/978-3-031-92037-0_4

therefore to its speciation). Indeed, the absorption rate by the human body is often directly linked to the chemical form and, in particular, to the solubility of the chemical species considered. These results contribute to the project to monitor the quality of tannery water in Mohammedia, whose work is continuing in order, in particular, to assess the environmental impact of tannery discharges into the environment.

Keywords Tannery wastewater · Metals speciation · Treatment

4.1 Introduction

Hide processing and treatment in the tanning industry generates liquid and solid effluents rich in toxic pollutants with significant environmental impacts. The wastewater discharged by these activities contains contaminants from the tanning process, particularly the chemicals used in the various stages of the process (Zhao et al. 2022). These liquid effluents are often accompanied by solid waste and atmospheric emissions, exacerbating the environmental impact, especially as they are generally not adequately treated at the source. In Morocco, this problem is exacerbated by the absence of strict regulations requiring effluent treatment before discharge (Saad and Youness 2024).

Heavy metals represent a category of micropollutants, such as metals that are non-biodegradable organic pollutants due to their toxicity and their ability to accumulate in living organisms (bioaccumulation). Although their discharge concentration is often low (in mg/L), the large volumes of wastewater compensate for these low proportions, making their harmful effects significant. Among these metals, chromium occupies a prominent position. When discharged in high quantities, it can disrupt the operation of wastewater treatment plants. Studies have shown that the toxicity of metals is highly dependent on their chemical form, with free forms being the most harmful (Egorova and Ananikov 2017; Ali et al. 2019; Sall et al. 2020).

Tanneries, particularly artisanal tanneries, are recognized worldwide as significant sources of chromium contamination of surface waters (Appiah-Brempong et al. 2022; Kapanga et al. 2024; Faye and Sibali 2025).

This study focuses on specifying the main heavy metals (Cr, Pb, Hg, Cd …) in liquid discharges from an industrial tannery, using ICP-AES as the analytical technique.

The problem of heavy metals is intrinsically linked to the diversity of chemical forms in which these elements appear in the environment. In addition to simple mineral forms (such as Pb^{2+}), heavy metals can exist in organic, complex, or chelated forms, each with specific properties influencing solubility and toxicity. These transformations, though often subtle, significantly alter their environmental behavior. The speciation of heavy metals remains an essential element in understanding and managing their impact on the receiving environment (Smith et al. 2022).

Speciation goes beyond simply measuring the total concentration of a metal in a sample by identifying the different chemical forms present in liquid discharges (soluble, colloidal, particulate, etc.). This approach is essential, as the toxicity of metals is closely linked to their chemical form.

4.2 Materials and Methods

4.2.1 Speciation of Metallic Elements in Tannery Wastewater

The speciation of heavy metals is of crucial importance because of the impacts they can have, even in trace amounts, on ecological and environmental systems. These impacts are determined mainly by the chemical forms in which these metals are found in the environment (Ali et al. 2019). In aquatic environments, trace metals are distributed between different phases: particulate, colloidal, and dissolved (Table 4.1).

Interactions between these phases play an essential role in the fate of metals and, consequently, in their environmental impact. For example, slow exchanges can occur between dissolved and particulate fractions, notably via the aggregation of colloids (Jensen et al. 2020). Furthermore, metals associated with low-molecular-weight ligands in the dissolved phase have long residence times in water, enabling them to be transported over long distances. Conversely, metals bound to small particles, such as colloids, are generally less mobile (Vardhan et al. 2019).

Metals associated with colloids often have lower bioavailability for aquatic organisms (Amde et al. 2017; Santschi 2018; Adams et al. 2020). Therefore, distinguishing between metals bound to colloids and those found in the truly dissolved (permeable) fraction is essential. The latter deserves particular attention for an accurate ecotoxicological assessment of discharges (Siddiqui and Brander 2024).

Assessing the environmental impact of metals from tannery effluents requires a thorough understanding of the physico-chemical characteristics of the receiving waters. These characteristics influence not only the uptake of metals by exposed aquatic organisms but also their speciation in particulate or dissolved form, and hence their bioavailability and toxicity (Saxena et al. 2017; Arti and Mehra 2023).

Geochemical processes play a key role in the distribution of metals between the dissolved and particulate phases in the effluent dispersion plume (Mkandawire 2019; Zeng et al. 2023).

Table 4.1 Definition of phase sizes

Fraction	Size
Particulate	$> 0.45\ \mu m$
Dissolved or filterable	$< 0.45\ \mu m$ (permeable + colloidal)
Colloidal	$< 0.45\ \mu m$ and > 10 kDa
Permeable or truly dissolved	< 10 kDa

The main objectives of this study are to analyze and determine the physical speciation of metallic elements in tanning process water and overall wastewater.

4.2.2 Sampling

This study's samples were taken from a tanning plant in Mohammedia, Morocco. Figure 4.1 shows the sampling points at the factory.

Wastewater is pre-treated by primary, secondary, and tertiary settling. The bulk sample obtained is carefully homogenized. In practice, this homogenization is facilitated by the use of a large container. The final samples for analysis are then taken using the quartering method.

Sampling is carried out manually over 3 days (7 samples per day):

- A at the end of five necessary process stages in the production cycle.
- From collection basin B3 (figure), which collects wastewater from the various stages directly into the sewage network at hourly intervals. Samples were taken at 7-h intervals (T1, T2, T3, T4, T5, T6, and T7). Samples were kept cold (+ 4 °C) before analysis.

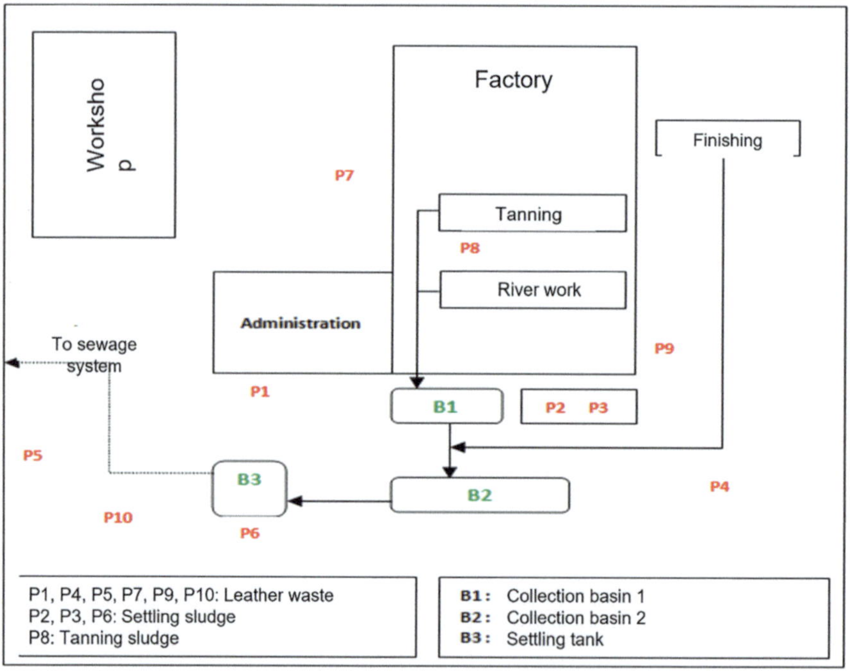

Fig. 4.1 Descriptive diagram of sampling points

4.2.3 Sample Preparation Techniques

To determine the percentage presence of heavy metals (chromium, cadmium, mercury, lead, copper, nickel, zinc, and iron) in the dissolved phases, wastewater samples were filtered using a Teflon membrane with 0.45 μm pores (47 mm diameter, Millipore™ FH type). This process separates particulate from dissolved phases before mineralization (see Table 4.1).

Trace elements in the various samples were determined by inductively coupled plasma atomic emission spectrometry (ICP-AES) after mineralization. This method was chosen for its high sensitivity, its ability to analyze several elements simultaneously, its limited matrix effects, its low memory requirements, its rapid execution, and its tolerance of high salt concentrations (Khan et al. 2022).

4.3 Results

The production unit generates wastewater of varying quality over time, with effluent characteristics differing from sample to sample. These variations can be explained by changes in the composition of the waste collected during the various stages of the process.

The analyses showed that average chromium concentrations varied significantly, ranging from 29.2 mg/L to 811.41 mg/L for samples T7 and T3, respectively. During the tanning water entertainment periods, maximum concentrations of up to 4325.8 mg/L were recorded. In contrast, as detailed in in this chapter, much lower concentrations were reported for cadmium (3–5 μg/L) and nickel (7–20 μg/L).

In this study, speciation refers to the physical fractionation of metals into three phases: particulate, colloidal, and dissolved. Speciation results for the particulate and dissolved phases of chromium, cadmium, lead, and mercury in effluent before discharge are presented in Figs. 4.2, 4.3, 4.4, and 4.5. These results are presented separately for (a) treated wastewater and (b) hourly samples taken from the pond.

Despite a limited database and a simplified definition of the dissolved fraction (< 0.45 μm), the results indicate that the particulate phase accounts for a significant proportion of heavy metals in almost all samples except for cadmium.

4.3.1 Chromium

Many health problems are linked to chromium, such as chronic ulceration and perforation of the nasal septum and allergic skin reactions (Hossini et al. 2022). The authors showed that chromium percentages in the particulate phase in the waters studied varied from 52.8% at time (T1) to 92.9% at a time (T3).

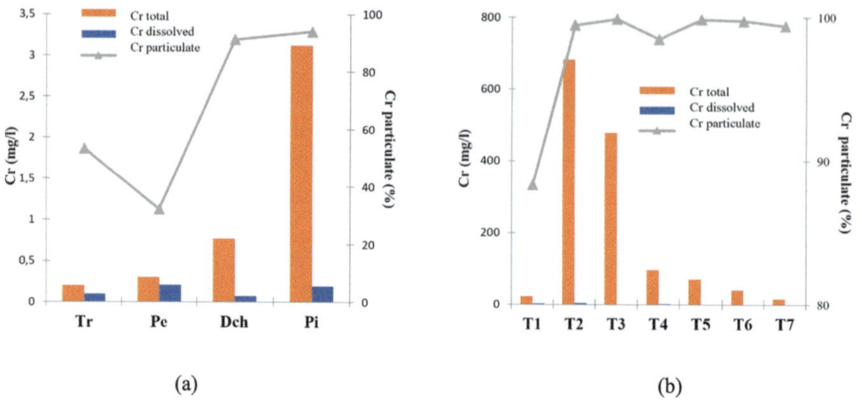

Fig. 4.2 Chromium speciation: **a** in process water, **b** in catchment water

Fig. 4.3 pH/pe dominance diagram for aqueous species

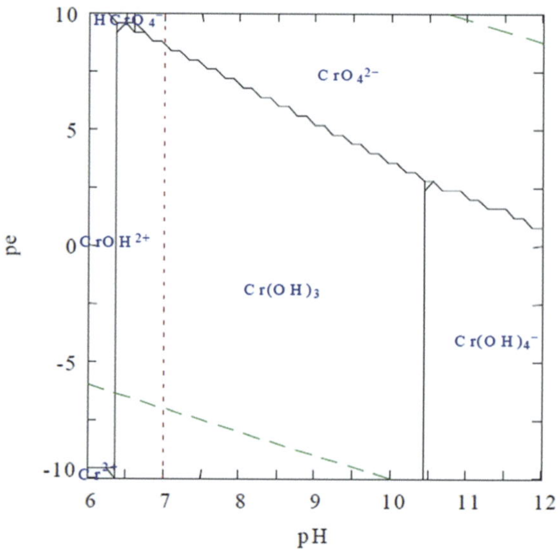

Chromium is acutely and chronically toxic to humans at high doses, with toxicity depending on chemical speciation. The chemical forms of exposure influence the associated health effects.

In the case of effluents from the quenching and dehairing stages, chromium is mainly in dissolved form (Fig. 4.2a). As for effluents from the deliming, pickling, and tanning stages, the particulate fraction of chromium varies around 4280 mg L^{-1}, with a particulate fraction exceeding 98%. Figure 4.2b shows that almost all chromium is particulate for the various samples.

Cr III salts are used effectively as tanning agents in the leather industry. Chromium is subjected to rapid sorption in the environment, so the concentration of soluble

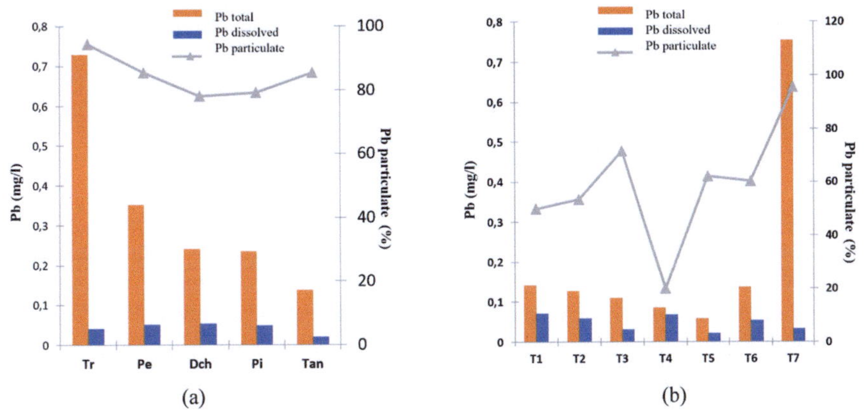

Fig. 4.4 Lead speciation: **a** in process water, **b** in catchment water

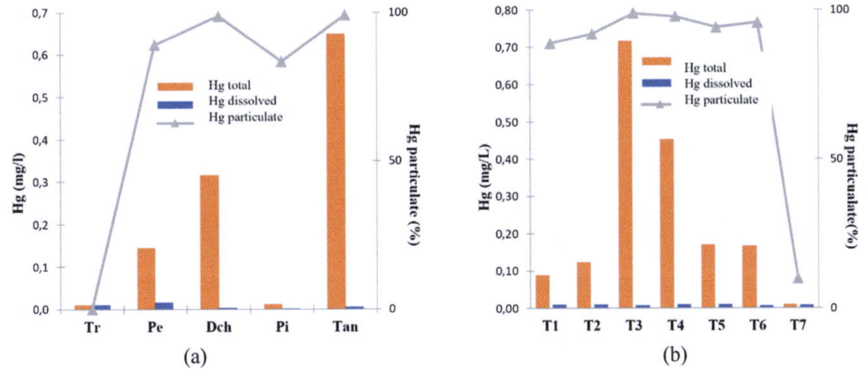

Fig. 4.5 Mercury speciation **a** in process water, **b** in collection basin water

chromium is low compared to adsorbed chromium (Prasad et al. 2021). However, it can be said that most of the chromium present at the end of the process is adsorbed by hair, skin fragments, etc. The toxicity, mobility, and bioavailability of Cr depend fundamentally on its chemical form.

The toxicity, mobility, and bioavailability of Cr depend fundamentally on its chemical form. Chromium in the environment may be present mainly as Cr^{3+} and Cr^{6+}. Cr^{6+} is highly soluble. It is around 300 times more toxic than Cr^{3+}. On the other hand, Cr^{3+} precipitates at the average pH of natural waters. Tannery wastewater mainly contains Cr^{3+}. The behavior of Cr in wastewater depends on the physico-chemical conditions of the effluent (Aniyikaiye et al. 2019).

The aqueous chemistry of chromium differs significantly according to redox conditions (Fig. 4.3). Under reducing to slightly oxidizing conditions, chromium concentration may be limited by the solubility of chromium III oxide. Under oxidizing conditions, aqueous chromium concentration is not limited by the solubility of chromium oxide III.

4.3.2 Lead

Lead ranks second to arsenic on the list of the 20 most toxic heavy metals. Its effects affect several target organs, including the bones, brain, blood, kidneys, reproductive system, cardiovascular system, and thyroid gland (Flora and Agrawal 2017; Mandal et al. 2022).

As shown in Fig. 4.4, particulate lead concentrations measured in the samples ranged from 50% at sampling point T1 to 95.62% at point T3. In process wastewater, the percentage of particulate lead systematically exceeds 80%.

In the aqueous phase, lead can be either free ions (Pb^{2+}) or complexes. It should be noted that lead solubility may increase due to the formation of complexes with organic matter.

4.3.3 Mercury

Mercury is widely recognized as one of the most toxic metals present in the environment (Wu et al. 2024). When it enters the food chain, it gradually accumulates as compounds in humans and animals. Mercury also causes genetic defects, such as chromosome breakage and disturbances in cell division, leading to abnormal chromosome distribution. It can cause impaired lung and kidney function, chest pain, and dyspnea (Zafar et al. 2024; Wu et al. 2024). Indeed, the results show that almost 99% of mercury is present in particulate form, both in samples taken from the collection basin (Fig. 4.5b) and in process water (Fig. 4.5a). This may suggest that this element is removed by simply settling in the collection tank.

Mercury is generally present in aquatic environments in three oxidation states: elemental (Hg^0), mercurous (Hg^+), and mercuric (Hg^{2+}) (Pavithra et al. 2023). In these environments, it rapidly binds to suspended matter, which explains why it represents only a small proportion of total mercury in the dissolved phase (Nittler et al. 2018; Pavithra et al. 2023).

The different chemical forms of mercury depend mainly on the presence or absence of organic matter, the redox potential, and the pH of the medium. In the presence of organic matter, mercury speciation becomes complex due to the diversity and unfamiliarity of organic ligands in solution. Mercury can form organic compounds such as R-Hg-X or R_2-Hg, where R represents a carbon group and X an anion (Cl^-, NO_3^-, etc.). In oxic environments, organic compounds generally dominate.

In a reducing environment, heterotrophic bacteria and phytoplankton are essential in reducing mercuric mercury (Hg^{2+}) to elemental mercury (Hg^0). In suspended matter, a large proportion of mercury settles there, whether on organic or inorganic particles. This suspended matter is crucial in mercury transport (Jędruch et al. 2017; Gonzalez-Raymat et al. 2017). In addition, according to Gonzalez-Raymat et al. (2017), trace metals can interact with suspended matter by two main mechanisms:

- As precipitates
- In adsorbed form.

4.3.4 *Cadmium*

Cadmium and some of its compounds are known carcinogens and can cause damage to various types of cells in the human body (Genchi et al. 2020b). Cadmium's main target organs are the kidneys and liver. This metal is of growing concern due to its presence in waste products, particularly sewage sludge, often disposed of on land. Human activities strongly influence its concentration in surface soils and sediments (Kubier et al. 2019). The results show that cadmium is found mainly in the particulate phase. However, its percentage remains below 60% in all wastewater samples, whether process water or water from the collection basin (Figs. 4.6a, b and 4.7).

Half the cadmium is transported in the dissolved phase. Generally, in solution, cadmium is mainly in the form of the divalent cation Cd^{2+} (Viala et al. 2021). Cadmium's behavior is similar to that of zinc. In freshwater at pH < 8.2 and in oxidizing environments, it is predominantly in Cd^{2+} form, but high chloride concentrations lead to $CdCl_2$ or $CdCl^+$ complexes forming. At pH values above 8, cadmium precipitates with carbonates.

Dissolved forms of this element in aqueous media are free species (Cd^{2+}) and formed by associations (complexation) of cadmium with inorganic or mineral compounds (ligands or complexing agents).

Tables 4.2 and 4.3 show metal speciation results: Cu, Fe, Ni, and Zn.

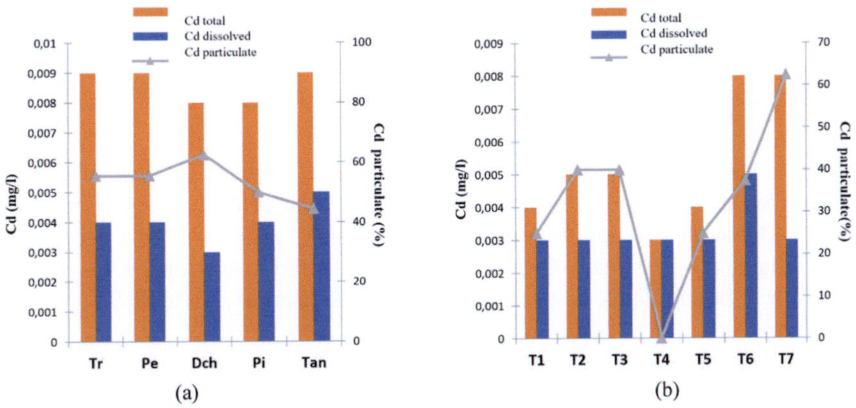

Fig. 4.6 Cadmium speciation **a** in process water, **b** in catchment water

Fig. 4.7 Predominance diagram pe/pH of aqueous cadmium species

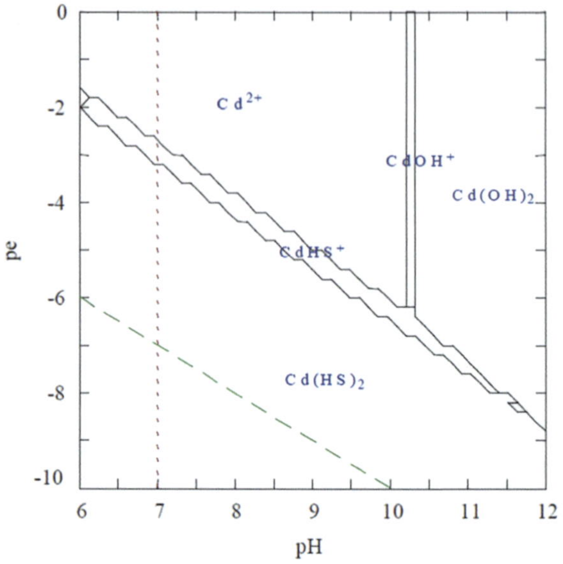

Table 4.2 Metal speciation in wastewater from the catchment area

	Cu		Fe		Ni		Zn	
	Dissolved	Particulate	Dissolved	Particulate	Dissolved	Particulate	Dissolved	Particulate
T1	47.2	52.8	46.7	53.3	64.3	35.7	80.8	19.2
T2	27.3	72.7	33.6	66.4	10.0	90.0	87.7	12.3
T3	7.1	92.9	13.2	86.8	0.0	100.0	54.6	45.4
T4	15.5	84.5	31.3	68.7	0.0	100.0	37.0	63.0
T5	9.3	90.7	58.1	41.9	7.9	92.1	34.7	65.3
T6	12.1	87.9	3.8	96.2	11.1	88.9	29.4	70.6
T7	12.2	87.8	29.2	70.8	14.3	85.7	1.1	98.9

4.3.5 *Zinc*

Zinc is a common metal in the human environment, although its toxic effects on humans are still poorly documented (Schoofs et al. 2024). However, high zinc levels can lead to serious complications such as pancreatitis, anemia, muscle pain, acute renal failure, and even death (Chasapis et al. 2020). In the waters studied, the proportion of zinc in particulate form ranged from 12.3% (sample T2) to 98.9% (sample T7) (Table 4.3).

Zn is only present in the divalent oxidized form Zn(II) in aqueous media. Due to its basic nature, Zn^{2+} hydrolysis is weak or non-existent under acidic or near-neutral conditions (Khalid et al. 2023).

Table 4.3 Metal speciation in process water

	Cu		Fe		Ni		Zn	
	Dissolved	Particulate	Dissolved	Particulate	Dissolved	Particulate	Dissolved	Particulate
Hardening	11.67	88.33	8.98	91.02	14.29	85.71	11.22	88.78
Hair removal/peeling	11.76	88.24	8.98	91.02	20.69	79.31	6.48	93.52
Deliming	7.46	92.54	57.34	42.66	50.00	50.00	0.14	99.86
Pickling	14.75	85.25	36.96	63.04	40.00	60.00	17.04	82.96
Tanning	37.50	62.50	40.85	59.15	21.88	78.13	24.40	75.60

4.3.6 Iron

Iron toxicity results from its rapid absorption by the body. The main sources of iron are drinking water, iron pipes, cooking utensils, and dietary supplements. Its target organs are the liver, cardiovascular system, and kidneys (Kobayashi et al. 2018; van Swelm et al. 2020). In this study, particulate iron proportions ranged from 41.9% (T5) to 96.2% (T6) (Table 4.3).

4.3.7 Nickel

Nickel is a carcinogenic metal, and any exposure can lead to adverse effects such as reduced body weight, skin irritation, and impacts on the heart and liver (Genchi et al. 2020a). In this study, the proportions of particulate nickel ranged from 35.7 to 100%, with the highest concentrations found in samples T3 and T4 and the lowest in T1 (Table 4.3).

Under oxidizing, acidic, or neutral conditions, nickel concentration is controlled by the solubility of the solid carbonate. $Ni(OH)_2$ hydroxide can precipitate at pH above 8, increasing its content in an alkaline medium.

4.3.8 Copper

Copper is an essential element for mammals, as it acts as an electron donor or acceptor in various metalloenzymes. However, excessive exposure can cause adverse health effects (Karim 2018; Taylor et al. 2020). In this study, particulate copper proportions ranged from 52.8% (T1) to 92.9% (T3) (Table 4.3).

In oxidizing environments, copper availability depends on the solubility of carbonate or hydroxyl phases, depending on the pH of the groundwater. Cu^0 or Cu_2S (weakly soluble) are readily formed in slightly reducing environments with sulfides.

4.4 Discussion

Environmental pollution by toxic heavy metals from industrial waste is a major problem, particularly in developing countries. These highly persistent metals profoundly alter terrestrial ecosystems' diversity, structure, and functions. The chemical speciation of elements in the environment is essential to understanding their behavior and effects. Indeed, two chemical forms of the same metal can present very different toxicity levels. Thus, measuring a metal's total concentration is insufficient to accurately assess its toxic impact (Egorova and Ananikov 2017; Ahmad et al. 2021).

Heavy metals like chromium are highly mobile in soil and can contaminate groundwater (Economou-Eliopoulos and Megremi 2021). Chromium (VI), in particular, migrates easily and deeply in soil (Wang et al. 2020). In this context, chemical speciation provides a better understanding of the environmental behavior of these metals.

In this study, the particulate fraction is dominant for all heavy metals except Cd. The dissolved metal fraction is defined as the concentration in solution (as a percentage) of the total concentration. Water samples taken from B3 and filtered through a 0.45 μm pore size membrane were analyzed.

As shown in Figs. 4.1 and 4.4, the Cr and Hg fractions are bound to particles, meaning that free metal ions make up less than 10% of the total metal concentration. On the other hand, for Pb, the dissolved and particulate fractions are approximately the same. Cd showed a higher fraction of dissolved species than the other metals, particularly at time T4.

Wastewater from tanning processes is a significant source of environmental pollutants, posing numerous risks to human health. Pressure from environmental protection agencies is becoming frequent, forcing manufacturers to close their tanneries in developed countries and developing ones.

Chromium III salts are the most widely used chemicals in tanning processes, with 60–70% of total chromium salts reacting with hides. In other words, around 30–40% of the chromium is discharged with solid waste and liquid rejects (mainly spent tanning solutions).

Conventional methods used to separate dissolved and particulate phases with a filter pore size of 0.45 μm, and colloids (< 0.45 μm and > 10 kDa) are included in the dissolved fraction itself. This lack of separation between the colloidal fraction and the permeable fraction may have certain consequences in studies on the fate and behavior of contaminants in the aquatic environment. Thus, we recognize the importance of the distinction between permeable and colloidal phases on the fate of biogeochemical trace metals in the aquatic environment (Cuss et al. 2020; Schindler et al. 2021).

The size distribution of trace metals is, therefore, a complementary characteristic of their chemical speciation and is essential for understanding their environmental fate and impact on aquatic organisms (Tutu 2017; Bjerregaard et al. 2022; Abdou and Tercier-Waeber 2022).

The particulate phase plays a significant role in the biogeochemical cycling of metals in aquatic systems. Through suspended solids, the particulate phase can transport and store metals. Depending on environmental conditions, metals can be easily transferred from the particulate phase to the dissolved phase. Studying metals in the particulate phase is also valuable and necessary.

The role of suspended solids (SS) in transporting metals in watercourses is generally much greater than that of the dissolved phase (Zhang et al. 2018).

At acid pH, most metal species are in the form of positively charged free ions, and particle surfaces are also positively charged, producing an electrostatic repulsion phenomenon between metal ions and particles. This prevents the adsorption of metals onto TSS. However, several authors believe that organic acids can promote the adsorption of metals onto particles under these acidic conditions (Peng et al. 2017; Amde et al. 2017).

Metal mobility generally takes the form of a transition from colloidal and particulate to dissolved form. pH, oxygen, salinity, hardness, and organic carbon content are the physico-chemical parameters that most influence this type of exchange in water. For example, when pH in water decreases, the concentration of metals in ionic form increases due to competition with (H_3O^+) ions for adsorption sites. As a result, metal mobility generally increases in the medium as pH decreases. Consequently, in watercourses affected by acid drainage, metals are preferentially found in solution (Neculita and Rosa 2019; Sandlin et al. 2021).

Dissolved metal in its ionic, or free, form is generally present in low proportions, as it forms complexes with numerous inorganic and organic ligands or adsorbs to submerged surfaces. Dissolved metals are in contact with sediments or suspended matter, which act as complex exchange surfaces.

Generally speaking, the behavior of a metal in an aquatic ecosystem depends on the physico-chemical conditions of the environment (Pinheiro et al. 2021; Miranda et al. 2021, 2022). For example, dissolved metals can precipitate when pH or redox conditions change. As the watercourse grows, the concentration of dissolved organic carbon increases. This allows the formation of stable complexes between metals and organic matter. Moreover, bacterial activity can cause conditions to become anoxic and reducing. Metals then tend to precipitate as metal sulfides. The interaction between living organisms and the metal depends, among other things, on the form of the metal in the environment. By analogy, the fixation of metal by organisms is also called complexation with the biotic ligand (Santore et al. 2018; Liang et al. 2021).

The most important chemical processes affecting the fate and mobility of trace metals in sediments are transfers from the solid phase to the liquid phase and vice versa (Brusseau and Chorover 2019).

In aquatic systems, trace metals (TMEs) are distributed between the dissolved and particulate phases. In the dissolved phase, TMEs may be free, i.e., not complexed with any ligand (apart from solvatation). However, a significant proportion is in the form of complexes with inorganic and organic ligands. Finally, many ETMs will be in the particulate phase, either adsorbed on particles with a strong affinity for metals or included in the crystalline structure of certain minerals in suspension. The metal distribution between these free or complex dissolved and particulate species constitutes its speciation. These different forms, which evolve according to the physico-chemical conditions of the environment, condition not only the residence time in the various compartments of the medium but also the bioavailability and even the toxicity of the metal to living organisms (Shahid et al. 2017; Peana et al. 2021).

In the receiving environment, physical, chemical, and biological processes interact to transform contaminants into more or less bioavailable forms for ecosystem organisms. Therefore, total concentrations in the environment only partially reflect the risk associated with a contaminant since bioavailability is very closely linked to the characteristics of the environment.

To assess the impact of a contaminant on the environment, we need to estimate the bioavailable concentrations and the environmental factors that affect them. Among the factors influencing metal speciation, it is easy to guess that organic matter plays a key role, as it contains numerous functional groups capable of complexing metal cations. Among the various physico-chemical parameters likely to influence metal bioavailability, the concentration and nature of dissolved organic matter are undoubtedly two fundamental parameters, as we shall see below.

Dissolved organic matter (DOM) is operationally distinguished from particulate organic matter (POM) by 0.45 μm membrane filtration. In industrial wastewater, DOM comprises a highly complex mixture, the majority of molecules of which are not structurally identified (Shi et al. 2021). The result is a wide variety of properties that are difficult to predict.

Tanneries are significant sources of chromium contamination of surface waters worldwide, particularly artisanal tanneries (Appiah-Brempong et al. 2022; Kapanga et al. 2024; Faye and Sibali 2025). Indeed, chromium speciation is becoming increasingly important, as Cr^{3+} can undergo transformation into Cr^{6+}, a more toxic element than the + 3 oxidation state. The environmental and ecotoxicological fate of chromium and the risks associated with emissions are poorly understood due to the complex chemistry of this element in water. Concentrations and proportions of trivalent and hexavalent chromium are not always easy to identify. Chromium in the aquatic environment can vary according to local conditions. The identification and quantification of Cr species is therefore essential in ecotoxicology from the point of view of assessing the impact on public health.

4.5 Conclusion

In this research, total concentrations of heavy metals, including total chromium, were successfully measured in tannery wastewater samples using the ICP-AES method. The high levels of metals detected reveal significant pollution of the water analyzed. This contamination exerts considerable negative impacts on the environment, which can affect human health throughout the food chain.

Speciation analyses show that most metals are mainly present in the particulate phase. The proportions of Cr, Cu, Pb, Hg, and Ni in the dissolved phase vary, respectively, between Cr: 0.05–8%; Cu: 7.1–47.2%; Pb: 4.4–50%; Hg: 2–11%, and Ni: 0–14%. Cadmium, on the other hand, has a much higher dissolved fraction, reaching 60–100%. The distribution of metals in particulate form follows a decreasing order: Cr > Ni > Hg > Cu > Fe > Pb > Zn > Cd.

The variations observed in the physical speciation of trace metals highlight the need to differentiate the truly dissolved phase from the colloidal phase to assess environmental discharges better. However, conventional laboratory methods cannot effectively separate these two fractions. This lack of distinction can lead to uncertainties in analyzing the behavior and fate of contaminants in aquatic environments. Particular attention needs to be paid to this colloidal fraction to improve the ecotoxicological assessment of discharges into these ecosystems.

These results are part of a project to monitor wastewater quality from tanneries in Mohammedia. Work continues to assess the environmental impact of tannery discharges on aquatic ecosystems.

References

Abdou M, Tercier-Waeber M-L (2022) New insights into trace metal speciation and interaction with phytoplankton in estuarine coastal waters. Mar Pollut Bull 181:113845. https://doi.org/10.1016/j.marpolbul.2022.113845

Adams W, Blust R, Dwyer R et al (2020) Bioavailability assessment of metals in freshwater environments: a historical review. Environ Toxicol Chem 39:48–59. https://doi.org/10.1002/etc.4558

Ahmad W, Alharthy RD, Zubair M et al (2021) Toxic and heavy metals contamination assessment in soil and water to evaluate human health risk. Sci Rep 11:17006. https://doi.org/10.1038/s41598-021-94616-4

Ali H, Khan E, Ilahi I (2019) Environmental chemistry and ecotoxicology of hazardous heavy metals: environmental persistence, toxicity, and bioaccumulation. J Chem 2019:6730305. https://doi.org/10.1155/2019/6730305

Amde M, Liu J, Tan Z-Q, Bekana D (2017) Transformation and bioavailability of metal oxide nanoparticles in aquatic and terrestrial environments. A review. Environ Pollut 230:250–267. https://doi.org/10.1016/j.envpol.2017.06.064

Aniyikaiye TE, Oluseyi T, Odiyo JO, Edokpayi JN (2019) Physico-chemical analysis of wastewater discharge from selected paint industries in Lagos, Nigeria. Int J Environ Res Public Health 16:1235. https://doi.org/10.3390/ijerph16071235

Appiah-Brempong M, Essandoh HMK, Asiedu NY et al (2022) Artisanal tannery wastewater: quantity and characteristics. Heliyon 8. https://doi.org/10.1016/j.heliyon.2021.e08680

Arti, Mehra R (2023) Analysis of heavy metals and toxicity level in the tannery effluent and the environs. Environ Monit Assess 195:554. https://doi.org/10.1007/s10661-023-11154-4

Bjerregaard P, Andersen Christian BI, Andersen O (2022) Ecotoxicology of metals—sources, transport, and effects on the ecosystem, chap 26. In: Nordberg GF, Costa M (eds) Handbook on the toxicology of metals, 5th edn. Academic Press, pp 593–627

Brusseau ML, Chorover J (2019) Chemical processes affecting contaminant transport and fate, chap 8. In: Brusseau ML, Pepper IL, Gerba CP (eds) Environmental and pollution science, 3rd edn. Academic Press, pp 113–130

Chasapis CT, Ntoupa P-SA, Spiliopoulou CA, Stefanidou ME (2020) Recent aspects of the effects of zinc on human health. Arch Toxicol 94:1443–1460. https://doi.org/10.1007/s00204-020-02702-9

Cuss CW, Glover CN, Javed MB et al (2020) Geochemical and biological controls on the ecological relevance of total, dissolved, and colloidal forms of trace elements in large boreal rivers: review and case studies. Environ Rev 28:138–163. https://doi.org/10.1139/er-2019-0014

Economou-Eliopoulos M, Megremi I (2021) Contamination of the soil–groundwater–crop system: environmental risk and opportunities. Minerals 11:775. https://doi.org/10.3390/min11070775

Egorova KS, Ananikov VP (2017) Toxicity of metal compounds: knowledge and myths. Organometallics 36:4071–4090. https://doi.org/10.1021/acs.organomet.7b00605

Faye G, Sibali LL (2025) A review on characterization, treatment, and impact of tannery wastewater in Ethiopia. Environ Qual Manage 34:e70015. https://doi.org/10.1002/tqem.70015

Flora SJS, Agrawal S (2017) Arsenic, cadmium, and lead, chap 31. In: Gupta RC (ed) Reproductive and developmental toxicology, 2nd edn. Academic Press, pp 537–566
Genchi G, Carocci A, Lauria G et al (2020a) Nickel: human health and environmental toxicology. Int J Environ Res Public Health 17:679. https://doi.org/10.3390/ijerph17030679
Genchi G, Sinicropi MS, Lauria G et al (2020b) The effects of cadmium toxicity. Int J Environ Res Public Health 17:3782. https://doi.org/10.3390/ijerph17113782
Gonzalez-Raymat H, Liu G, Liriano C et al (2017) Elemental mercury: its unique properties affect its behavior and fate in the environment. Environ Pollut 229:69–86. https://doi.org/10.1016/j.envpol.2017.04.101
Hossini H, Shafie B, Niri AD et al (2022) A comprehensive review on human health effects of chromium: insights on induced toxicity. Environ Sci Pollut Res 29:70686–70705. https://doi.org/10.1007/s11356-022-22705-6
Jędruch A, Kwasigroch U, Bełdowska M, Kuliński K (2017) Mercury in suspended matter of the Gulf of Gdańsk: origin, distribution and transport at the land–sea interface. Mar Pollut Bull 118:354–367. https://doi.org/10.1016/j.marpolbul.2017.03.019
Jensen LT, Wyatt NJ, Landing WM, Fitzsimmons JN (2020) Assessment of the stability, sorption, and exchangeability of marine dissolved and colloidal metals. Mar Chem 220:103754. https://doi.org/10.1016/j.marchem.2020.103754
Kapanga PM, Nyakairu GWA, Nkanga CI et al (2024) A review of dye effluents polluting African surface water: sources, impacts, physicochemical properties, and treatment methods. Discov Water 4:85. https://doi.org/10.1007/s43832-024-00129-2
Karim N (2018) Copper and human health—a review. J Bahria Univ Med Dent Coll 8:117–122. https://doi.org/10.51985/JBUMDC2018046
Khalid M, Park Y, Karri RR, Walvekar R (2023) Advanced oxidation processes for micropollutant remediation. CRC Press
Khan SR, Sharma B, Chawla PA, Bhatia R (2022) Inductively coupled plasma optical emission spectrometry (ICP-OES): a powerful analytical technique for elemental analysis. Food Anal Methods 15:666–688. https://doi.org/10.1007/s12161-021-02148-4
Kobayashi M, Suhara T, Baba Y et al (2018) Pathological roles of iron in cardiovascular disease. Curr Drug Targets 19:1068–1076. https://doi.org/10.2174/1389450119666180605112235
Kubier A, Wilkin RT, Pichler T (2019) Cadmium in soils and groundwater: a review. Appl Geochem 108:104388. https://doi.org/10.1016/j.apgeochem.2019.104388
Liang W-Q, Xie M, Tan Q-G (2021) Making the biotic ligand model kinetic, easier to develop, and more flexible for deriving water quality criteria. Water Res 188:116548. https://doi.org/10.1016/j.watres.2020.116548
Mandal GC, Mandal A, Chakraborty A (2022) The toxic effect of lead on human health: a review. Hum Biol Public Health 3. https://doi.org/10.52905/hbph2022.3.45
Miranda LS, Ayoko GA, Egodawatta P et al (2021) Physico-chemical properties of sediments governing the bioavailability of heavy metals in urban waterways. Sci Total Environ 763:142984. https://doi.org/10.1016/j.scitotenv.2020.142984
Miranda LS, Ayoko GA, Egodawatta P, Goonetilleke A (2022) Adsorption-desorption behavior of heavy metals in aquatic environments: influence of sediment, water and metal ionic properties. J Hazard Mater 421:126743. https://doi.org/10.1016/j.jhazmat.2021.126743
Mkandawire M (2019) Geochemical prediction of metal dispersion in surface and groundwater systems. In: Nano and bio-based technologies for wastewater treatment. Wiley, pp 1–40
Neculita CM, Rosa E (2019) A review of the implications and challenges of manganese removal from mine drainage. Chemosphere 214:491–510. https://doi.org/10.1016/j.chemosphere.2018.09.106
Nittler LR, Chabot NL, Grove TL, Peplowski PN (2018) The chemical composition of mercury, pp 30–51
Pavithra KG, SundarRajan P, Kumar PS, Rangasamy G (2023) Mercury sources, contaminations, mercury cycle, detection and treatment techniques: a review. Chemosphere 312:137314. https://doi.org/10.1016/j.chemosphere.2022.137314

Peana M, Pelucelli A, Medici S et al (2021) Metal toxicity and speciation: a review. Curr Med Chem 28:7190–7208. https://doi.org/10.2174/0929867328666210324161205

Peng H, Gao P, Chu G et al (2017) Enhanced adsorption of Cu(II) and Cd(II) by phosphoric acid-modified biochars. Environ Pollut 229:846–853. https://doi.org/10.1016/j.envpol.2017.07.004

Pinheiro JPS, Windsor FM, Wilson RW, Tyler CR (2021) Global variation in freshwater physico-chemistry and its influence on chemical toxicity in aquatic wildlife. Biol Rev 96:1528–1546. https://doi.org/10.1111/brv.12711

Prasad S, Yadav KK, Kumar S et al (2021) Chromium contamination and effect on environmental health and its remediation: a sustainable approaches. J Environ Manage 285:112174. https://doi.org/10.1016/j.jenvman.2021.112174

Saad AM, Youness F (2024) The Moroccan leather sector between modernity and authenticity: what problems? What solutions? Discov Sustain 5:306. https://doi.org/10.1007/s43621-024-00496-9

Sall ML, Diaw AKD, Gningue-Sall D et al (2020) Toxic heavy metals: impact on the environment and human health, and treatment with conducting organic polymers, a review. Environ Sci Pollut Res 27:29927–29942. https://doi.org/10.1007/s11356-020-09354-3

Sandlin W, Langman J, Moberly J (2021) A review of acid rock drainage, seasonal flux of discharge and metal concentrations, and passive treatment system limitations. Int J Min Reclam Environ 35:34–47. https://doi.org/10.1080/17480930.2020.1728035

Santore RC, Ryan AC, Kroglund F et al (2018) Development and application of a biotic ligand model for predicting the chronic toxicity of dissolved and precipitated aluminum to aquatic organisms. Environ Toxicol Chem 37:70–79. https://doi.org/10.1002/etc.4020

Santschi PH (2018) Marine colloids, agents of the self-cleansing capacity of aquatic systems: historical perspective and new discoveries. Mar Chem 207:124–135. https://doi.org/10.1016/j.marchem.2018.11.003

Saxena G, Chandra R, Bharagava RN (2017) Environmental pollution, toxicity profile and treatment approaches for tannery wastewater and its chemical pollutants. In: de Voogt P (ed) Reviews of environmental contamination and toxicology, vol 240. Springer International Publishing, Cham, pp 31–69

Schindler RJ, Comber SDW, Manning AJ (2021) Metal pollutant pathways in cohesive coastal catchments: influence of flocculation and biopolymers on partitioning and flux. Sci Total Environ 795:148800. https://doi.org/10.1016/j.scitotenv.2021.148800

Schoofs H, Schmit J, Rink L (2024) Zinc toxicity: understanding the limits. Molecules 29:3130. https://doi.org/10.3390/molecules29133130

Shahid M, Shamshad S, Rafiq M et al (2017) Chromium speciation, bioavailability, uptake, toxicity and detoxification in soil-plant system: a review. Chemosphere 178:513–533. https://doi.org/10.1016/j.chemosphere.2017.03.074

Shi W, Zhuang W-E, Hur J, Yang L (2021) Monitoring dissolved organic matter in wastewater and drinking water treatments using spectroscopic analysis and ultra-high resolution mass spectrometry. Water Res 188:116406. https://doi.org/10.1016/j.watres.2020.116406

Siddiqui S, Brander SM (2024) Aquatic ecotoxicology: understanding pollutants, aquatic organisms, and their environments. Springer Nature

Smith DR, Maroney MJ, Monica N, Tyson JF (2022) General chemistry of metals, sampling, analytical methods, and speciation. In: Nordberg GF, Costa M (eds) Handbook on the toxicology of metals, 5th edn. Academic Press, pp 15–54

Taylor AA, Tsuji JS, Garry MR et al (2020) Critical review of exposure and effects: implications for setting regulatory health criteria for ingested copper. Environ Manage 65:131–159. https://doi.org/10.1007/s00267-019-01234-y

Tutu H (2017) Water quality. BoD—Books on Demand

van Swelm RPL, Wetzels JFM, Swinkels DW (2020) The multifaceted role of iron in renal health and disease. Nat Rev Nephrol 16:77–98. https://doi.org/10.1038/s41581-019-0197-5

Vardhan KH, Kumar PS, Panda RC (2019) A review on heavy metal pollution, toxicity and remedial measures: current trends and future perspectives. J Mol Liq 290:111197. https://doi.org/10.1016/j.molliq.2019.111197

Viala Y, Sappin-Didier V, Bussière S et al (2021) Simple models efficiently predict free cadmium Cd^{2+} in the solutions of low-contaminated agricultural soils. Sci Total Environ 778:146428. https://doi.org/10.1016/j.scitotenv.2021.146428

Wang X, Li L, Yan X et al (2020) Processes of chromium (VI) migration and transformation in chromate production site: a case study from the middle of China. Chemosphere 257:127282. https://doi.org/10.1016/j.chemosphere.2020.127282

Wu Y-S, Osman AI, Hosny M et al (2024) The toxicity of mercury and its chemical compounds: molecular mechanisms and environmental and human health implications: a comprehensive review. ACS Omega 9:5100–5126. https://doi.org/10.1021/acsomega.3c07047

Zafar A, Javed S, Akram N, Naqvi SAR (2024) Health risks of mercury. In: Kumar N (ed) Mercury toxicity mitigation: sustainable nexus approach. Springer Nature Switzerland, Cham, pp 67–92

Zeng J, Tabelin CB, Gao W et al (2023) Heterogeneous distributions of heavy metals in the soil-groundwater system empowers the knowledge of the pollution migration at a smelting site. Chem Eng J 454:140307. https://doi.org/10.1016/j.cej.2022.140307

Zhang J, Zhou F, Chen C et al (2018) Spatial distribution and correlation characteristics of heavy metals in the seawater, suspended particulate matter and sediments in Zhanjiang Bay, China. PLOS ONE 13:e0201414. https://doi.org/10.1371/journal.pone.0201414

Zhao J, Wu Q, Tang Y et al (2022) Tannery wastewater treatment: conventional and promising processes, an updated 20-year review. J Leather Sci Eng 4:10. https://doi.org/10.1186/s42825-022-00082-7

Part II
Advanced Approaches in Tannery Wastewater Treatment

With growing awareness of the environmental impacts of tannery effluents, the industry is turning to more advanced methods of wastewater treatment. Part 2 of this book introduces innovative technologies that can help mitigate the environmental damage caused by tannery effluents while also offering practical solutions to the industry's long-standing pollution problems.

Chapter 5 provides an overview of tannery wastewater treatment, covering both conventional methods and emerging technologies. It examines the principles behind chemical, biological, and physical treatment processes, discussing their effectiveness and limitations in the context of tannery operations. Chapter 6 highlights adsorption strategies for environmental remediation, using a case study of tannery operations in Morocco to showcase how adsorption methods can reduce harmful pollutants in wastewater. In Chapter 7, electrochemical treatment techniques are explored as a sustainable approach to tannery effluent management. This chapter discusses the potential benefits of electrochemical treatment for breaking down toxic chemicals and offering an eco-friendlier alternative to traditional methods.

Chapter 5
Treatment of Tannery Wastewater

Wafa Hassen, Bilel Hassen, and Abdennaceur Hassen

Abstract The leather tanning industry is among the oldest in the world, having evolved from ancient times to the present day. This sector is notable for its ability to convert animal by-products, particularly from cattle, goats, and sheep, into leather through tanning processes. These processes, which date back to ancient history, have enabled humans to wear leather, primarily utilizing two methods: vegetable tanning, which employs natural plant extracts, and chemical tanning, often based on chromium. The effluents produced by tanneries are complex and rich in organic and mineral substances, including hexavalent chromium, a highly toxic compound for the environment. These discharges, which exhibit varying bio-physico-chemical characteristics over time, are often challenging to biologically degrade and, if not properly treated, can lead to significant pollution of waterways and surrounding ecosystems. The odor nuisances generated by tanning industries pose a significant barrier to their establishment and expansion in urban areas. Additionally, these industries are substantial consumers of water, with requirements ranging from less than 25 to over 80 L/kg of rawhide. To mitigate this pollution and prevent environmental impacts, it is crucial to enhance production technologies to optimize ecological performance, which includes improving the efficiency of chemical use, reducing water and energy consumption, and recovering or recycling generated waste. The optimization of water usage can be achieved through various methods, including strict monitoring of wastewater volumes, batch washing instead of using running water, minor modifications to existing equipment, the adoption of low-drift techniques with upgraded machinery, the reuse of wastewater in less critical processes, and the recycling of specific process liquors. Traditional soaking and liming processes account for over

W. Hassen · A. Hassen (✉)
Laboratory of Treatment and Wastewater Valorization, Center of Research and Water Technologies (CERTE), Soliman, Tunisia
e-mail: abdohas@gmail.com

W. Hassen
e-mail: hassen.wafa@gmail.com

B. Hassen
Group of Aquatic Bacteriology and Biotechnology (GAB2), LR16INSTM05, The National Institute of Science and Technology of the Sea, University of Carthage, Salammbô, Tunisia
e-mail: bilel.hassen@fst.utm.tnv

M. A. Aboulhassan and S. Souabi (eds.), *Sustainable Practices in the Tannery Industry*,
https://doi.org/10.1007/978-3-031-92037-0_5

50% of the biochemical oxygen demand (BOD) and chemical oxygen demand (COD) in tannery effluents. Several alternatives can be explored to replace sulfide, recycle lime/sulfide liquors, and implement techniques to reduce hair consumption. To mitigate chrome pollution, it is feasible to increase the levels of chrome fixed in the tanning bath while minimizing losses in subsequent stages. Additional strategies include recycling used chrome liquors, which also helps lower effluent salinity, and treating chrome-containing liquors with an alkali to precipitate chrome as hydroxide for potential reuse. Current regulations about the treatment and quality of tannery discharges are particularly stringent.

Keywords Leather · Tanning processes and industries · Tannery effluents · Hexavalent chromium · Odor nuisance · Regulations

5.1 Introduction: Leather Tanning

Tannery effluents are complex discharges that are heavily laden with organic and mineral substances, particularly chromium. Their characteristics can vary over time and are often difficult to biodegrade. The tanning industry has a long-standing history, utilizing animal by-products, especially from cattle, goats, and sheep sourced from slaughterhouses, to produce leather through a well-established process known as tanning. This ancient technique has been recognized since antiquity as a means for humans to create leather garments. To facilitate tanning, tanneries employ many chemical agents, resulting in significant volumes of wastewater and solid waste. Approximately, 35–40 L of water are consumed for every kilogram of treated hide, making this industry one of the most water-intensive. The leather tanning sector can range from small-scale artisanal operations to large-scale industrial enterprises. The tanning process involves various operations, including the use of tannin, a natural product derived from plants, which chemically neutralizes the biological properties of the hides, enhancing their durability, flexibility, and resistance to decay.

This process primarily occurs in facilities known as tanneries. Tanning involves various chemical substances, including sodium hydroxide, sodium hypochlorite, potassium dichromate, lime, chlorides, sulfuric acid, formic acid, surfactants, sodium sulfide, and sodium and ammonium salts, among others. These often-oxidizing agents ultimately contribute to wastewater with a significant increase in pollutant load. The tanning process is categorized into two types: vegetable and mineral, the latter primarily using chromium.

Vegetable tanning, as previously mentioned, utilizes tannins, which are natural organic compounds derived from plant sources such as sap, leaves, and bark, available in various concentrations. Tannins are recognized for their anti-parasitic properties, providing natural protection against pests. Each type of tannin possesses unique and complex chemical characteristics, resulting in different effects on leather, which influences the choice of specific plant species based on the desired outcomes. Depending on the intended results regarding flexibility, color, or other attributes, vegetable tanning primarily employs extracts from fruits, woods, leaves, barks, and roots, including sources like acorns, myrobalan, rhubarb, chestnut, quebracho, oak, sumac, and canaigre, among others.

The artisanal vegetable tanning method consists of several stages. The initial phases, called river work, focus on preparing the hides that arrive at tanneries, whether they are fresh, salted, or dried. Initially, these hides are submerged in water, often directly in a river, to soften them, a process known as reverdissage. They are then placed on frames to remove hair, followed by fleshing, where workers use large, sharp knives on zinc frames to eliminate residual flesh and fat. Finally, during the liming process, the hides are immersed in vats containing lime milk for three to four weeks, after which the lime is removed through maceration for about a week. Throughout the skin treatment process, decomposing organic waste is directly discharged into the watercourse. Fleshing occurs under shelters with sloped, paved floors, facilitating the drainage of residue-laden water into the river. The liming vats produce a "thick liquid" that flows into the stream, and repeated immersions in tannin baths result in effluents that stain the river a reddish hue.

The red-tinted waters are consistently linked to pollution and cause various nuisances, particularly olfactory, by saturating the air with both harmful and repugnant emissions resulting from the decomposition of organic materials and the chemicals used in tanning. Furthermore, allowing animals to drink from these contaminated waters can have health implications for livestock, nearby pastures, and humans, posing risks of epizootic infections such as anthrax, which is caused by the bacterium *Bacillus anthracis*. As a precaution against this disease, anthrax vaccine is often administered to tannery workers.

Tanning is primarily divided into three main stages: soaking, re-tanning, and pit processing. During the soaking phase, hides are stacked in vats filled with a tanning solution derived from the leaching of plant barks, with oak bark being the most commonly used due to its abundance and high tannin content. The use of quebracho, a hard bark from a Latin American species, gained popularity with the importation of hides from Buenos Aires in the late nineteenth century. Hides are immersed successively in several vats containing fresh tanning liquid, with the immersion time gradually increasing. The re-tanning process involves alternating the hides in buried vats with layers of bark, all submerged in a tanning solution. After several months in increasingly tannin-rich pits, the tanned hides undergo a finishing process that includes beating to soften the leather and remove creases, followed by shaving to refine the surface and adjust thickness. The hides are then air-dried in large, specialized buildings. Depending on their final use, the surface may be treated with oil or grease. Traditional vegetable tanning, which spans twelve to eighteen months, is regarded as a slow process, and tanners emphasize the quality of their product by contrasting it with the rapid chrome tanning method.

During the twentieth century, tanners experienced significant technical advancements alongside economic setbacks. Despite improvements in manufacturing processes and increased productivity, many companies faced market declines, leading to numerous closures. This context set the stage for a shift toward sustainable development at the century's end and the beginning of the twenty-first century, characterized by a renewed interest in vegetable tanning. Concurrently, over the past two decades, mineral tanning, primarily using chromium salts, has been applied to the hides of thicker and fattier animals, such as calves and kids, resulting in heavy leather. The rapid production rates of modern industry have established this tanning method as the standard, with most tanned hides on the market utilizing chromium or aluminum salts. While this process has enhanced productivity and yield, it has compromised product quality and environmental standards, as chromium tanning generates highly concentrated sludge containing chromium salts. This heavy metal is significantly polluting and harmful when released into the natural environment without proper treatment and neutralization. Therefore, it is crucial to chemically "neutralize" these substances using magnesium oxide, lime, or soda.

Chemical neutralization is essential to decrease the solubility of chromium and limit its mobility within aquatic environments. Once dried, the residues from chromium neutralization cannot be utilized in agricultural applications, unlike those from vegetable tanning, and must be stored in Class I landfills designated for hazardous waste. The significant advantage of vegetable tanning over chrome tanning lies in the by-products generated, which can serve as restructuring agents and fertilizers for agricultural soils. Although chrome tanning is notably faster and more productive on an industrial scale, it has become the most prevalent technology, despite the known risks of allergic reactions associated with chromium, as highlighted by consumer advocacy groups. This chapter outlines the various artisanal and industrial tanning processes for leather, as well as the average quality of effluents produced by commonly recognized tanning methods worldwide (Aboulhassan et al. 2008).

5.2 The Different Types of Tannery Discharges

Tanning is a process that transforms hides into leather through the application of tannins, rendering the material decay-resistant, flexible, and water-resistant. Tannins replace the water molecules found between the collagen fibers of raw hides, ensuring their stability. Until the late nineteenth century, vegetable tanning was the dominant method, but it was largely overtaken in the twentieth century by chemical tanning, primarily using chromium. A significant advancement in this field was the introduction of chrome tanning at the end of the nineteenth century, which involves binding the fibers of hides with chromium sulfate salts during dehydration. After the initial processing, the hides are rinsed and treated with sea salt. The tanning process occurs in two stages, gradually increasing the concentration of chrome salts, starting with low pH salts to facilitate penetration and prevent excessive surface tanning. This method employs more astringent salts that adhere quickly, allowing the hides to undergo a single bath. This more efficient approach enhances the productivity of small enterprises with minimal investment while utilizing existing facilities.

The time required for tanning a hide has been significantly reduced, allowing each pit to process a greater number of hides. Productivity gains affect all facets of production, including capital and labor. However, the chemical tanning technique using chromium poses higher pollution risks compared to traditional methods, and the leniency of regulatory authorities encourages industries to overlook pollution control measures. The speed of chromium tanning is a major advantage over the high costs associated with traditional oak bark tanning, which can take twelve to eighteen months. This process is essential to the productivity model of the Fordist era, which faced criticism during the post-Fordist period.

The transition to chromium tanning faces some resistance, yet economic efficiency currently outweighs environmental concerns. Nonetheless, chromium has a significant drawback: its high pollution levels. As a heavy metal, it contributes to substantial wastewater pollution generated by the manufacturing process. These issues are well recognized and create hesitance among both companies and public authorities. However, this does not hinder the widespread adoption of this method. Economic considerations overshadow environmental issues, relegating sustainability to a secondary priority while productivity remains paramount. In the recent years, the leather industry has faced intense competition from the textile and chemical fiber sectors, which are also highly polluting.

Environmental impacts arise from liquid, solid, and gaseous emissions, as well as the consumption of resources such as raw hides, energy, chemicals, and water. Discharges into the wastewater system primarily result from wet processes associated with river work, tanning, and leather finishing. Atmospheric emissions are mainly produced during dry finishing processes, although gases may also be released from heat production facilities. Solid waste generated from the process largely comes from fleshing, splitting, and trimming, while sludge from effluent treatment represents another potential source of solid waste, particularly when treatment occurs on-site. However, some of these wastes can be classified as by-products, as they may be marketed as raw materials for other industries.

5.3 The Different Tanning Processes Recommended Worldwide

The production and trade of leather have significantly increased in recent years. This rise in activity has led to a corresponding increase in wastewater generated by leather manufacturing facilities, necessitating the implementation of effective systems for treating the effluents produced by this industry. Tanning is treating and transforming the hides of various animals into leather, which has become essential for communities as clothing, bags, and various tools. This process helps prevent deterioration caused by environmental conditions such as sunlight, humidity, winds, and the degrading actions of various microorganisms, fungi, parasites, insects, and other microscopic life forms.

The process involves the application of a series of tanning agents to animal hides, primarily consisting of chemical salts such as chromium and/or plant-based tannins. These agents penetrate the leather and bond with its structure, inhibiting the chemical and biological degradation processes that lead to unpleasant odors, which are exacerbated by moisture and temperature. This treatment results in a stable and durable leather suitable for commercial use. Once the hide is removed from the slaughtered animal, it is typically treated with sodium chloride (NaCl) to prevent microbial growth and decay, preserving it until further processing. Tanning occurs in several stages, which are heavily water-intensive, requiring between 10 and 20 L of water per kilogram of treated leather, and generates significant gaseous, particulate, and liquid waste pollution.

Considering the wide range of pollutants found in effluents generated by leather production, treatment processes typically integrate various technologies and procedures to effectively manage heavily contaminated discharges, particularly those containing metals such as chromium, which is essential in chemical tanning and complex residual sludges. The most common procedure begins with a homogenization step, followed by a pretreatment phase involving clarification through dissolved air flotation (DAF) systems. Dissolved air flotation is one of the most widely employed methods due to its efficiency and versatility, applicable in both industrial and municipal wastewater treatment, desalination plants, and water reuse systems. The clarified water then undergoes biological treatment to separate contaminants, culminating in a sludge management process that offers multiple alternatives.

Furthermore, membrane processes are utilized when the treated water is intended for reuse in the production process (Suthanthararajan et al. 2003). The four primary stages of tanning include soaking, hair removal, fleshing, and splitting; chrome tanning and liming; a repetition of the tanning process to enhance treatment; and finally, dyeing, drying, and packaging. The waste produced from tanning processes primarily consists of solid waste and liquid effluents (Córdova et al. 2014).

5.4 Characterization of Tannery Residual Products and Effluents

The various residual products generated by tanneries, whether in liquid or solid form, consist of a range of waste materials including salts, shavings, and scraps of tanned hides, sludges, chemical residues, and packaging, among others. These materials are non-recoverable and must be transported to controlled landfill sites, which presents an additional environmental challenge. Untanned by-products, such as raw hide trimmings, hair, wool, fats, tallow, and flesh, can sometimes be recovered and redirected for industrial applications, particularly in the production of gelatin, adhesives, and collagen.

Regarding wastewater, the average water consumption ranges from 25 to 60 m^3/ton of fresh leather, yielding approximately 500 kg of finished leather per ton. Depending on the specific processes used in the industry, over 440 kg of chemicals may be utilized per ton of leather. These chemicals dissolve in the treatment water, resulting

Table 5.1 Average composition of homogenized effluents from the tanning industry

Parameter (mg/L)	General effluents	Effluents previously subjected to hair recovery, chromium and desulfurization
COD	7000–8000	5000–5500
BOD_5	4000–4500	3000–3500
NH_3	200–250	< 200
SS	3500–4000	2500–3000
Cr^{3+}	200–300	80–100
S^{2-}	200–250	< 2
SO_4^{2-}	1800	1800
Cl^-	5000–6000	5000–6000

Adapted from Artiga (2005)

in effluent with a high organic load and inorganic pollutants such as chromium, chlorides, ammonium, sulfides, and sulfates.

Two options are available for the collection and treatment of wastewater. The first option involves treating wastewater separately at each stage of the production process, with each facility specifically designed according to the source of the wastewater, whether from riverbanks, tanning, or finishing. The second approach entails mixing and homogenizing all effluents generated at each stage, followed by a common treatment for the entire wastewater. This method is more prevalent as it allows for the installation of a single treatment plant rather than separate facilities for each type of discharge. Typical wastewater from tanneries exhibits a high concentration of biodegradable organic matter (BOD) and significant nitrogen content due to the protein composition, keratin, and fats present in leather. Additionally, various organic compounds such as tanning agents, synthetic substances, fats, and dyes are utilized during the tanning process. The wastewater also contains inorganic elements, including chromium, chlorides, ammonium, sulfides, and sulfates, with chromium being the most concentrated and problematic. The concentration of salts is notably high, as salts are used for leather preservation, including sulfides and chromium salts. The presence of oils and fats is generally minimal. The typical composition of homogenized wastewater from the tanning industry is presented in Table 5.1 (Alvarez et al. 2004).

5.5 Tannery Wastewater Treatment

The characteristics of various pollutants found in wastewater can be described as follows: The presence of salts leads to high conductivity, complicating removal processes and biological treatment due to their excess. Organic matter is present in high concentrations and is primarily biodegradable; a well-designed biological treatment, combined with effective pretreatment, can eliminate up to 99% of the chemical oxygen demand (COD). Sulfides can be removed through an oxidation process during pretreatment. Total nitrogen and ammoniacal nitrogen originate from leather and ammonium-based products used in the tanning process, and they are eliminated during biological treatment through nitrification–denitrification. Chromium typically exists in the form of Cr^{3+}; while it rarely affects biological treatment, it is advisable to recover it during pretreatment. The main challenge in designing pretreatment lies in selecting an appropriate technology to reduce chromium in the effluent. Establishing specific treatment plants for each phase of tanning allows for better adaptation to the typical characteristics of each effluent, thereby facilitating treatment and enhancing the recovery of valuable components like chromium. However, this approach is still uncommon due to its high management costs compared to uniform wastewater management.

The characteristics of various pollutants found in wastewater can be summarized as follows: The most commonly used methods for treating wastewater from tanneries include physico-chemical processes and biological treatments. Additionally, membrane technologies are increasingly adopted as they enhance the reuse of treated water. Inorganic compounds present in this wastewater can have toxic and inhibitory effects on the microorganisms involved in biological treatment, necessitating a physico-chemical pretreatment for their removal. When effluents are not separated and wastewater must be mixed and homogenized for treatment, several factors must be considered in the design of the treatment process. A high chromium content, primarily in the form of Cr^{3+} but occasionally as Cr^{6+}, limits the potential for reusing and valorizing the sludge generated from biological treatment. A high concentration of organic load may require the addition of a post-treatment step, typically utilizing membrane technologies, if biological treatment does not meet performance standards. A high sulfate level makes anaerobic biological treatment inadvisable, prompting the use of desulfurization technologies during pretreatment. Lastly, significant solid matter can be effectively reduced through well-designed pretreatment processes, including coagulation–flocculation and primary clarification methods. The recent processes, including decanters and dissolved air flotation (DAF) systems, are efficient in eliminating solids. Adequate design of the pretreatment phase can protect the equipment from possible harm.

An appropriate design of pretreatment can safeguard equipment from solids and significantly reduce the volume of influent directed toward biological treatment, leading to savings in equipment, space, and energy. When considering the reuse of treated water in addition to wastewater treatment, several factors must be addressed. The recovery of chromium during pretreatment is crucial; a well-designed pretreatment system can remove between 95 and 100% of chromium from the water. By eliminating chromium, the sludge produced during biological treatment can be repurposed as fertilizer. The installation of anaerobic biological treatment promotes biogas production. If this treatment is implemented, it is essential to first remove sulfides and sulfates. Following anaerobic treatment, aeration must be introduced to ensure that the organic load is reduced under discharge standards. The physico-chemical process involves treating the homogenized effluent by adding precipitating agents for chromium removal (sodium hydroxide, NaOH), coagulants (ferric chloride $FeCl_3$ or aluminum salts), and flocculants (polyelectrolytes) (Fabregas 2021).

5.6 Physico-Chemical Treatments

5.6.1 Pretreatment Physico-Chemical Treatments

The pretreatment of wastewater from the tanning industry is carried out using physicochemical processes that effectively remove solids, sulfides, chromium, and other contaminants, thereby enhancing the efficiency of subsequent biological treatment. Commonly employed operations in this pretreatment phase include various techniques designed to optimize the overall treatment process.

5.6.1.1 Homogenization

Homogenization is defined as a process aimed at reducing a substance to achieve a uniform mixture by standardizing particle sizes. Products subjected to this treatment may include fat globules from tannery effluent, which are transformed into extremely small particles. This results in an even distribution of fat within the effluent. The process involves emulsifying the fat droplets present in the suspension of leather and reagents, thereby fragmenting the fat globules into much smaller sizes. This creates a uniform mixture of the various wastewater streams generated at each stage of the tanning process, facilitating consistent flow rates and concentrations of influent during treatment. In contrast, the roughing process ensures the physical separation of coarse solids using screens and sieves with varying mesh sizes.

5.6.1.2 Sulfide Removal

The removal of sulfides is essential in the treatment of tannery effluents, as these substances are significant contributors to unpleasant odors. Typically, the environment surrounding tanneries is filled with noxious smells resulting from leather tanning processes. Therefore, it is crucial to eliminate these odors, even if only partially. Ozonation emerges as an advanced treatment technology that effectively addresses these effluents. This process is adept at degrading various pollutants due to the strong oxidizing nature of ozone, which can break down a wide range of contaminants found in tanning wastewater, including organic compounds. Ozone effectively degrades complex organic substances such as fats and proteins into simpler molecules, thereby reducing the organic load in water. Additionally, it can convert hexavalent chromium, a toxic metal, into the less harmful trivalent chromium, facilitating its precipitation and removal from the effluents. Ozone can oxidize other residual compounds into more resistant chemical substances, including sulfides and nitrogenous compounds.

Air oxidation or other stronger oxidants, particularly those derived from chlorine, are utilized. Air oxidation, which occurs at a pH of approximately 11, is a slow process that requires a catalyst to enhance its rate. Effective removal of sulfides can be achieved with concentrations below 1 mg/L in the effluent. Additionally, other powerful oxidizing agents include hypochlorous acid (HClO), chlorine (Cl_2), and hydrogen peroxide (H_2O_2). Ozonation treatment not only reduces pollutants, particularly odor-causing sulfides during tanning, but also improves water quality parameters, such as chemical oxygen demand (COD) and biological oxygen demand (BOD). Specifically, ozone helps lower COD and BOD levels, making the water more compliant with discharge standards. Ozone neutralizes odor-causing substances, including sulfides and their derivatives in the effluent, while providing disinfectant properties that eliminate pathogens present in the wastewater (American Public Health Association 1995).

5.6.1.3 Coagulation–Flocculation

In addition to naturally settling solid particles, there are colloidal particles and those that do not easily settle due to their electrostatic charge, which causes them to repel one another and remain suspended. In such instances, coagulants are utilized to destabilize this charge, allowing colloidal and suspended particles to precipitate. Occasionally, a polyelectrolyte, such as aluminum and iron chlorides, may be employed as a flocculant to aggregate these precipitated particles, forming larger flocs that enhance sedimentation or flotation, thereby optimizing the subsequent clarification process. This precipitation process aids in reducing chromium, sulfates, and organic matter associated with the separated solids, which is a slowly biodegradable material that hinders biological treatment.

Within the category of "polyelectrolyte" polymers, there are distinct types, including polyamine polymers utilized as organic coagulants and copolymer acrylamide polymers employed as flocculants. Polyamines, which are low molecular weight cationic polymers, are available commercially in solutions that range from low to high viscosity (50–10,000 centipoises, a unit of dynamic viscosity). These products can be directly pumped from a container without prior preparation, and a proper secondary dilution aids in the discharge from the dosing pump. In contrast, acrylamide copolymers, or organic flocculants, are characterized by their high molecular weight and are primarily found in powder or emulsion form. They can be anionic or cationic, with some available as highly viscous solutions (5000–10,000 centipoises) that can be pumped directly with secondary dilution during the discharge process. As previously mentioned, flocculation is a separation process for particles dispersed in a given fluid, such as leather tanning residues, achieved by adding a chemical flocculating agent, often polyelectrolytes like aluminum and iron salts, to promote the agglomeration of particles into larger aggregates that can be more easily separated through physical sedimentation (Haydar and Aziz 2009; Song et al. 2003).

Coagulation and flocculation are essential processes in various applications, particularly in wastewater treatment facilities for the leather, food, and drinking water industries, where they facilitate the removal of suspended particles and enhance water quality. This process typically occurs in two stages: coagulation, which involves aggregating suspended particles by altering their electrostatic charges by adding a chemical agent, and flocculation, where the charged particles clump together to form larger, more easily separable flocs (Aboulhassan et al. 2006).

In this context, the Jar Test is a small-scale laboratory experiment that simulates the coagulation–flocculation process found in wastewater treatment plants. This test enables operators to determine the optimal dosages of coagulants and flocculants required to effectively remove suspended particles and pollutants while minimizing treatment costs. In certain industries, such as tanning, chromium in the tanning process necessitates an additional test known as the general leaching test for metals, primarily focusing on chromium in the treatment of tannery waste. This leaching test assesses the amount of solute that can be released from a solid into a solvent by agitating the solid in the solvent for a specified duration and then measuring the concentration of the dissolved solute.

The leaching test is useful for identifying harmful substances, such as chromium, in residual aqueous solutions by employing acetic acid. This method assesses the release of pollutants from solidified waste into water at a reference temperature of approximately 20 °C, as well as at elevated temperatures that enhance the leaching process compared to results obtained at the reference temperature. The test can be conducted under conditions that simulate specific disposal environments, such as using groundwater, to yield representative values. Average removal rates that can be achieved include over 60% COD, over 55% BOD_5, over 30% sulfates, and over 95% chromium. Commonly used coagulants and precipitants include aluminum and iron salts, primarily ferric chloride ($FeCl_3$), aluminum chloride ($AlCl_3$), and aluminum sulfate ($Al_2[SO_4]_3$), along with precipitating bases like calcium hydroxide ($Ca[OH]_2$) or sodium hydroxide (NaOH). The removal of chromium using these agents requires a controlled pH of around 7.5. In certain tanning processes, chrome is replaced with tannins, which are less biodegradable and are removed during pretreatment through precipitation with aluminum or iron salts, resulting in a reduction of COD by approximately 50%.

5.6.1.4 Solids Separation by Flotation

Dissolved Air Flotation (DAF) systems are widely utilized in the pretreatment of effluents from the tanning industry. In a DAF installation, the flotation and sedimentation of particles, along with the flocs produced during the coagulation–flocculation process, occur simultaneously. The clarified water is pressurized and saturated with dissolved air. Upon entering the flotation chamber, this saturated flow experiences depressurization, resulting in the formation of microbubbles that lift the particles and flocs that cannot settle or float to the surface. Particles with adequate sedimentation capacity settle at the bottom of the DAF unit. This equipment effectively removes

Table 5.2 Characterization of effluents from each stage of the tanning process

Parameter	Process	Bronzage	Post-polymerization	Finition	Total
Volume (m^3/ton)	20–25	1–3	3–8	1	24–37
COD (kg/ton)	120–160	10–20	15–40	10	145–230
BOD_5 (kg/ton)	40–60	3–7	5–15	4	48–86
SS (kg/ton)	70–120	5–10	10–20	5	85–155
Cr^{3+} (kg/ton)	–	2–5	1–2	–	3–7
S^{2-} (kg/ton)	8–12	–	–	–	8–12
Nitrogen (kg/ton)	10–20	–	1–2	–	11–22
Chlorides (kg/ton)	120–200	50–60	5–10	–	175–270
Sulfates (kg/ton)	5–20	30–50	10–40	–	45–110

Values are given based on tons of raw bovine leather. Adapted from Artiga (2005)

both floating and settled sludge with a high solids concentration (up to 5%) and produces high-quality clarified sludge. The sludge generated in a DAF station is subsequently directed to dehydration treatment, employing various techniques such as centrifugation, pressure filtration, thermal drying, or vacuum filtration. Additionally, there are sedimentation units that can separate settleable solids; however, they are less effective at separating non-sedimentable solids and require more space than DAF clarifiers (Table 5.2).

5.6.1.5 Alternatives Pretreatments

There are alternative technologies capable of achieving significant removal efficiencies for chrome, sulfides, and solids. However, their application is often restricted due to high costs, as seen with ion exchange and activated carbon adsorption. The method involves separating pollutants through coagulation–flocculation, followed by the removal of flocs and floating solids using a dissolved air flotation (DAF) system. The flocs produced during coagulation–flocculation are optimally sized for separation from water in a DAF unit. This coagulation–flocculation process can also be conducted continuously in a stirred tank. DAF technology can be developed and integrated with the principles of dissolved air flotation and sedimentation, featuring optimal equipment design, thus providing an effective and robust separation process for oils, fats, colloids, ions, macromolecules, microorganisms, and fibers. The coagulation–flocculation followed by DAF sequence is commonly used in treating effluents from the tanning industry, demonstrating both operational efficiency and cost-effectiveness in terms of chemical and energy consumption. The analysis of effluents during coagulation–flocculation is conducted using Jar Test assessments to identify the best combination of chemicals (polyelectrolytes) for optimizing the process and adjusting pH levels.

In a DAF system, compressed air is introduced into a recirculation flow of clarified product. This air dissolves in the liquid medium and subsequently generates bubbles of approximately 30–50 μm when released through a dispersion head in the flotation chamber. Coagulated and flocculated particles adhere to these bubbles and rise to the top of the DAF unit, where they are mechanically removed. Settled materials flow into the sediment compartment at the bottom of the DAF unit and are discharged by a sludge extraction system, typically a screw conveyor. The clarified water exits the DAF unit through an adjustable overflow system, with a portion of this clarified effluent being redirected by the recirculation pump back into the air compression and saturation system. A wide range of DAF flotation equipment is available, customizable based on the flow rate to be treated and the available space, capable of handling flows from 5 to 1000 m^3/h. The treatment capacity of the equipment can accommodate pollutant load ranges of up to 40 kg of solids per flotation surface. The advantages of DAF systems include high-quality treated water, rapid startup, high-speed operation, thicker sludge resulting in lower sludge volume production, reduced ecological footprint, user-friendly design, and robust, well-known technology with simple, adaptable, and efficient control systems.

5.6.2 *Biological Treatment*

Bacteria and other microorganisms play a vital role in the decomposition of organic matter found in effluents, including those from tanneries. The most common biological treatment techniques are activated sludge and aerobic oxidation processes. Most industrial facilities utilize biological treatment methods, which follow the stages of homogenization and physicochemical pretreatment. This treatment type is particularly effective, as nearly all-organic matter in wastewater is biodegradable. Wastewater from tanning activities contains readily biodegradable organic matter, making biological treatment highly efficient, if the pretreatment is properly designed to remove previously mentioned inhibitory agents. Both aerobic and anaerobic methods can be employed.

5.6.2.1 Aerobic Systems

A commonly used wastewater treatment system is the activated sludge process, which operates through a combination of anoxic and aerobic reactors. This method, facilitated by biological reactions involving microorganisms, effectively decreases chemical oxygen demand (COD) and nitrogen levels. Treatment can be conducted continuously in one or more reactors or sequentially, as seen in sequencing batch reactor (SBR) systems (Fig. 5.1).

The SBR process consists of five distinct stages. The initial stage, filling, involves introducing wastewater into the reactor without mixing or aeration. Water is injected

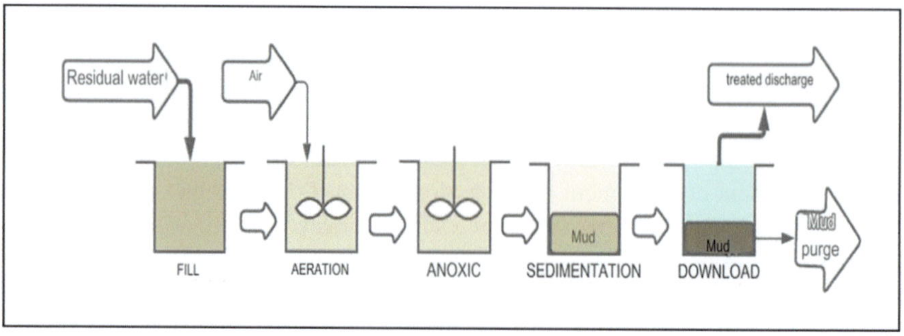

Fig. 5.1 Operating stages of an SBR biological reactor for COD and nitrogen removal

near the bottom of the tank, facilitating contact with the sludge and effective homogenization of the contents. The second stage, aeration, represents aerobic biological treatment, where air is supplied to provide the necessary oxygen for aerobic bacteria to degrade organic matter. This phase is intermittent, with its duration varying based on the characteristics of the wastewater and the pollutant removal objectives. The third stage, anoxia, occurs without air supply, creating an anoxic environment conducive to denitrification, where nitrates are converted into nitrogen gas. The fourth stage, sedimentation, involves halting aeration and agitation to allow activated sludge to settle at the bottom of the basin. Finally, the fifth stage, discharge, entails pumping the upper portion of the clarified effluent to a minimal reserve volume while removing excess sludge rich in solids. Recirculation is unnecessary, as the biomass concentration in the reactor maintains the desired sludge age through operational and extraction cycles. The SBR treatment process is notable for its efficiency in space and energy, as it allows the entire process to be conducted within a single tank.

5.6.2.2 Anaerobic Systems

Anaerobic systems can tolerate high levels of COD and BOD_5 while promoting the production of biogas, an energy-rich gas generated by the biological decomposition of organic matter in the absence of oxygen. In the treatment of wastewater from tanneries, Up flow Anaerobic Sludge Blanket (UASB) reactors are commonly utilized. These compact and efficient reactors facilitate COD removal, biogas collection, and solid separation within a single unit. However, anaerobic systems have notable drawbacks, including sensitivity to toxic substances and a COD reduction that rarely exceeds 60%, necessitating adequate pretreatment and complementary aerobic biological treatment. Table 5.3 illustrates the overall removal rates achieved through various technologies for wastewater treatment in the tanning industry. Anaerobic digestion is a complex process in which organic pollutants in liquid effluents are decomposed into methane, carbon dioxide, and water. This degradation process occurs in two main phases: acidogenesis, which includes hydrolysis and acidification, and methanogenesis, leading to biogas production. In the first phase, organic complexes such as polysaccharides and proteins are hydrolyzed into soluble monomers like sugars and amino acids, aided by exo-enzymes produced by fermentation bacteria. In the subsequent phase, the hydrolysates undergo fermentation, transforming into volatile fatty acids, hydrogen, carbon dioxide, ethanol, other alcohols, and lactate, characterizing acidogenesis (Fig. 5.2).

The fermentation products are converted into acetate, hydrogen, and carbon dioxide by bacteria known as acetogens. These bacteria closely coexist with methanogenic bacteria, forming a nearly symbiotic relationship. This association is crucial, as the conversion of fermentation products is only thermodynamically feasible if hydrogen concentration remains low, a condition maintained by the activity of methanogenic bacteria during the final phase of the reaction. The conversion of acetate, hydrogen, and carbon dioxide into methane by methanogenic bacteria represents the final stage of anaerobic digestion. The chemical oxygen demand (COD) present in wastewater is thus transformed into methane and a reduced amount of biomass. Due to the low solubility of methane in water, this gas easily escapes, facilitating the removal of COD from the effluent. This degradation process can be summarized as follows: both mechanisms occur within a single Up flow Anaerobic Sludge Blanket (UASB) reactor. The specific design of the process allows the UASB reactor to create a set of convection cells and an upward flow that, in interaction with the biogas, leads to the flotation of the granular sludge layer. At the base, the reactor is equipped with a system for distributing raw tannery effluents. A three-phase separator (water-sludge-biogas) is installed at the top of the reactor, enabling

Table 5.3 Estimated wastewater yields from the tanning industry, depending on the treatment applied

	COD	BOD_5	SS	Chromium	Sulfides	NTK
Pretreatments						
Fat flotation	20–40	–	–	–	–	–
Sulfide oxidation	10	–	–	–	90	–
Chromium precipitation	–	–	–	90–99	–	–
Pretreatment combinations						
Homogenization + sedimentation	25–35	25–35	50–70	20–30	–	25–35
Homog. + physical–chemical + sedimentation	50–65	50–65	80–90	20–50	2–10	40–50
Homog. + physical–chemical + flotation	55–75	55–75	80–95	20–50	2–5	40–50
Biological treatment (pretreatment + …)						
… aeration	85–95	90–97	90–98	–	–	50
… anoxia—aeration	85–95	90–97	90–98	–	–	80–90
… anaerobic UASB	65–75	60–70	50–80	–	–	20–30

Adapted from Black et al. (2013)

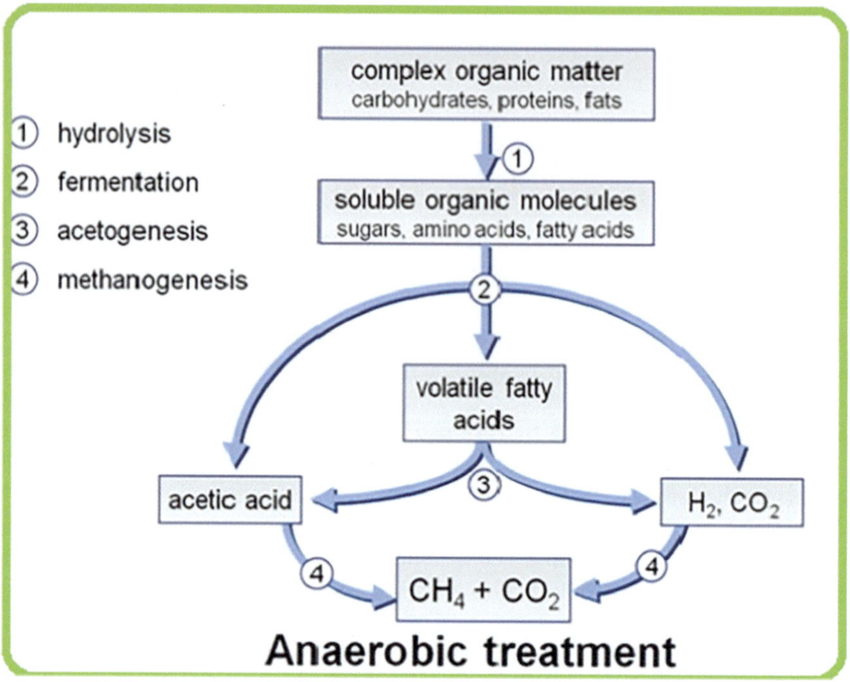

Fig. 5.2 Primary phases of anaerobic methanization process

the collection of biogas through inverted chutes and the overflow of treated effluent, while preventing the escape of granular sludge through a deflector system.

The advantages of the UASB reactor with medium loading are numerous. It produces minimal sludge and requires low energy for operation. Additionally, it generates biogas, which serves as a renewable energy source. Its high volumetric loading capacity allows for a reduction in reactor size. This system can also facilitate the acquisition of green certificates, thereby promoting carbon credit management. The proposed reactor is characterized by its simple and robust design, proven reliability, remarkable stability, as well as operational flexibility and reduced supervision and maintenance requirements (Fig. 5.3) (Bernardino 2019).

The sludge derived from wastewater in tanning processes contains approximately 60–70% organic matter and 3–5% nitrogen, while its potassium content is minimal. Before treatment, it is essential to dehydrate this sludge to reduce its volume and facilitate transportation. Various equipment, such as belt filters, filter presses, and centrifuges, are employed to condition the sludge, often with salts or polyelectrolytes. For final treatment, several methods can be utilized, including anaerobic digestion that produces biogas, agricultural application subjected to contaminant regulations, aerobic composting to generate legally compliant organic compost, and thermal treatment through incineration, gasification, or pyrolysis, which allows for energy recovery.

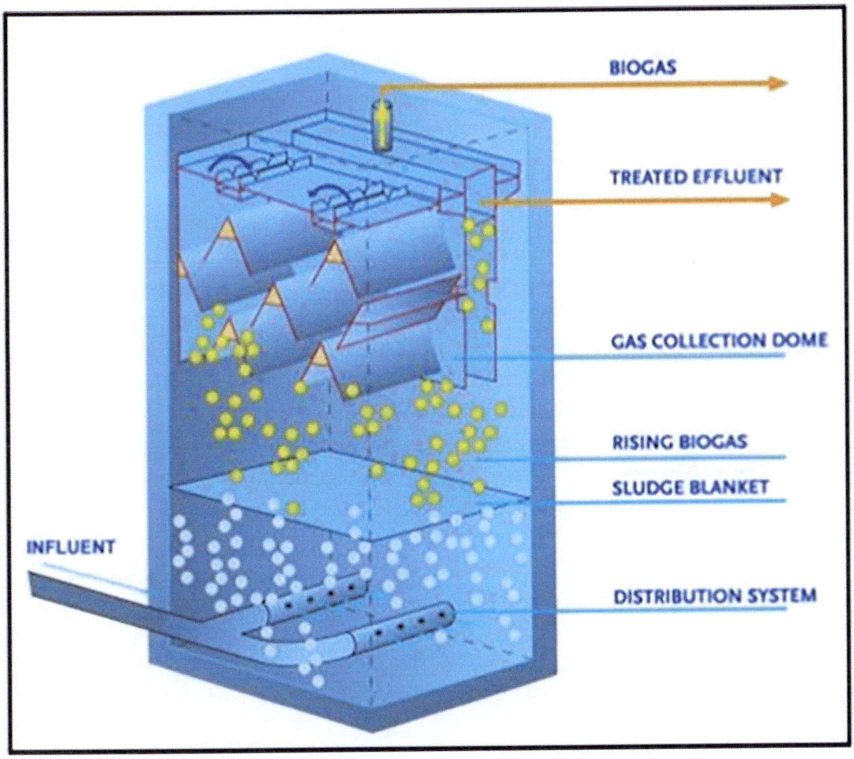

Fig. 5.3 Schematic internal view of a UASB reactor

Traditional methods of pretreatment and biological treatment are effective in removing sulfides, chromium, organic load, nitrogen, and suspended solids. However, the resulting effluent still contains significant concentrations of salts and dissolved solids, such as sodium, chloride, sulfate, calcium, and magnesium, along with challenging impurities that hinder the reuse of water without further treatment of these contaminants. In water-intensive industrial sectors, adopting a circular model is crucial to facilitate the reuse of treated wastewater. Membrane filtration technologies have proven effective in eliminating these pollutants and promoting water reuse. This technology can be integrated into the wastewater treatment process in the tanning industry either as post-treatment using filtration membranes or through the application of membrane bioreactors (MBR).

5.6.2.3 Filtration Membranes as Post-treatment

There are four categories of membranes classified according to the size of their pores: microfiltration (MF), ultrafiltration (UF), nanofiltration (NF), and reverse osmosis

(RO). This classification is illustrated in Fig. 5.4, which details the pore dimensions (in nm) and the pollutants each membrane can retain.

Microfiltration (MF) effectively removes colloids, suspended solids, bacteria, and viruses. It is often employed as a pretreatment step for other membranes such as UF, NF, or RO; however, its standalone application is limited due to insufficient effectiveness in treating wastewater from tanneries.

Ultrafiltration (UF) is a selective filtration process that operates under pressures of up to approximately 10 bars. It can eliminate macromolecules and high molecular weight organic particles, thereby facilitating the fractionation of organic matter. The performance of UF membranes largely depends on the materials used in their construction. They are commonly utilized for prefiltration to enhance reverse osmosis processes and prevent fouling; making them, the most frequently installed membranes in membrane bioreactor (MBR) systems.

Nanofiltration (NF) employs membranes specifically designed to remove persistent organic substances and heavy metals while generating minimal concentrate volume. Close monitoring of these membranes is crucial to prevent any risk of fouling and clogging.

Reverse osmosis (RO) is an effective wastewater purification method that utilizes pressure to enhance its efficiency. This process concentrates dissolved solids, trace organic compounds, heavy metals, and monovalent ions. By integrating reverse osmosis membranes with techniques such as ultrafiltration or nanofiltration, it is possible to remove approximately 95% of dissolved solids, 94% of Na^+ and Cl ions, 98% of sulfates, 65% of Mg^{2+}, and 55% of Ca^{2+}.

However, reverse osmosis has notable drawbacks, including a high tendency for fouling, which necessitates pretreatment, and the generation of significant concentrate that requires proper management. The primary challenge in employing

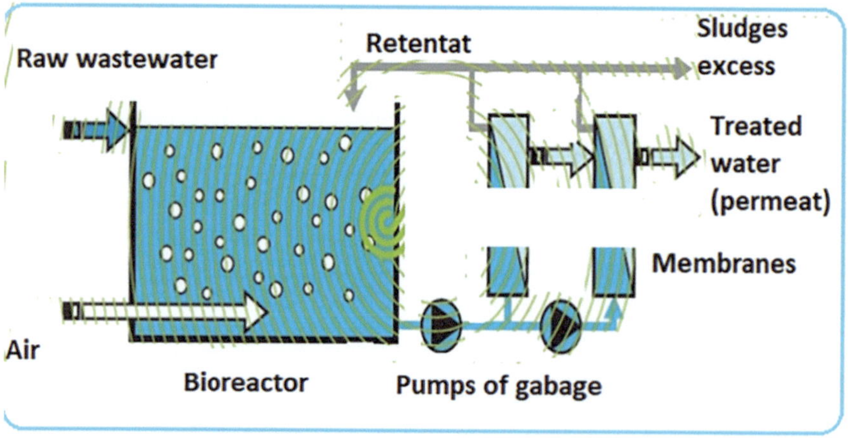

Fig. 5.4 External membrane bioreactors

membrane technologies lies in achieving an optimal balance between cost and efficiency while addressing fouling issues. It is essential to carefully select cleaning products and methods during the design and installation of a membrane system to optimize both operational costs and performance.

5.6.2.4 MBR Membrane Bioreactors

This technology incorporates ultrafiltration membranes and biological treatment, creating a unique system that retains biomass. The membranes act as a separator for the sludge produced in the reactor, thereby eliminating the need for an additional clarifier. They also preserve biomass, including nitrifying microorganisms and those capable of breaking down difficult-to-degrade organic matter, enabling high biomass concentrations within the reactor. These reactors are designed to generate significant biomass concentration, minimize sludge production, and maximize pollutant removal, while providing flexibility in their design and operation. It is common to add reverse osmosis units as post-treatment following an MBR reactor. However, membrane fouling remains the primary drawback of these systems, necessitating a carefully designed cleaning mechanism. Membrane bioreactors are mainly categorized into two types: external membrane bioreactors, which are located outside the aeration tank, and submerged membrane bioreactors, which are integrated directly within the aeration tank.

In external membrane bioreactors, the membranes are housed within a casing that can be arranged in either series or parallel configurations. These membranes, which can be tubular or flat, may consist of organic or inorganic materials and operate through internal filtration, where the flow moves from the inside to the outside. Due to the high concentration of suspended solids, direct filtration is not feasible; therefore, tangential filtration is employed. This process relies on a pressure differential across the membrane, utilizing high shear rates to prevent the buildup of solids. Chemical cleaning is required two to four times a month.

These data example illustrates the findings of a case study on implementing a dedicated wastewater pretreatment station in the tanning sector through coagulation and flocculation. Filtration rates range from 60 to 130 L h^{-1} m^{-2}. However, the energy consumption associated with tangential filtration, which is between 4 and 8 kWh m^{-3}, along with the need for extensive pretreatment in the presence of fibrous materials, renders this process unsuitable for urban wastewater treatment.

In submerged membrane bioreactors, membranes, whether hollow fiber or flat, are directly integrated into the activated sludge. These external skin membranes facilitate filtration from the outside to the inside, through either hydrostatic pressure or vacuum. The accumulation of materials on the membrane surface is managed through specific aeration, supplemented by automatic cleaning phases such as backwashing. Part of the aeration may also meet the oxygen requirements of the biomass. When membranes are installed in a dedicated tank, recirculation to the aeration tank is necessary, while their direct placement in the latter depends on the chosen configuration. The modular design of this system allows for the rehabilitation of existing

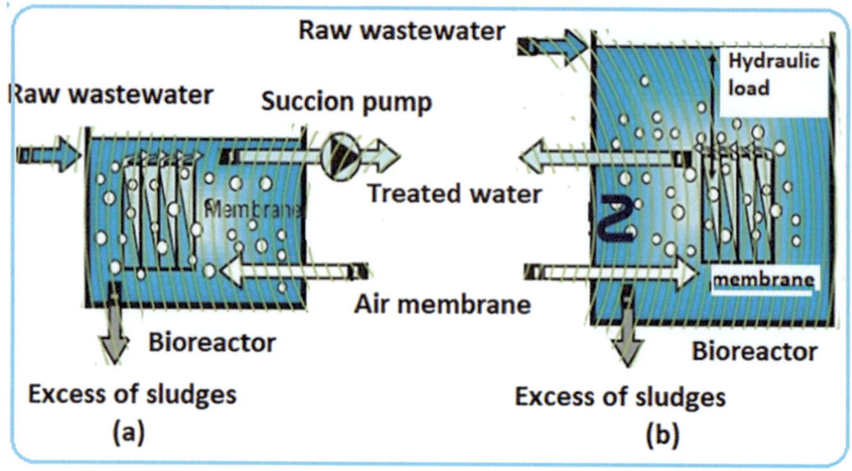

Fig. 5.5 Submerged membrane bioreactors

facilities by progressively integrating membranes into an already operational aeration tank.

Filtration rates range from 10 to 40 L h^{-1} m^{-2}, with a relatively low operating pressure between 2 and 5 m CE. Chemical cleaning, which requires partial interruption of the filtration stage, is conducted two to three times annually. For the treatment of urban and industrial wastewater, there are systems that utilize high-performance hollow fiber organic membranes specifically designed for the clarification of tannery effluents (Figs. 5.4 and 5.5) (Table 5.4).

5.7 Quality of Tanning Effluents

5.7.1 Tannery Wastewater Quality

The process of transforming rawhide into leather involves treatment through multiple baths of chemical solutions. This procedure primarily removes excess materials and imparts essential leather characteristics, such as resistance to decay, flexibility, and coloration. The pollution generated arises from both the hide itself and the reagents that are not fully absorbed. Effluents from tanneries typically contain organic loads related to preparation stages and mineral loads associated with tanning and re-tanning phases that utilize chromium sulfate as a tanning agent.

The regulatory framework governing the leather tanning industry falls under classified installations for environmental protection. Monitoring the quality of wastewater discharges is a significant concern for this sector. Pollution parameters to be monitored are outlined in the prefectural authorization orders for companies, under

Table 5.4 Results from the pretreatment facilities of tanning industry wastewater

Origin of water	Mixed wastewater from the shoe tanning process
Cible	Compliance with legal discharge limits into the sewage system, which will be directed to a WWTP, and the possibility of reusing effluent. Full sludges treatment
Flow	200 m^3/day
Influent characteristics	
	• pH = 3
	• TSS = 1000 mg/L
	• COD = 3000 mg/L
	• BOD_5 = 1500 mg/L
	• Total nitrogen = 100 mg/L
	• Chromium Cr^{3+} = 140 mg/L
Installed equipment	
	• Roughing screens
	• Homogenization tank
	• Preparation and dosing system for coagulant ($FeCl_3$), flocculant (poly-electrolyte), and pH control (NaOH)
	• Coagulation, flocculation and pH control tank system
	• Clarifier
	• Complete sludge treatment system: conditioning and dewatering of sludge by filter press, preparation and dosing of milk of lime ($Ca(OH)_2$)
Performance	
	> 90% in SST
	> 67% in DCO
	> 67% of BOD_5
	> 50% of total nitrogen
	> 98% Cr^{3+} chrome
	Sludge dry matter = 35%

specific ministerial regulations. Most facilities, even if they have a pretreatment system for effluents, are connected to the municipal sewage network. In this context, additional monitoring parameters may be established through a discharge agreement between the company and the municipality. There is no standardized list of parameters to monitor, except for a few commonly observed ones such as BOD_5, COD, suspended solids, total chromium, total metals, pH, sulfides, total nitrogen, and total phosphorus.

It is essential to monitor the occupational risks associated with chrome handling in tanneries. It is widely recognized that processing one ton of rawhide requires an average of 500 kg of chemicals, with chrome, tannins, and other tanning agents constituting 23% of this total. Chronic exposure to chrome can lead to significant health issues, particularly respiratory problems, for workers involved in chrome handling within tanneries. Recent cross-sectional epidemiological studies conducted in tanneries across various countries have indicated that nearly 71% of both male smokers and non-smokers exhibit respiratory symptoms. These symptoms typically manifest only after more than ten years in the profession ($p < 0.001$), with those performing tasks away from the tanning process being the least affected. The concentration of chrome in the final effluents studied ranged from 1.4 to 9 mg L^{-1}. The study highlights human, material, and organizational constraints, necessitating the involvement of all responsible authorities to enhance the tanning sector and safeguard both public health and water quality in local waterways.

As environmental protection standards become increasingly stringent, the leather industry often faces significant criticism. In response, chemical industry stakeholders supplying tanneries are implementing various research and development initiatives to modernize their processes. However, environmental preservation should not be the sole focus; consumers also demand improvements in leather quality and the development of new applications to meet their growing needs.

The concept of "sustainable development" is rapidly emerging as a compromise between economic growth and environmental protection. Its ambiguous definition allows for diverse interpretations, alleviating tensions between economic and environmental objectives, although it may also obscure their contradictions. This raises concerns that sustainable development could become merely a catchphrase. Water is increasingly viewed as a precious global resource, necessitating effective management, particularly within the leather industry. The transformation of raw hides into finished leather requires substantial water usage across various stages. To address this waste, leather industry players are altering their practices by adopting new cleaning techniques and replacing multiple rinses with discontinuous processes, enabling more precise water management. Over the past two decades, companies in the sector have successfully reduced their water consumption by approximately 60%.

Hexavalent chromium, widely utilized in the leather industry, is recognized as a significant pollutant and allergenic agent. Criticism directed at leather manufacturers primarily concerns the use of this highly toxic compound, which infiltrates the leather and partially ends up in rinse waters. In addition to causing considerable pollution, hexavalent chromium is associated with the onset of allergies. Currently, it is employed in the tanning process of approximately 85% of leathers due to its ability to impart flexibility and softness while enhancing dye fixation. However, efforts have been made to minimize the use of these salts to the essential minimum, resulting in a 90% reduction of chromium discharge into wastewater while maintaining the required quality. Initiatives have been established to valorize the waste and by-products generated from leather production, acknowledging that a significant amount of waste is produced from the slaughterhouse to the final tanned leather, which was previously disposed of untreated. Only 20% of the initial weight of the hide is converted into leather. Recently, several methods have been developed to repurpose this waste. Untanned hides can be transformed into organic products such as glue and food gelatin. Fats and other materials from leather processing are converted into biogas through fermentation or biomethanization. During the initial stage of tanning, the hair from the hides, which was sent directly to the treatment plant with water, has seen its volume reduced by 50% and is now collected for use as fertilizer in agriculture.

It is important to emphasize those new solvents, which are also regarded as pollutants originating from tanning workshops, play a vital role in environmental protection. These solvents, utilized during the finishing stages of production, have contributed to a 90% reduction in environmental pollution through the adoption of more eco-friendly compounds.

5.7.2 *Treatment of Tanning Water Effluents by Ozonation*

Tanning is a critical process, both artisanal and industrial, for transforming animal hides into leather, a vital material used in various sectors such as fashion, automotive, and furniture. However, this industry generates wastewater that contains complex pollutants, including chrome salts, fats, proteins, and other chemicals. Ozonation emerges as an advanced treatment technology, providing an effective solution for managing these effluents by treating wastewater produced during the tanning process. The wastewater from tanneries contains a range of pollutants, including organic residues like skin fragments, hair, fats, and proteins, as well as dissolved chemicals such as chrome salts, sulfates, sulfides, acids, and bases utilized in tanning. Additionally, these effluents include suspended solids, comprising leather particles and other solid materials, along with heavy metals, primarily chrome, which is used in chemical tanning.

The ozonation treatment operates on a mechanism that is both straightforward and effective. Initially, ozone is produced using a specialized device that converts molecular oxygen (O_2) into ozone, either through electrical discharge or corona effect. This ozone is then injected into the effluent stream via diffusers, where it mixes with water. The subsequent phase involves an oxidation reaction, during which ozone interacts with existing contaminants; this process can be enhanced by an agitator to improve contact. After the reaction is complete, any residual ozone is removed, either by decomposing it into oxygen or through a destruction system. Finally, the by-products of this reaction are separated from the treated effluents using filtration or sedimentation methods.

The ozonation of tannery effluents offers significant advantages, primarily due to its remarkable effectiveness in degrading pollutants. Ozone (O_3) is recognized as the most potent oxidant, capable of breaking down a wide range of contaminants found in tannery wastewater. This process converts complex organic compounds, such as fats and proteins, into simpler molecules, thereby reducing the organic load in the water. Ozone is essential in the transformation of heavy metals, particularly chromium, by converting highly toxic hexavalent chromium into less harmful trivalent chromium. This reaction facilitates the precipitation and removal of chromium from the effluents. Additionally, ozone oxidizes certain residues into recalcitrant chemical substances, including sulfides and nitrogen compounds.

The notable improvement in water quality from tanneries is evident through ozonation, which diminishes pollutants and enhances quality indicators such as chemical oxygen demand (COD) and biological oxygen demand, ensuring that the treated water meets discharge standards. Moreover, ozone, with its oxidizing power, aids in neutralizing substances that cause unpleasant odors in the effluents. It also serves a disinfecting role by eliminating pathogens and reducing accompanying flora.

Considering its numerous advantages, ozonation produces less sludge and solid waste compared to other methods, thereby simplifying the management of by-products. In summary, ozonation serves as an innovative solution to treat tanning effluents, presenting promising prospects for more sustainable and effective environmental management in the leather industry. The treatment of tanning wastewater remains a complex process, requiring an integrated approach that combines physicochemical and biological techniques to meet environmental standards and safeguard aquatic ecosystems.

5.8 The Principle of Pollution Control

The rapid growth of industrialization, particularly in the leather and textile sectors, has led to detrimental environmental impacts, evidenced by numerous soil, air, and water contamination from toxic substances and other pollutants. This has severely affected human health and the integrity of ecosystems. Additionally, the increased and intensive use of material and energy resources has placed cumulative pressures on ecosystem quality at local, regional, and global levels. Before the implementation of coordinated efforts to mitigate pollution effects, environmental management primarily adopted a laissez-faire approach, focusing on waste disposal to prevent severe local nuisances, often viewed in the short term. Corrective measures were only considered when damage was deemed unacceptable. As industrial activities intensified and understanding of cumulative effects improved, the principle of pollution control gradually became central to environmental management.

This approach is based on two fundamental concepts: the first, the capacity for assimilation, asserts that a threshold of emissions does not disrupt environmental balance or human health; the second, the principle of prevention, argues that environmental damage can be avoided by modifying the methods, pace, and rate of pollutant introduction. Preventing pollution and minimizing discharges is achievable through the analysis of industrial effluents at their source within manufacturing operations and processes, which often helps avert the release of significant pollutant quantities. Recycling techniques are essential to prevent and treat pollution associated with tannery effluents. For instance, the wastewater-recycling program in tanneries, as demonstrated by Preul (1981), emphasizes the recovery and reuse of chromium and the complete recycling of wastewater, ensuring that no effluent is discharged into waterways except in emergencies. Integrating recycling into the treatment of tannery effluents presents significant challenges to enhance productivity while minimizing environmental and health risks associated with discharges. Effluent treatment can be accomplished through various methods, including thermal, physicochemical, biological, or membrane processes, which also allow for the recovery of valuable materials such as certain rare metals, oils, and fats, as well as the reuse of treated wastewater for diverse applications, including irrigation of green spaces or industrial crops like cotton.

Advanced wastewater treatment methods have been increasingly implemented in the recent years to enhance the removal of pollutants from industrial effluents, tailored to specific requirements. These methods include sand and multi-media filtration, chemical precipitation, activated carbon adsorption, electrodialysis, distillation, nitrification, algal harvesting, effluent valorization, micro-screening, ammonia stripping, reverse osmosis, ion exchange, land application, denitrification, and the use of wetlands. The selection of the most appropriate process for a given situation should be based on the quality and quantity of the raw wastewater, the requirements of the receiving waters, and, importantly, the associated costs (Tables 5.5 and 5.6).

5.9 Conclusion

The processing of animal hides significantly affects the environment. The wastewater generated contains pollutants from the decomposition of hides, along with chemical residues from preparation and tanning processes. This can also lead to solid waste and air emissions, resulting in unpleasant odors. Public concerns regarding tanneries primarily focus on water pollution from untreated discharges and offensive smells. Additionally, the increasing use of synthetic chemicals, such as dyes and pesticides, raises alarms about their toxicity and persistence in the environment. Pollution control measures may also have secondary effects, such as soil and groundwater contamination. Although modern tanning technology, which utilizes fewer chemicals and less water, has a reduced environmental impact compared to traditional methods, its widespread adoption still faces numerous challenges.

The management of pollution caused by liquid and solid effluents, particularly untreated tannery waste, can lead to a rapid decline in the physical, chemical, and biological characteristics of receiving environments, especially surface waters. Implementing simple treatment methods for effluents from tannery facilities can reduce suspended solids and biochemical oxygen demand (BOD) by over 50%. More advanced techniques can achieve even higher treatment levels. Given the variety of chemical components in tannery effluents, it is essential to adopt a sequence of treatment processes. Separating the streams is beneficial to ensure distinct treatment of concentrated waste and liquid or solid discharges, while also protecting the natural environment. The use of specific solvents during the tanning process has further helped to mitigate atmospheric pollution often produced by tannery operations.

The biological decomposition of organic matter, along with the emissions of sulfide and ammonia from wastewater, contributes to the unpleasant odors commonly associated with tanneries. The location of these facilities raises concerns due to the odor nuisances they generate. Addressing these odors relies more on proper

Table 5.5 List of common parameters to be monitored in leather tanning industry effluents

Sigle	Parameter	Unit	Sigle	Parameter	Unit
Ag	Silver	μg/L	DS550	Dissolved solids at 550 °C	mg/L
Al	Aluminum	μg/L	EC	Electrical conductivity	μmhos/cm
ALG	Algae	Nb/100 mL	ENTR	Enterococci	Nb/100 mL
ALKM	Basicity expressed by $CaCO_3$ arsenic	mg/L	F^-	Fluorides	mg/L
As	Arsenic boron barium bromides	μg/L	FCOL	Fecal coliforms	Nb/100 mL
B	Boron	mg/L	Fe	Ferrous	μg/L
Ba	Barium	μg/L	HARD	Hardness as $CaCO_3$	mg/L
Br	Bromides	mg/L	HCO_3^-	Bicarbonates in the form of HCO_3^-	mg/L
Ca	Calcium	mg/L	Hg	Mercury	μg/L
Cd	Cadmium	μg/L	K	Potassium	mg/L
Cl	Chlorides	mg/L	Li	Lithium	μg/L
CLDE	Chlorine demand	mg/L	MBAS	Detergents	μg/L
CLRL	Chlorophyll	μg/L	Mg	Magnesium	mg/L
CN	Cyanides	μg/L	Mn	Manganese	μg/L
Co	Cobalt	μg/L	Mo	Molybdenum	μg/L
COLR	Color (cobalt platinum)		Na	Sodium	mg/L
Cr	Chromium	μg/L	NH_4^+	Ammonia in NH_4^+ form	mg/L
Cu	Copper	μg/L	DS105	Dissolved solids at 105 °C	mg/L
DBO	Biochemical oxygen demand	mg/L	DS105	Dissolved solids at 105 °C	mg/L
DCO	Chemical oxygen demand	mg/L	Ni	Nickel	μg/L
DO	Dissolved oxygen (as O_2)	mg/L	NKJT	Total nitrogen Kjeldahl	mg/L
DOC	Dissolved organic carbon	mg/L	NO_2^-	Nitrite as NO_2^-	mg/L
			NO_3^-	Nitrates as NO_3^-	mg/L
PTOT	Total phosphorus in the form of P	mg/L	ODOR	Odor threshold number	
RSCL	Free chlorine residual	mg/L	OG	Oils and greases	μg/L
SAR	Sodium adsorption ratio		Pb	Lead	μg/L
Ni	Nickel	μg/L	PHEN	Phenols	μg/L

(continued)

Table 5.5 (continued)

Sigle	Parameter	Unit	Sigle	Parameter	Unit
NKJT	Total nitrogen Kjeldahl	mg/L	PHFD	pH measured in situ	
NO_2^-	Nitrite as NO_2^-	mg/L	PO_4	Phosphates in the form of PO_4^{-2}	mg/L
NO_3^-	Nitrates as NO_3^-	mg/L	TS550	Total solids at 550 °C	mg/L
Se	Selenium	μg/L	TURB	Turbidity	NTU
Si	Silica in the form of H_2SiO_3	mg/L	UV	UV (absorption at 254 nm)(/cm × 10)	
Sn	Tin	μg/L	Zn	Zinc	μg/L
SO_4	Sulfates	mg/L	STRP	Streptococci	Nb/ 100 mL
Sr	Strontium	μg/L	T	Temperature	°C
SS100	Suspended solids at 100 °C	mg/L	TCOL	Total coliforms	Nb/ 100 mL
SS550	Suspended solids at 550 °C	mg/L	TOTB	Total bacteria	Nb/ 100 mL

Table 5.6 Technological choices for tannery effluent treatment

Settling before treatment	Mechanical screening to remove coarse materials Flow equalization (balancing)
Primary treatment	Sulfide removal from greenhouse effluents Dechroming of tanning effluents Physico-chemical treatment for BOD removal and neutralization secondary treatment
Secondary treatment	Biological treatment Activated sludge (oxidation pit) Activated sludge (conventional) Lagooning or *Lagoon-based treatment systems* (aerated, optional or anaerobic)
Tertiary treatment	Nitrification and denitrification
Sedimentation and sludge treatment	Various shapes and sizes of tanks and basins

maintenance than on technological advancements. The solvents and other emissions resulting from finishing operations vary based on the chemicals used and the techniques employed to minimize their production and release. Approximately 30% of the solvents utilized may be lost through emissions, whereas modern processes can reduce this figure to around 3% in many instances. Lastly, the practice of incinerating solid waste and scraps by certain tanneries highlights the necessity for well-designed incinerators and implementing stringent operational practices.

The management of waste and the treatment of sludge from tanning processes pose significant disposal challenges, aside from effluents. Organic sludges, which are free from chrome or sulfides, can be utilized as soil conditioners and have a mild fertilizing effect due to their nitrogenous compounds. To enhance these benefits, immediate incorporation into the soil is recommended. The agricultural use of soils contaminated with chrome is a contentious issue in various regions, where guidelines delineate permissible applications.

There are several markets for the valorization of trimmings and flesh into by-products, including gelatin, glue, leather board, tallow, and proteins for animal feed. Treated process effluents, following appropriate treatment and quality control, may sometimes be used for irrigation in areas facing water scarcity or where effluent disposal is heavily regulated.

To mitigate leaching and odor issues, it is crucial that only solids and dehydrated sludges are sent to landfills. Measures must be implemented to ensure that tanning waste does not react with other industrial residues, such as acidic waste, which could generate toxic hydrogen sulfide gas. Uncontrolled incineration can lead to unacceptable emissions and is therefore discouraged.

Combating pollution from tanning facilities relies significantly on optimizing production technologies to enhance environmental performance and achieve various objectives, including improving chemical usage efficiency, reducing water and energy consumption, and recovering or recycling waste materials.

References

Aboulhassan AM, Souabi S, Yaacoubi A, Baudu M (2006) Improvement of paint effluents coagulation using natural and synthetic coagulant aid. Hazard Mater 138:40–45

Aboulhassan AM, Souabi S, Yaacoubi A, Zaim N, Bouthir FZ (2008) Tannery effluents characterization and impacts on the marine environment. J Water Sci 21(4):463–473. ISSN 1718-8598

Alvarez SG, Maldonado M, Gerth AY, Kuschk P (2004) Characterization of tannery wastewater and study of water lily in chromium recovery. Inf Technol 15(3):75–80. ISSN 0718-0764. https://doi.org/10.4067/S0718-07642004000300012

American Public Health Association (APHA) (1995) Standard methods for the examination of water and wastewater. Water Environment Federation, Alexandria, Virginia

Artiga P (2005) Contribution to the improvement of biological wastewater treatment in the tanning industry. Thèse de doctorat Université de Saint-Jacques-de-Compostelle, Département d'ingénierie chimique

Bernardino S (2019) Production of biogas/bioSNG from anaerobic pretreatment of milk processing wastewater. In: Substitute natural gas from waste, pp 397–424

Black M, Canova M, Rydin S, Scalet BM, Roudier S, Sancho LD (2013) Best available techniques (BAT) reference document for the tanning of hides and skins. Eur Comm Database 46:2013. http://eippcb.jrc.es/pages/FActivities.htm

Córdova HM, Vargas R, Cesare MF, Flores L, Visitación L (2014) Treatment of wastewater from traditional and alternative tanning processes with chromium complexing agents. J Chem Soc Peru 80(3):183–191

Fabregas J (2021) Technology blog/wastewater treatment in the tanning industry. Newsletter Sigma DAF, France

Haydar S, Aziz AJ (2009) Coagulation/flocculation studies of tannery wastewater using combination of alum with cationic and anionic polymers. Hazard Mater 168(15):10

Jallouli S, Wali A, Buonerba A, Zarra T, Belgiorno V, Naddeo V, Ksibi M (2020) Efficient and sustainable treatment of tannery wastewater by a sequential electrocoagulation-UV photolytic process. J Water Process Eng 38:101642. ISSN 2214-7144. https://doi.org/10.1016/j.jwpe.2020.101642

Khanal KS, Huang J (2003) ORP-based oxygenation for sulfide control in anaerobic treatment of high-sulfate wastewater. Water Res 37:2053–2062

Metarfi Y, El Rhazi K, El Madani FZ, Omor A, Rais Z (2018) Risques professionnels liés à la manipulation du chrome au sein des tanneries modernes de la ville de Fès, Maroc. Rev d'Épidémiol Santé Publ 66(Supplément 3):S187. ISSN 0398-7620. https://doi.org/10.1016/j.respe.2018.03.319

Park D, Yun SY, Park MJ (2004) Use of dead fungal biomass for the detoxification of hexavalent chromium screening and kinetics. Process Biochem 40:2559–2565

Perrin C (2014) Sustainable development in historical perspective: the example of tanneries. Man Soc 193:37–55. ff10.3917/lhs.193.0037ff. Ffhal-01470967

Preul HC (1981) Application of fluidic devices for control of wastewater treatment operation. Pascal-Francis Water Sci Technol 13(9):139–146, ISSN 0273-1223; GBR; DA

Song Z, Williams CJ, Edyvean RGJ (2003) Treatment of tannery wastewater by chemical coagulation. Dessalement 164:249–259

Suthanthararajan R, Ravindranath E, Chitra K, Umamaheswari B, Ramesh T, Rajamani S (2003) Application of membranes to recover and reuse water from treated tannery wastewater. Dessalement 164:151–156

Chapter 6
Adsorption Strategies for Environmental Remediation: A Case Study of Tanneries in Morocco

Saliha Elabbas, Amina Lissaneddine, and Faissal Aziz

Abstract Adsorption is emerging as an innovative and promising method for treating wastewater from tanneries in Morocco. This sector is crucial to the local economy yet contributes significantly to environmental pollution. The tanning process produces effluents rich in organic and inorganic contaminants, including heavy metals, dyes, and toxic chemicals. Consequently, there is an urgent need to develop sustainable solutions for managing these wastewater streams to safeguard environmental and water resource integrity. This chapter investigates the utilization of calcium carbonate-based materials as adsorbents for treating tannery effluents. Given their availability and relatively low cost, these materials present a viable alternative to more expensive and energy-intensive treatment technologies. Adsorption, as a physical process, facilitates the attachment of pollutants to the surface of adsorbents, thereby enhancing their removal from wastewater. Experimental studies were conducted to evaluate the adsorption efficiency of various forms of calcium carbonate, including calcite and aragonite. The findings indicate a significant adsorption capacity, evidenced by marked reductions in the concentrations of heavy metals and organic contaminants in the treated effluents. Furthermore, the research explores the optimal conditions influencing adsorption efficacy, including pH, temperature, contact time, and initial pollutant concentrations. The chapter also elucidates the underlying mechanisms of adsorption, such as electrostatic interactions and chemical bonding, which enhance the effectiveness of the process. Incorporating calcium

S. Elabbas (✉) · A. Lissaneddine · F. Aziz (✉)
National Center for Research and Studies on Water and Energy (CNEREE), Cadi Ayyad University, Marrakech, Morocco
e-mail: salihaelabbas@gmail.com

F. Aziz
e-mail: f.aziz@uca.ac.ma

S. Elabbas
Faculté d'Ingénierie et d'Innovation, Agriculture, Environnement et Gestion de Ressources, Université Privée de Marrakech, Maroc, Marrakech, Morocco

A. Lissaneddine · F. Aziz
Water Sciences, Microbial Biotechnologies and Natural Resources Sustainability Laboratory, Cadi Ayyad University, Marrakech, Morocco

M. A. Aboulhassan and S. Souabi (eds.), *Sustainable Practices in the Tannery Industry*,
https://doi.org/10.1007/978-3-031-92037-0_6

carbonates into an adsorption framework addresses wastewater treatment and aligns with waste management strategies that convert byproducts from the leather industry into valuable resources. In conclusion, the implications of adopting this innovative approach for the Moroccan leather industry are significant. By implementing adsorption as a wastewater treatment method, tanneries can mitigate their environmental footprint while enhancing compliance with regulatory standards. This strategy fosters a transition toward more sustainable industrial practices, contributing to a circular economy that prioritizes resource optimization and waste reduction. The chapter advocates for continued technological innovation and research in wastewater treatment methodologies, emphasizing the potential of utilizing natural and cost-effective materials as sustainable solutions to the environmental challenges confronting the tanning sector in Morocco.

Keywords Adsorption · Tannery wastewater · Chromium · Organic contaminants

6.1 Introduction

6.1.1 Overview of Environmental Pollution and the Role of Tanneries

Pollution is the introduction of harmful substances into the environment. It can come in many forms and from various sources, including industrial, agricultural, and urban activities, each contributing uniquely to the overall environmental burden. The impacts of pollution are diverse and complex, encompassing air, water, and soil contamination, as well as waste accumulation, each necessitating specific strategies for effective reduction and management (Elabbas et al. 2016).

Environmentally sensitive raw materials and inputs are used in leather tanning industries. Thus, tannery wastewater contains considerable amounts of degradable organic waste such as fats, proteins, and nitrogen-containing compounds and a major quantity of non-degradable toxic heavy metals, especially chromium ions and other inorganic contaminants like sulfides. These chlorides increase the BOD and COD. Globally, about 90% of the tanneries discharge their effluents into water resources, thus making the resources highly toxic and unfit to use (Chowdhury et al. 2015). To tackle water pollution effectively, it is essential to implement advanced wastewater treatment technologies, enforce strict regulatory measures, and raise public awareness about sustainable practices. These actions are vital in mitigating the harmful effects of water pollution, thereby preserving water quality and safeguarding the health of ecosystems and human populations (Li et al. 2021).

6.1.2 Importance of Sustainable Remediation Strategies

Tanneries are notable for their substantial environmental impact due to the discharge of effluents containing high concentrations of toxic substances such as heavy metals, dyes, and hazardous compounds. Sustainable remediation strategies are essential to addressing these issues by providing long-term solutions to reduce pollution and restore environmental balance. These approaches involve minimizing waste generation, recovering valuable resources, and ensuring cleaner discharge into natural systems (Ines Del Giudice 2024).

Moreover, adopting such strategies supports global efforts to combat climate change by reducing emissions and preserving natural ecosystems that act as carbon sinks (Singh and Babu 2024). They also help ensure compliance with increasingly stringent environmental regulations, enhance the reputation of the tannery sector, and promote social responsibility by improving living conditions for nearby communities (D'Arcangelo et al. 2023). By prioritizing sustainability, tanneries can mitigate their environmental footprint, foster resilience, and contribute positively to broader ecological and societal goals (FAO 2022).

6.1.3 Objectives and Scope of the Chapter

This chapter explores the potential of adsorption, particularly using calcium carbonate-based materials, as an innovative and sustainable solution for treating wastewater from tanneries in Morocco. It aims to examine the environmental challenges posed by tannery effluents, which contain harmful organic and inorganic pollutants, including heavy metals and dyes. The chapter emphasizes the need for effective, low-cost, and energy-efficient treatment technologies to mitigate the environmental impact of this crucial yet pollutant-generating sector.

The scope of the chapter includes a detailed investigation of the efficiency of calcium carbonate-based adsorbents, such as eggshells and powdered marble, in removing contaminants from tannery wastewater. Experimental studies are discussed, focusing on the adsorption capacity of these materials under varying conditions (e.g., pH, temperature, contact time, and adsorbent dose). The chapter further explores the underlying adsorption mechanisms that enhance chromium removal and COD, including electrostatic interactions and chemical bonding.

In addition to evaluating the technical effectiveness of adsorption, the chapter also addresses the broader implications of using these natural materials for wastewater treatment. It connects the practice of adsorption to waste management strategies that contribute to resource recovery and a circular economy, where byproducts from the tanning process are repurposed into valuable resources.

Ultimately, the chapter advocates for adopting sustainable wastewater treatment solutions in the Moroccan leather industry, highlighting the importance of continued research and innovation in this area. It calls for integrating natural, cost-effective materials into industrial practices to reduce environmental pollution, comply with regulatory standards, and promote a more sustainable future for the sector.

6.2 Environmental Impact of Tanneries

6.2.1 *Tanneries in Morocco: Industry Overview and Economic Significance*

The tanning industry in Morocco, with deep historical roots in cities like Marrakech and Fez, plays a vital role in the national economy. It contributes significantly to the country's export market, representing over 58% of formal production intended for export. Footwear is the primary export product, accounting for 80% of the leather sector's exports. Other products, including leather and hides, comprise 12% of exports, while leather goods and clothing products account for 8% (FEDIC 2015). The industry contributes to economic growth and supports tourism, particularly through the traditional tanneries that attract visitors with their rich heritage.

However, despite its economic importance, the tanning industry in Morocco faces significant environmental challenges. The tanning process generates large quantities of pollutants, including toxic substances like chromium (Cr) and sulfides, organic pollutants (such as high chemical oxygen demand (COD) and biological oxygen demand (BOD_5)), and inorganic pollutants like sulfates and chlorides. Additionally, atmospheric pollutants such as hydrogen sulfide (H_2S) and solid waste, including epidermis and subcutaneous tissue, contribute to the environmental burden.

In cities like Marrakech, many tanning operations are situated in the historic Medina, where they discharge untreated effluents directly into local water bodies. This practice has led to the accumulation of pollution in the region, with annual waste loads comprising 99 tons of BOD_5, 302 tons of COD, over 370 tons of total suspended solids (TSS), 270 tons of chlorides, 138 tons of sulfides, and 40 tons of chromium. The high biodegradability index (COD/BOD_5 ratio of 3) indicates that tannery wastewater is not easily biodegradable, further complicating treatment efforts (Ministry of Energy Transition and Sustainable Development 2023) (Fig. 6.1).

6.2.2 *Pollutants Generated by Tanneries*

Tannery wastewater is characterized by strong alkaline, high salt, and chromium content, with a dark brown color and offensive odor. It has a high content of organic substances that varies according to the type of process and chemical used (Jahan et al. 2014). It is highly contaminated with Cr, S^{2-}, $SO_4{}^{2-}$, suspended solid, BOD_5 and COD content, oil, grease, and chlorides (Suganthi et al. 2013).

Fig. 6.1 Tannage au chrome

Impacts of Organic Pollutants

When high-load BOD_5 is discharged to the river, the much-populated bacteria will grow and require proportional free DO to degrade organic matter. It leads to stress on aquatic organisms, including fish, due to the lack of free available DO. Likewise, BOD_5 and TSS also have a significant impact on aquatic life. The concentrated TSS may cause scatters of sunlight, impaired photosynthetic activities of algae and stress aquatic animals.

Impacts of Heavy Metals

The concentrated levels of heavy metals, including Cr, Zn, Cd, and Pb, are highly toxic to plants. Chromium is one of the major pollutants in the tannery industry, and it is a transitional metal that exists in different oxidation states (from Cr^{+3} to Cr^{+6}). The trivalent chromium (Cr^{+3}) and hexavalent chromium (Cr^{+6}) are the most abundant form. Except for Cr^{+3}, the remains are unstable (Terfie and Asfaw 2015). Cr is removed from the solution as $Cr(OH)_3$ aq. The effect of chromium depends on the concentration of chromium, the organ at risk and exposure. It may cause acute and chronic effects based on concentration and duration (Achmad et al. 2017).

6.2.3 *Current Wastewater Treatment Practices and Limitations*

Several treatment methods have been explored in recent years, such as precipitation/flocculation using sodium hydroxide, calcium hydroxide, and magnesium oxide for Cr(III) removal (Misganaw et al. 2024), ion exchange (Ahmad et al. 2014), membrane processes (Luján-Facundo et al. 2019), chromium adsorption on activated carbon (Nure et al. 2023), and phytoremediation (Fernandez et al. 2023).

However, many of these methods face challenges, including high costs and limited effectiveness (Ahmed ElSidig and Gasmelseed 2020). The need for cost-effective and scalable treatment solutions remains urgent to address the environmental impact of the tannery industry. Finding alternative, sustainable solutions is key to ensuring that Morocco's tannery sector can continue contributing to the economy while minimizing its ecological footprint.

6.3 Adsorption as a Sustainable Remediation Strategy

6.3.1 Fundamentals of Adsorption: Definition and Principles

The adsorption phenomenon involves a mass transfer of solutes (adsorbates) to solid surfaces (adsorbents) in a solution containing both the adsorbates and adsorbents. In the process, the adsorbates become physically and or chemically bound to the surface of the adsorbent (Adekeye et al. 2019a, b). This process is based on intermolecular attractions, which remain by residuals directed outwards. These forces represent surface energy per unit area, comparable to the tension surface of liquids. They are neutralized when mobile particles (a gas or a solute) are fixed on the surface, and we say that they adsorb (Talidi 2006). The adsorption process continues until an equilibrium is obtained, corresponding to the solute's equilibrium concentration.

Various low-cost adsorbents have been synthesized from clay materials, polymers, agricultural wastes, Industrial by-products and so on and applied to remove heavy metals from their aqueous solutions (Maharana et al. 2020a, b). Activated carbons and biochar have also been used to remove various inorganics, including heavy metals (Braghiroli et al. 2018). Other adsorbents that have been used for the removal of heavy metals include photocatalyst beads, red mud, coal, biomass, fertilizer industrial waste, algae, fly ash, waste iron, iron slags, hydrous titanium oxide, activated sludge biomass, etc., has generated increasing excitement. The adsorption materials can be mechanically, physically, and or chemically modified for improved adsorption of metals (Zaimee et al. 2021).

6.3.2 Advantages of Adsorption Over Other Remediation Techniques

Adsorption is a widely recognized method for environmental remediation due to its efficiency and versatility. It offers several key advantages over other techniques:

1. High Efficiency: Adsorption effectively removes many pollutants, including heavy metals, organic compounds, and dyes, even at low concentrations (Zou et al. 2021).
2. Simplicity and Flexibility: The process is straightforward, requiring simple equipment. It can be applied in batch or continuous systems and is easily scalable for different applications (Manna et al. 2023).

3. Cost-Effectiveness: Adsorbents such as activated carbon, agricultural residues, and industrial by-products are affordable and readily available. Their regeneration and reuse further reduce costs (Zaimee et al. 2021).
4. Environmental Friendliness: Adsorption produces minimal secondary waste and often uses natural or bio-based adsorbents, making it an eco-friendly option (Raji et al. 2023).
5. Selectivity: Adsorbents can be customized to target specific contaminants, enhancing treatment efficiency (Gupta and Nayak 2022).
6. Low Energy Demand: Compared to methods like incineration or electrocoagulation, adsorption requires significantly less energy, making it suitable for resource-limited areas (Karnwal and Malik 2024).

6.3.3 Adsorbent Materials: Types and Selection Criteria

Adsorbent materials are widely used to treat tannery wastewater, which contains high concentrations of heavy metals, organic pollutants, and other harmful substances. These materials are selected based on their efficiency, cost, and environmental impact. Activated carbons, derived from carbonaceous sources such as coconut shells and wood, effectively remove organic compounds and heavy metals like chromium, a major contaminant in tannery wastewater (Zhao et al. 2022).

Natural and synthetic clays, including bentonite and montmorillonite, are low-cost and readily available materials with a strong affinity for cationic pollutants. Research by Bhattacharyya and Gupta (2008) highlights their effectiveness in adsorbing dyes and heavy metals from tannery effluents. Agricultural and industrial waste-based adsorbents, such as rice husks, banana peels, and fly ash, offer a sustainable alternative by utilizing waste products. According to Chand et al. (2009), these adsorbents show significant potential for removing chromium and other pollutants when pre-treated or chemically activated.

Emerging materials like biochar and nanomaterials provide advanced solutions for tannery wastewater treatment. Biochar, produced via pyrolysis of organic residues, has been shown to effectively adsorb heavy metals, as noted by Inyang et al. (2016). Nanomaterials, such as graphene and carbon nanotubes, offer exceptional adsorption capacity due to their large surface area and reactivity, as discussed in the work of Qu et al. (2013).

These adsorbent materials are critical for addressing the environmental challenges of tannery wastewater. Their application improves water quality and aligns with sustainable waste management practices.

6.4 Case Study: Adsorption for Chromium Tannery in Marrakesh

The present study aims to evaluate and compare the adsorption capacity of two low-cost carbonaceous adsorbents, eggshell and powdered marble, on chromium removal from highly concentrated chrome tanning wastewater. The effects of adsorbent dose, contact times, effluent pH and temperature were studied. Batch kinetics and adsorption isotherms have also been analyzed.

In this work, the wastewater samples were obtained from a chromium tanning tank in a small semi-traditional tannery in Marrakech city, with a treatment capacity of about 500 skin/day. Generally, the main inorganic contaminant of chrome tanning wastewater is chromium, with a concentration range of 0.5–5 g/L (Sahu et al. 2009). Table 6.1 lists the characteristics of the raw wastewater tested.

The eggshell used in this study came from local cafeteria waste. The collected eggshells were washed with distilled water to remove the adhering membranes and dried in the ambient atmosphere for several days. The dried eggshells were ground, sieved, and stored in a desiccator for further use. The powdered marble came from a large private marble processing unit and was used in the experiment without prior treatment.

The study investigated the removal of chromium ions (Cr(III)) from wastewater using eggshell and powdered marble as cost-effective adsorbents. The results demonstrated that the adsorption efficiency was significantly influenced by adsorbent dose, contact time, and pH.

Effect of Adsorbent Dose

The effect of eggshell and powdered marble doses, ranging from 4 to 50 g and from 1 to 20 g/L, respectively, was examined for a contact time of 24 h (Fig. 6.2). The chromium removal increases as the amount of the adsorbent increases and reaches a

Table 6.1 Main characteristics of the raw wastewater tested

Parameter	Value	SD
pH (at 25 °C)	3.81	0.10
EC (mS/cm) (at 25 °C)	60.0	3.0
Cl^- (g/L)	11.6	0.5
COD (g/L)	3.75	0.10
SO_4^{2-} (g/L)	1.6	0.2
NH_4^+ (mg/L)	33.2	0.4
Total P (mg/L)	7.0	0.2
PO_4^{3-} (mg/L)	2.1	0.1
Cr(III) (g/L)	3.21	0.05
CO_3^{2-} (g/L)	0.41	0.01

SD standard deviation

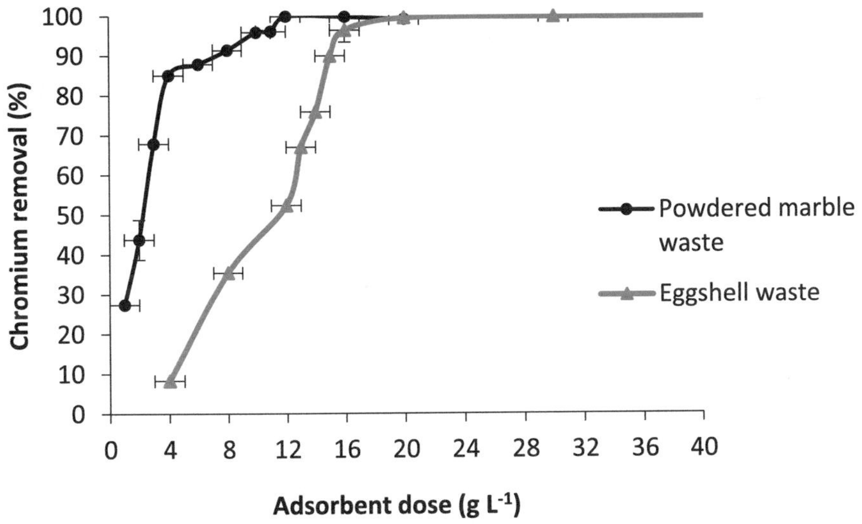

Fig. 6.2 Evolution of the Cr removed by **a** eggshell and **b** powdered marble versus adsorbent dose

maximum value (ca. 100%) at 20 g/L and 12 g/L for eggshell and powdered marble respectively. The adsorbent dose strongly affects chromium removal efficiency. The results could be attributed to increased adsorbent surface areas, augmenting the number of available adsorption sites (Pérez Marín et al. 2009). However, any further increase in an adsorbent concentration greater than the maximum values obtained does not significantly improve the percentage of metal removal because the equilibrium between chromium ions and the solid phase was reached. The difference in their surface areas can explain the difference between the equilibrium doses obtained for the two adsorbents. Therefore, the obtained results may suggest that the number of activate sites is more concentrated on the surface of powdered marble than on the surface of the eggshell, requiring larger dosing.

Effect of Contact Time

The change of chromium capacity adsorption onto eggshell and powdered marble versus time is presented in Fig. 6.3. The adsorption capacity and the time required to reach the equilibrium are indicators of the absorbate transfer kinetics from the liquid phase to the adsorbent surface. The adsorption equilibrium of 99% chromium was obtained within 30 min for powdered marble, and marble (262 mg/g) was much faster than eggshell (159 mg/g). During the first stage, the adsorption rate increases more rapidly in the case of powdered marble due to the availability of a higher number of active sites at this adsorbent surface than for eggshells. The reaction decreases when the remaining active sites are less available, and the equilibrium phase is reached (Ankareddy and Chandra 2023).

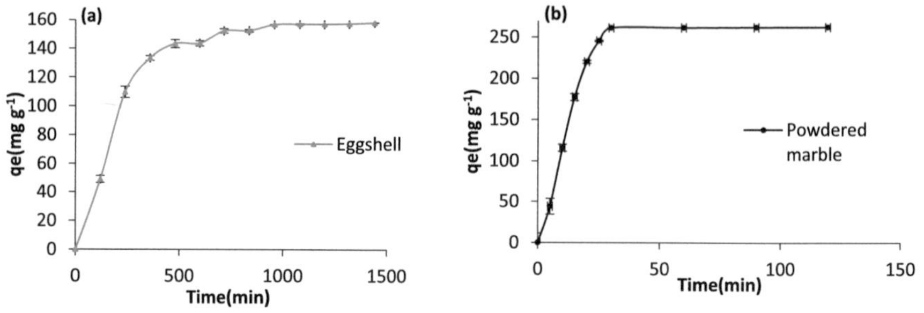

Fig. 6.3 Effect of contact time on the adsorption of Cr(III) ions by eggshell (20 g) and powdered marble (12 g) adsorbents

Effect of pH

The pH of the solution has a vital role in the sorption process since it is responsible for protonation of metal binding sites, calcium carbonate solubility and metal speciation in the solution (Elzinga and Arai 2021). In these experiences, the initial concentration of Cr(III) (3.21 g/L), the amount of eggshell and powdered marble (20 g/L and 12 g/L, respectively) and contact time (24 h) were kept constant. Figure 6.4 represents Cr(III) sorption capacity on eggshell and powdered marble versus pH. The removal efficiency of chromium ions increased when the pH was increased. Still, the eggshell sorbent presents a relatively slight sorption increase, probably indicative of its poor pore rate and accessibility compared to powdered marble absorbent. At lower pH (2 and 3), the low chromium removal may be attributed to (i) the elevated concentration of protons that will compete with chromium cations on the surface sorbent active sites and hence reduce their adsorption; (ii) the dissolution of $CaCO_3$, the principal constituent of the two adsorbents which lowers the adsorption. At higher pH (4 and 5), the important removal of chromium in the range can be attributed to (i) the competitive effect (Hþ vs. Cr^3), which becomes weaker, increasing the adsorption of Cr(III) (Elzinga and Arai 2021). (ii) The adsorption of hydrolysis products and precipitation of the metal as colloidal insoluble hydroxides, $Cr(OH)_3$. Similar results were also reported for Cu(II) adsorption onto eggshells (Rais et al. 2012) and for Al(III) adsorption onto powdered marble (Ghazy and Gad 2014). On the other hand, even though calcite is the leading principal component of both sorbents, there is a difference between their Cr(III) sorption capacities versus pH. This may be explained by the fact that eggshells are composed of calcite ($CaCO_3$) layers, containing embedded proteins that make eggshells less prone to dissolution than calcitealone (Nys and Gautron 2007).

Effect of Temperature

The values obtained for thermodynamic parameters are shown in Table 6.2. The negative value of DG indicates that the adsorption of Cr(III) on the eggshell and powdered marble is feasible and spontaneous. In addition, the absolute DG values increased with increasing temperature, from 298 to 328 K. Similar results were also

Fig. 6.4 Evolution of the adsorption capacity of Cr(III) on eggshell (20 g/L) and powdered marble (12 g/L) versus pH and C_i of 3.21 g/L

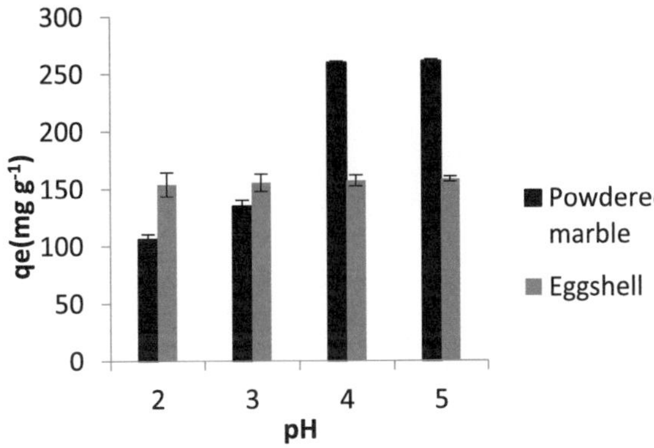

reported for Cr(III) adsorption by natural diatomite (Smith et al. 2022). The positive value of DH shows that the nature of adsorption is endothermic. According to Sepehr et al. (2014), the values of enthalpy changes (102–121 kJ/mol) suggested that the adsorption of Cr(III) onto powdered marble and eggshells may be dominated by chemical adsorption. The positive value of DS reveals the increased randomness at the solid solution interface during the fixation of the Cr(III) ion on the activities of the studied adsorbents. Similar results were also observed by Torab-Mostaedi et al. (2013), who studied the adsorption of cadmium and nickel on grapefruit peels.

The present study shows that the two carbonaceous materials tested (eggshell and powdered marble) are very efficient adsorbents for chromium removal from tanning effluent. Maximum chromium removal was found to beat pH5, and an adsorbent dose of 12 g/L of powdered marble and 20 g/L of eggshell was used for the initial concentration of chromium up to 3.21 g/L. The maximum sorption capacities and the adsorption kinetic rate noticed for powdered marble are higher than those of eggshells. The equilibrium adsorption of chromium by the two studied adsorbents was best represented by the Langmuir other than the pseudo-second-order kinetic

Table 6.2 Thermodynamic parameters for Cr(III) adsorption using eggshell and powdered marble ($C_0 = 3.21$ g L^{-1} and pH = 3.81)

Adsorbent	T (K)	$\Delta G°$ (kJ mol^{-1})	$\Delta H°$ (kJ mol^{-1})	$\Delta S°$ (J mol^{-1} K^{-1})
Eggshell	298	– 5.66 (0.44)[a]		
	308	– 7.34 (0.18)[a]	121.35 (2.90)[a]	423.1 (3.23)[a]
	318	– 12.66 (0.57)[a]		
	328	– 18.28 (0.55)[a]		
Powdered marble	298	– 6.99 (0.27)[a]		
	308	– 10.25 (0.59)[a]	102.83 (3.11)[a]	368.23 (4.21)[a]
	318	– 14.99 (0.94)[a]		
	328	– 17.66 (1.17)[a]		

[a] indicates the standard deviation value

model, indicating the dynamical behavior of the adsorption of chromium onto the two adsorbents studied (Elabbas et al. 2016).

6.5 Mechanisms of Adsorption in Tannery Wastewater Treatment

6.5.1 Physical Adsorption Versus Chemical Adsorption

The treatment of tannery wastewater in Morocco can benefit from applying both physical and chemical adsorption techniques, each offering unique advantages tailored to the diverse composition of effluents. These methods have been extensively studied for their potential to mitigate environmental impacts and meet stringent wastewater discharge standards (Husain et al. 2024; Lissaneddine et al. 2023).

Physical adsorption mainly utilizes adsorbents like activated carbon and bone char, which rely on surface location and pore length to lure pollutants (Fig. 6.5). Activated carbon, derived from tannery strong waste, has tested powerful in getting rid of azo dyes, achieving elimination efficiencies of 51.94 and 49.73% for specific dyes (Mella et al. 2019). The big floor vicinity and porous shape of activated carbon permit the adsorption of large molecular contaminants, making it a flexible solution for tannery wastewater treatment. Similarly, bone char has tested great efficacy in eliminating COD and color from vegetable tannery effluents, with 77.4% and 98.01% removal quotes, respectively (Appiah-Brempong et al. 2024). In those instances, the adsorption system is ruled via film diffusion, and the Freundlich isotherm model successfully describes the adsorption conduct.

Chemical adsorption, alternatively, leverages molecular interactions and the presence of precise purposeful agencies for pollutant elimination (Fig. 6.5). For instance, activated carbon derived from cattle hair waste contains functional groups like –CH=CH–, which play a critical role in the adsorption of toxic compounds (Mella et al. 2019). Another notable chemical adsorption method is adsorbing colloid flotation, which employs chemical collectors to bind pollutants effectively. This technique

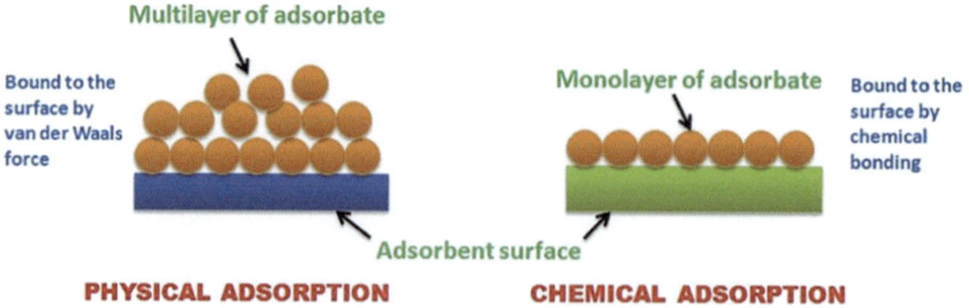

Fig. 6.5 Schematic representation of the physical and chemical adsorption (Cuong 2020)

is particularly feasible for the primary treatment of tannery wastewater, although it often requires integration with other processes for comprehensive treatment (Wang et al. 2007).

While physical adsorption provides a straightforward and adaptable approach using materials like activated carbon and bone char, chemical adsorption offers targeted removal of specific contaminants through molecular-level interactions. The nature of the pollutants and the desired treatment outcomes should guide the choice between these methods.

Integrating physical and chemical adsorption in a hybrid treatment system could further enhance the efficiency of tannery wastewater treatment in Morocco. Combining these methods with advanced processes such as sequential batch reactors and phycoremediation has the potential to address the diverse and complex composition of tannery effluents, paving the way for sustainable and effective wastewater management ("Innovative Strategies for Tannery Wastewater Remediation with Sequential Batch Reactor and Phycoremediation" 2024).

6.5.2 Adsorption Isotherms (Langmuir, Freundlich)

Adsorption isotherms, specifically Langmuir and Freundlich fashions, play an important function in the expertise of the adsorption approaches in tannery wastewater treatment in Morocco. These fashions assist in comparing the efficiency and capability of diverse adsorbents utilized in disposing of contaminants from wastewater. The study papers help discover unique adsorbents and their effectiveness in treating tannery wastewater, highlighting the application of those isotherms (Lissaneddine et al. 2023; Nayak et al. 2021; Mia 2020; Mella et al. 2019).

The Langmuir isotherm equation (Eq. 6.1) shows an adsorption model of a homogeneous system with a single layer, and it is expressed as follows:

$$\frac{C_e}{q_e} = \frac{1}{K_L \times q_m} + \frac{C_e}{q_m} \tag{6.1}$$

where q_e (mg g^{-1}) is the pollutant amount adsorbed at equilibrium, K_L is the Langmuir constant related to the energy of adsorption (L g^{-1}), and q_m (mg g^{-1}) is the maximum pollutant adsorption capacity.

The Freundlich adsorption isotherm model is applied for multiple-layer adsorption and is founded on a heterogeneous process. The Freundlich adsorption isotherm equation (Eq. 6.2) is given as follows:

$$\ln q_e = \ln K_f + \frac{1}{n} \times \ln C_e \tag{6.2}$$

where K_f is a Freundlich constant that presents the adsorption capacity of the adsorbent, and n is a constant that presents the greatness of the relationship between pollutant and adsorbent.

The Freundlich isotherm model was found to describe best the adsorption of COD and color onto bone char in a study, indicating a heterogeneous surface adsorption process. The maximum monolayer adsorption capacities were 559.61 mg/g for COD and 2011.76 Pt–Co/g for color, demonstrating the effectiveness of bone char as an adsorbent (Appiah-Brempong et al. 2024). This model is particularly useful for systems with non-uniform surface energies, which is common in real-world wastewater treatment scenarios.

Although no longer explicitly highlighted inside the provided papers, the Langmuir isotherm generally assumes monolayer adsorption onto a surface with a finite range of identical websites. This model is frequently used to describe adsorption methods wherein the adsorbent surface is homogeneous.

While the Freundlich model was most effective in the context, this highlights the importance of selecting appropriate isotherm models based on the specific adsorbent and contaminants involved. This underscores the complexity and variability in adsorption processes for tannery wastewater treatment.

6.5.3 Kinetics and Thermodynamics of Adsorption Processes

The kinetics and thermodynamics of adsorption tactics in tannery wastewater treatment in Morocco contain various advanced strategies aimed at efficaciously disposing of pollution, particularly chromium and organic compounds. Recent studies spotlight the combination of advanced oxidation procedures with adsorption strategies, showcasing extensive improvements in pollutant elimination performance.

A simple kinetic analysis of adsorption is the pseudo-first-order equation (Lagergren 1898):

$$\log(q_e - q_e) = \log(q_e) - \frac{K_1}{2.303} \times t \tag{6.3}$$

On the other hand, the pseudo-second-order kinetic model has been expressed in the form (Ho and McKay 1999):

$$\frac{t}{q_t} = \frac{1}{K_2 \times q_e^2} + \frac{t}{q_e} \tag{6.4}$$

Adsorption techniques have additionally shown enormous promise, with substances like kaolin and kaolin/ZnO nanocomposites displaying great elimination capabilities. These substances completed 100% elimination of Cr(VI) and verified excessive efficiencies for other pollutants, consisting of COD and BOD (Mustapha et al. 2020). In addition, unconventional adsorbents like tea waste had been explored for their effectiveness in chromium removal, achieving a 97% reduction in Cr(VI) underneath optimized conditions. The adsorption kinetics in this research accompanied a pseudo-second-order model, emphasizing the chemical interactions between adsorbent and adsorbate (Nigam et al. 2019).

The thermodynamic parameters are the Gibbs energy (ΔG in kJ mol^{-1}), standard entropy (ΔS in J mol^{-1} K^{-1}) change, and standard enthalpy (ΔH in kJ mol^{-1}) change of the adsorption process (Eqs. 6.5–6.7) (Lissaneddine et al. 2021):

$$\Delta G = -R \times T \times \ln K_c \tag{6.5}$$

$$\Delta G = \Delta H - T \times \Delta S \tag{6.6}$$

$$\ln K_c = -\frac{\Delta H}{R \times T} + \frac{\Delta S}{R} \tag{6.7}$$

where K_c is the thermodynamic distribution coefficient calculated using Eq. 6.8:

$$K_c = \frac{q_e}{C_e} \tag{6.8}$$

Thermodynamic analyses screen that most adsorption approaches are endothermic, indicating stronger pollutant removal at higher temperatures (Mustapha et al. 2020; Nigam et al. 2019). The mechanisms underlying those tactics involve both intra-particle and film diffusion, reflecting the complexity of adsorption dynamics (Mustapha et al. 2020). These findings underscore the importance of optimizing operational parameters to maximize the performance of adsorption-primarily based remedy systems (Fig. 6.6).

Aggregating advanced oxidation tactics and adsorption techniques offers a flexible and effective technique for treating commercial wastewater. By addressing each organic and inorganic pollutant, those techniques provide an incorporated answer to fulfill stringent environmental requirements while selling sustainable wastewater control practices.

While these superior techniques show promise, challenges remain, including managing residual chromium in sludge and effluent salinity, necessitating research into optimizing treatment mixtures and value effectiveness (Zhao et al. 2022).

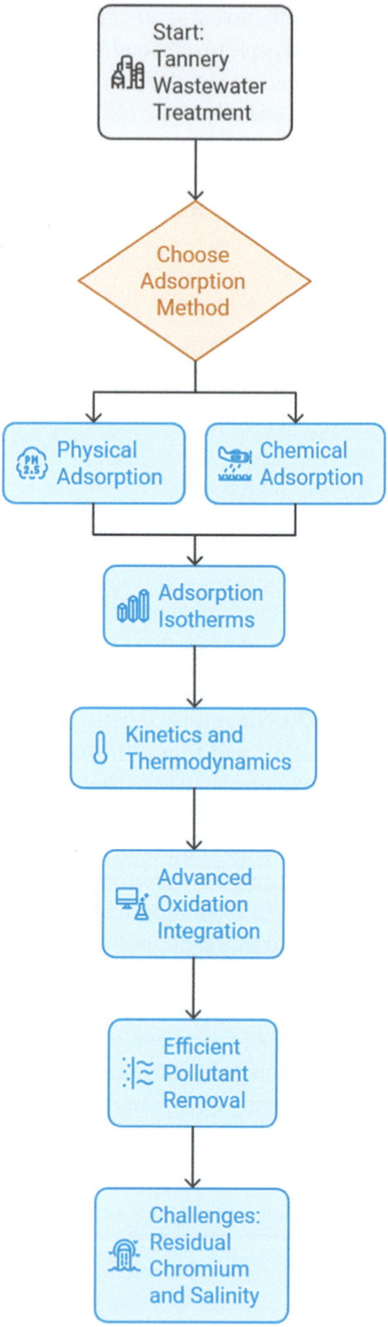

Fig. 6.6 Kinetics and thermodynamics of adsorption processes

6.6 Economic and Environmental Considerations

6.6.1 Cost–Benefit Analysis of Adsorption-Based Solutions

The cost-advantage evaluation of adsorption-primarily based answers for tannery wastewater treatment in Morocco shows tremendous blessings over traditional methods. Adsorption techniques, especially whilst blended with flotation strategies, exhibit high performance in doing away with pollutants, along with heavy metals like chromium, which might be generic in tannery effluents. This evaluation will explore the economic feasibility, effectiveness, and capacity for resource restoration related to these strategies.

Economic feasibility, treatment efficiency, and resource restoration are essential in assessing adsorption-based total wastewater remedy structures for tannery effluents. These factors no longer most effectively determine such structures' financial viability; additionally, their environmental and operational benefits over traditional strategies.

An adsorption-based treatment plant demonstrates strong economic potential, with a projected Net Present Value (NPV) of $31 million over a ten-year operational period under optimal conditions (Mak 2019). In contrast, conventional wastewater treatment methods often yield negative NPVs due to higher operational costs and limited financial returns, highlighting their lower economic viability (Mak 2019). These findings underscore the economic advantage of advanced adsorption techniques in tannery wastewater management.

From an efficiency perspective, adsorption processes, particularly when combined with techniques like flotation, achieve significant removal of both dissolved and suspended pollutants (Wang et al. 2007). Recent advancements in nanotechnology have further enhanced treatment outcomes. For instance, nanostructured chitosan and iron-oxide nanoparticles have demonstrated superior removal efficiencies for hazardous metals and dyes, making them promising candidates for integration into adsorption-based systems.

In addition to pollution mitigation, adsorption systems offer resource recovery potential, which can contribute to offsetting treatment costs. By-products such as steam generated during treatment can be monetized, with an estimated revenue potential of $6.2 million (Mak 2019). Furthermore, recovering solid wastes from tannery operations for composting or extracting valuable materials presents additional economic opportunities (Sawalha et al. 2020).

Combining high treatment efficiency, economic feasibility, and resource restoration competencies makes adsorption-based totally systems a compelling opportunity for conventional wastewater treatment techniques. These structures align with sustainable development dreams by lowering environmental impact, promoting useful resource performance, and ensuring long-term economic viability.

While adsorption-based total solutions gift a promising alternative for tannery wastewater remedy, demanding situations stay in the initial investment and era adoption, mainly in developing regions like Morocco. Balancing these factors is crucial for sustainable implementation.

6.6.2 Life Cycle Assessment (LCA) of Adsorbents: From Production to Disposal

The lifecycle evaluation of adsorbents used in tannery wastewater treatment in Morocco encompasses various ranges, from manufacturing to disposal. This evaluation highlights the environmental blessings of using low-value, natural adsorbents derived from waste substances, which could effectively eliminate harmful contaminants like chromium from tannery effluents. The following sections detail these adsorbents' production, utility, and disposal aspects (Fig. 6.7).

Manufacturing adsorbents from herbal and waste-derived materials offers an environmentally pleasant and fee-effective solution for wastewater treatment. Materials consisting of fish scales and wooden biochar, sourced from comfortably available waste streams, lessen the environmental burden of disposal even by supplying high-overall performance adsorbents for pollutant elimination (Akter et al. 2024; Singh et al. 2024). These natural materials are further enhanced through chemical treatments, such as sodium hydroxide (NaOH) or orthophosphoric acid, which significantly increase their surface area and porosity, improving adsorption capacity (Akter et al. 2024).

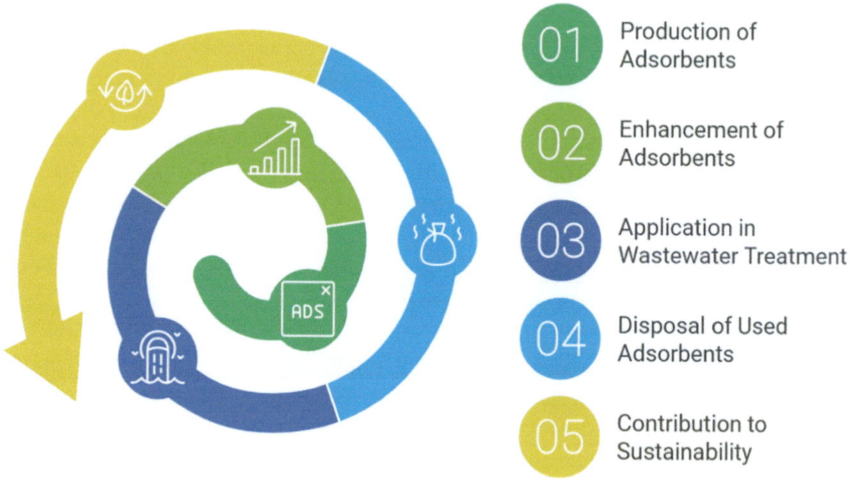

Fig. 6.7 Lifecycle assessment of natural adsorbents

Studies have suggested elimination efficiencies as high as 96% for pollution like Cr(VI) and different contaminants, highlighting their capability for addressing business effluents (Singh et al. 2024; Bhardwaj et al. 2023). Additionally, waste-derived adsorbents give a value-effective alternative to traditional treatment methods, making them specifically attractive for aid-confined settings (Nashy et al. 2022).

The environmental effect and disposal of used adsorbents also align with sustainability standards. By using waste substances to create adsorbents, these systems reduce the environmental footprint of wastewater treatment and cope with broader waste control demanding situations, contributing to a round economic system (Nashy et al. 2022). Moreover, the treated effluents regularly observe national discharge standards, thereby minimizing environmental dangers and ensuring regulatory compliance (Nashy et al. 2022).

Incorporating adsorbents derived from natural and waste materials into wastewater treatment techniques represents a sustainable and effective approach. Combining excessive removal efficiencies, value-effectiveness, and environmental advantages, those substances help the transition towards greater sustainable wastewater management practices.

While using natural adsorbents affords numerous blessings, demanding situations continue to ensure steady overall performance and scalability in various operational contexts. Further studies are needed to optimize these tactics for broader application in Morocco's tannery enterprise.

6.6.3 *Challenges in Scalability and Integration into Existing Systems*

The challenges in the scalability and integration of tannery wastewater treatment systems in Morocco are multifaceted, mostly stemming from technological, financial, and regulatory factors. Despite improvements in remedy technologies, the transition from pilot to business scale remains a good-sized hurdle, compounded by the excessive pollutant concentrations in tannery effluents and the reluctance of industries to undertake eco-friendly practices. The following sections outline key challenges confronted in this context.

Advancements in wastewater remedy technologies, including systems with ahead osmosis and nanofiltration, offer promising answers for tannery effluents. However, those systems regularly contain state-of-the-art designs and operations, posing sizeable challenges to scalability and practical implementation (Pal et al. 2017; Nayak et al. 2021). One crucial difficulty is membrane fouling, which reduces the efficiency and longevity of these systems, necessitating frequent preservation and increasing operational complexity (Pal et al. 2017; Moktadir et al. 2023).

The economic constraints associated with advanced treatment technologies further hinder their adoption. High initial capital investments required to install these systems often deter tannery operators, especially those focused on maximizing short-term profits (Nayak et al. 2021; Elkarrach et al. 2021). In addition, the non-stop operation of these complex structures incurs substantial preservation and power charges, challenging their long-term financial feasibility (Saravana Kumar et al. 2023).

Regulatory and environmental elements also play a pivotal role in adopting wastewater treatment technologies. While tanneries are under increasing strain to comply with stringent environmental standards, many industries are sluggish to conform due to earnings-pushed reasons and a loss of instant economic incentives (Nayak et al. 2021; Elkarrach et al. 2021). Moreover, inadequate governmental help, along with subsidies or tax incentives for sustainable practices, limits the willingness of tannery operators to put money into powerful wastewater treatment answers (Nayak et al. 2021).

Addressing these technological, financial, and regulatory limitations is essential to advancing sustainable wastewater remedies in the tannery sector. Solutions with stronger governmental aid, revolutionary cost-reduction techniques, and improved sturdy and coffee-preservation technologies may want to bridge the distance between promising innovations and realistic programs.

In assessment, whilst the challenges are full-size, there is potential for revolutionary answers through hybrid technologies and collaborative efforts among industries and regulatory bodies that could beautify the scalability and integration of wastewater remedy systems in Morocco.

6.7 Policy and Regulatory Framework

6.7.1 Environmental Standards for Tannery Effluent in Morocco

In Morocco, environmental standards for tannery effluent generally aim to reduce the discharge of hazardous pollutants, particularly chromium and sulfides, commonly found in tannery wastewater. The Ministry of Energy, Mines, and Environment, together with local government, oversees the control of those effluents. Regulations are designed to mitigate the environmental impact of tannery operations, focusing on water pollutants. Recent studies imply that superior treatment techniques can help comply with these requirements, ensuring that handled effluents are safe for discharge into the surroundings. Key criminal texts guide those efforts to defend water resources efficiently (Elkarrach et al. 2020; Nunes et al. 2014).

- Law No. 11-03: This regulation addresses protecting and enhancing the environment, placing the muse for environmental standards in diverse industries, including tanning.
- Decree No. 2-07-253: This decree specifies the bounds for discharges into water our bodies, along with those from tanneries.

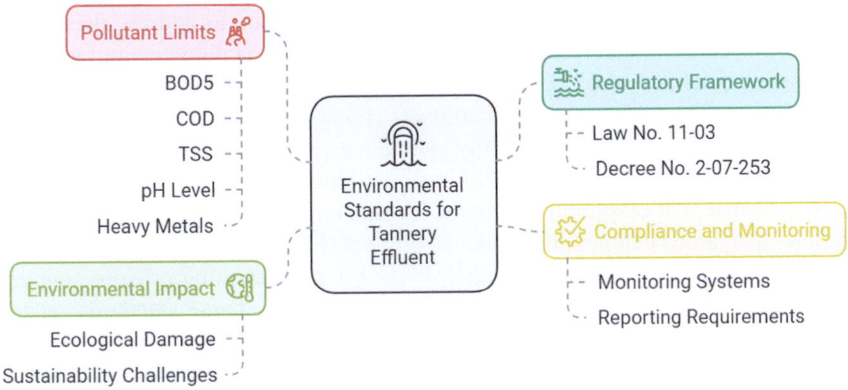

Fig. 6.8 Environmental standards for tannery effluent in Morocco

The effluent from tanneries should follow unique limits for diverse pollution to maintain water-friendliness (Fig. 6.8). The following are a number of the crucial parameters and their respective limits:

- BOD_5: Maximum of 300 mg/L
- COD: Maximum of 600 mg/L
- TSS: Maximum of 200 mg/L
- pH Level: Must be among 6 and 9
- Heavy Metals: Specific limits for metals along with chromium, lead, and cadmium, which have not to exceed 0.1 mg/L for chromium and 0.05 mg/L for lead.

Tanneries in Morocco must establish tracking systems to ensure adherence to environmental requirements. This consists of normal sampling and analysis of effluent, with the results pronounced to the applicable government. Failure to conform can also result in penalties, such as fines or the suspension of operations. The tanning enterprise has a massive environmental footprint, mainly in phrases of water pollutants. The release of untreated or insufficiently treated effluent can cause tremendous ecological harm, which includes harm to aquatic ecosystems and contamination of water resources. Therefore, compliance with environmental regulations is crucial to shield herbal resources and promote the industry's sustainability.

Tannery effluent is characterized by high concentrations of hazardous pollution, mainly chromium and sulfides, which pose substantial environmental and fitness dangers. Chromium, particularly in its hexavalent shape (Cr(VI)), is toxic and a primary issue in wastewater management. Studies have verified the efficacy of bioaugmentation strategies, attaining up to 96% elimination of Cr(VI) through specialized bacterial traces (Elkarrach et al. 2020). Sulfides are another critical pollutant commonly found in tannery effluents. Their effective removal has been reported to reach rates as high as 99.98% when employing combined chemical and biological treatment processes (Elkarrach et al. 2021).

Advanced treatment technologies have been developed to address these challenges. Combined structures, such as ferric chloride precipitation incorporated with sequencing batch reactors (SBR), have established over 99% removal efficiency for chromium, sulfides, and other contaminants (Elkarrach et al. 2021). Additionally, bioaugmentation gives an eco-friendly alternative via leveraging going on microorganisms to enhance the degradation of precise pollution. This method is now not the most effective goal chromium; however, it also contributes to the general discount of other dangerous compounds inside the effluent (Elkarrach et al. 2020).

Regulatory compliance is a crucial driver for implementing effective tannery wastewater treatment systems. In Morocco, effluent discharge standards require treated wastewater to meet stringent COD, chromium, and sulfide limits. Studies have shown that advanced treatment methods can consistently meet these standards, ensuring environmental protection and aligning with national policies (Elkarrach et al. 2021; Ilou et al. 2014).

Integrating modern technologies and eco-friendly practices, which include bioaugmentation, into tannery wastewater management techniques is vital for mitigating pollution. These techniques no longer best decorate pollutant removal efficiency but additionally help compliance with regulatory requirements, contributing to sustainable business operations.

While considerable progress has been made in treating tannery effluents to fulfill environmental requirements, demanding situations constantly achieve these outcomes throughout all facilities. Continuous monitoring and development of treatment technologies are crucial for sustainable practices in the tannery industry.

6.7.2 Role of Government and Stakeholders in Promoting Sustainable Practices

The function of government and stakeholders in promoting sustainable practices in tannery wastewater treatment in Morocco is important for addressing environmental challenges and ensuring compliance with rules. Effective collaboration among numerous entities can result in the adopting of progressive treatment technologies and sustainable practices.

The Moroccan authorities have carried out stringent discharge requirements for tannery wastewater to mitigate its environmental impact. These rules mandate the usage of superior remedy techniques to ensure compliance with permissible degrees of pollutants, consisting of COD, chromium, and sulfides (Zhao and Chen 2019). In addition to promoting sustainability, adopting Zero Liquid Discharge (ZLD) systems is encouraged. ZLD systems facilitate the recycling and reuse of treated wastewater, significantly lowering the release of dangerous effluents into the environment (Ricky et al. 2022).

Effective tannery wastewater control calls for collaboration among government bodies, industry stakeholders, and researchers. Such partnerships are essential for transferring innovative technologies and promoting sustainable practices (Nayak et al. 2021). Financial incentives and subsidies are critical to supporting local tanneries in adopting eco-friendly technologies. For instance, bioelectrochemical systems, which generate bioelectricity while simultaneously treating wastewater, offer a dual benefit of energy production and pollution control, making them a promising solution for Moroccan tanneries (Elabed et al. 2019).

Advancements in treatment technologies have enabled compliance with Moroccan discharge standards for tannery wastewater. Combined systems, such as ferric chloride precipitation paired with sequencing batch reactors (SBR), have effectively reduced pollutant loads. These integrated approaches ensure regulatory compliance and enhance the overall sustainability of tannery operations (Elkarrach et al. 2021).

Advancements in treatment technology have enabled compliance with Moroccan discharge requirements for tannery wastewater. Combined structures and ferric chloride precipitation paired with sequencing batch reactors (SBR) have proven excellent efficacy in reducing pollutant masses. These included processes ensure regulatory compliance and decorate the overall sustainability of tannery operations (Elkarrach et al. 2021).

Through sturdy regulatory frameworks, financial assistance, and the advertising of innovative solutions, Morocco is fostering a transition towards environmentally sustainable practices within the tannery enterprise. These measures align with international efforts to reduce business pollution and guard water assets.

Stakeholders must invest in and sell these revolutionary answers to beautify sustainability within the tannery quarter.

While authorities' regulations are pivotal, the energetic participation of stakeholders, industry gamers, and researchers is equally critical in fostering a subculture of sustainability in tannery wastewater management.

6.7.3 Recommendations for Integrating Adsorption Technologies into National Policies

Integrating adsorption technology into national guidelines requires a multifaceted technique that addresses environmental, monetary, and technological factors. The destiny prospects for those technologies are promising, especially concerning carbon seizure, water remedy, and electricity efficiency. The following sections outline key recommendations for policy integration and future study instructions.

Governments play a pivotal position in developing adsorption technologies by helping with research and improvement tasks. Funding should be directed toward looking at superior adsorbents, especially biochar and nanomaterials, which demonstrate big capacity in CO_2 seize and pollutant elimination applications (Dicko et al. 2018; Das et al. 2023). Promoting interdisciplinary collaboration between academia, industry, and government is similarly vital to foster innovation and facilitate the realistic utility of this technology in numerous sectors (Dicko et al. 2018).

Establishing robust regulatory frameworks is another essential degree of coverage. Such rules must encourage the adoption of adsorption technology, mainly in industries with excessive environmental footprints, along with power generation and wastewater treatment (Das et al. 2023; El-Baz et al. 2020). Incentives for sectors to transition to sustainable practices can similarly boost the mixing of adsorption structures into mainstream applications.

Destiny studies should focus on improving novel adsorbents with excessive overall performance and adaptability to varying operational situations to cope with the evolving demands of adsorption generation. Such advancements can enhance performance and expand the applicability of adsorption processes (Das et al. 2023; Brauer 1985).

Another promising route includes integrating adsorption structures with renewable strength sources. For example, coupling adsorption methods with sun thermal systems should beautify energy efficiency, decreasing the environmental effect of treatment systems (Sawalha et al. 2020).

Scalability is likewise an urgent situation. Large-scale implementation of adsorption approaches requires devoted studies to ensure that laboratory successes can be correctly translated into business packages without compromising efficiency or fee effectiveness (Dicko et al. 2018).

Through strategic coverage assistance and centered research efforts, adsorption technologies can become a cornerstone of sustainable business practices, addressing worldwide environmental demanding situations whilst selling monetary growth.

While combining adsorption technology into the country's broad guidelines provides giant possibilities, demanding situations such as fees, public popularity, and technological readiness must be addressed to ensure a successful implementation.

6.8 Future Prospects and Research Directions

6.8.1 Innovations in Adsorbent Materials

Innovations in adsorbent substances for tannery wastewater remedies in Morocco have seen sizeable improvements, specifically in using natural and engineered materials. These improvements focus on enhancing elimination efficiencies of harmful contaminants, which include chromium and natural pollutants while ensuring fee effectiveness and sustainability.

Natural substances, including gravel, clay, lime rock, and seashells, have been investigated as powerful and green adsorbents for removing pollutants from wastewater. Among those, lime rock tested the best chromium (VI) elimination efficiency, reaching a removal rate of 63.59% and an adsorption capacity of 65.5 mg/g under top-rated conditions (Fseha et al. 2024). The number one mechanism using this elimination is electrostatic appeal, stimulated through parameters including pH, contact time, and adsorbent dosage (Fseha et al. 2024).

While those innovations gift promising solutions, challenges continue to be in scaling those technologies for tremendous application in Morocco's tannery industry, particularly concerning monetary feasibility and regulatory compliance.

6.8.2 Hybrid Systems Combining Adsorption with Other Treatment Techniques

Hybrid systems that integrate adsorption with other treatment techniques have proven full-size promise in the remedy of tannery wastewater in Morocco. These integrated procedures enhance the removal performance of pollution, specifically chromium and natural compounds, which are widely widespread in tannery effluents (Fig. 6.9).

The aggregate of UV/TiO_2/H_2O_2 superior oxidation with adsorption has been powerful, reducing COD from 7600 to 504 mg/L within nine hours. This method additionally converts Cr(III) to Cr(VI), which is then effectively adsorbed with the use of anion alternate resin, resulting in a total chromium removal efficiency of 99.5% (Yu et al. 2024).

A hybrid treatment involving ferric chloride precipitation and a sequencing batch reactor (SBR) has demonstrated remarkable results, with removals of 99.89% for COD and 99.99% for total chromium. This method is effective and economically viable for treating tannery effluents (Elkarrach et al. 2021).

Using co-immobilized microbial biochar beads in hybrid biofilters has reduced Cr(VI) by 92.5% and COD by 89.5%. This method utilizes electrostatic attraction and complexation to enhance pollutant removal (Deepa et al. 2022).

While these hybrid systems present effective solutions, challenges such as operational costs and the need for further optimization remain. Future studies should focus on refining those approaches to maximize performance and decrease fees.

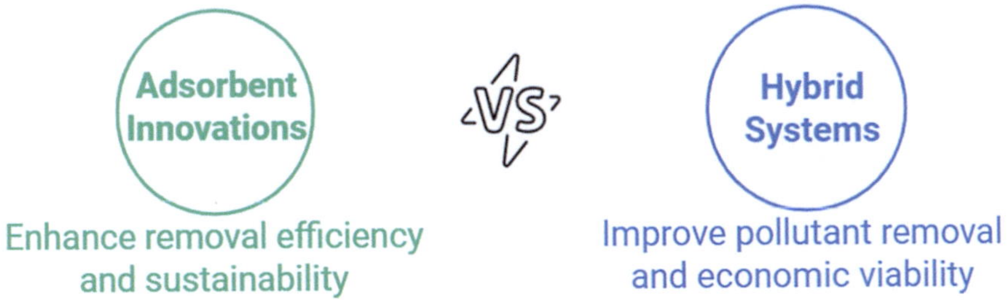

Fig. 6.9 Which wastewater treatment approach should be prioritized for tannery effluents in Morocco?

6.8.3 Role of Artificial Intelligence and Machine Learning in Optimizing Adsorption Processes

Integrating artificial intelligence (AI) and system mastering (ML) in optimizing adsorption strategies for tannery wastewater treatment in Morocco has proven widespread promise. These technologies enhance the performance and effectiveness of treatment strategies, especially when dealing with the complicated pollution observed in tannery effluents.

Artificial Intelligence (AI) has been instrumental in optimizing wastewater treatment adsorption approaches, particularly via advanced modeling strategies. Artificial Neural Networks (ANN) and Adaptive Neuro-Fuzzy Inference Systems (ANFIS) have been hired to expect and optimize COD elimination. Comparative studies indicate that while both methods exhibit high accuracy, ANFIS outperforms ANN in terms of precision and computational efficiency, delivering low error rates in predictions ("Application of Artificial Neural Network (ANN) and Adaptive Neuro-Fuzzy (ANFIS) Techniques for the Modelling and Optimization of COD Adsorption Process" 2022).

Hybrid models, such as integrating ANN with Genetic Algorithms (GA), have enhanced prediction capabilities in tannery wastewater treatment. These models have achieved exceptional accuracy, with maximum prediction errors as low as 0.28%, demonstrating their reliability in optimizing treatment parameters and outcomes (Yu et al. 2024). Such hybrid approaches offer a robust framework for managing complex adsorption systems by combining the strengths of multiple AI techniques.

AI applications extend beyond optimization to improve the efficiency of advanced treatment methods. Integrating Advanced Oxidation Processes (AOPs) and UV/TiO_2/H_2O_2 with adsorption has achieved good-sized discounts in COD levels. These blended methods exhibit AI's potential to beautify the performance of conventional strategies through identifying ultimate operational situations and remedy sequences (Yu et al. 2024).

One of the vital advantages of AI in adsorption technology is its capability to lessen operational charges. AI contributes to more excellent, efficient, and helpful resource usage by minimizing the need for trial-and-error experiments and streamlining process design. Moreover, AI-driven optimization guarantees specific manipulation over treatment parameters, essential for attaining regular outcomes in industrial packages (Vijayan et al. 2024).

Through the software of AI strategies, adsorption-based wastewater remedy approaches may be appreciably advanced in phrases of accuracy, performance, and fee effectiveness, paving the manner for sustainable and superior solutions to environmentally demanding situations.

While AI and ML gift vast blessings in optimizing adsorption processes, demanding situations stay in the scalability and flexibility of this technology to diverse wastewater compositions, necessitating further research and development.

6.9 Conclusion

6.9.1 *Summary of Key Findings and Implications*

The treatment of tannery wastewater in Morocco has been extensively studied, focusing on progressive adsorption strategies. The findings spotlight the effectiveness of combining superior oxidation approaches with adsorption to reduce harmful pollution, such as chromium and natural compounds, establishing extensive discounts in COD and BOD.

Similarly, hyphenated Fenton oxidation paired with adsorption yielded excessive elimination fees for COD and Cr(VI), while photocatalytic oxidation mixed with adsorption performed an exquisite 99.5% overall chromium elimination. These effects underscore the capability of coupling adsorption with advanced treatment technologies for enhanced performance.

In addition to high-tech solutions, cost-effective natural adsorbents, such as fish scales, have shown promise for addressing tannery effluent treatment. These low-cost, eco-friendly materials provide dual benefits by effectively removing pollutants and repurposing waste products (Fig. 6.10).

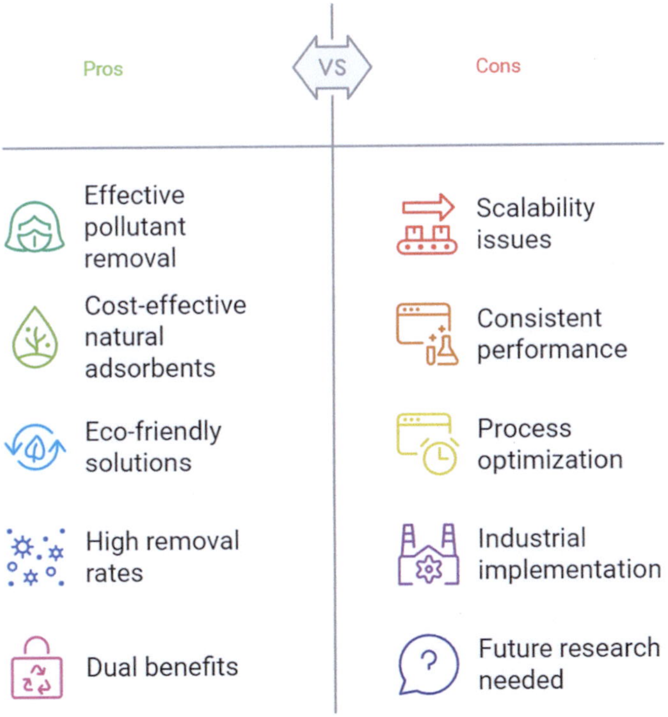

Fig. 6.10 Tannery wastewater treatment in Morocco

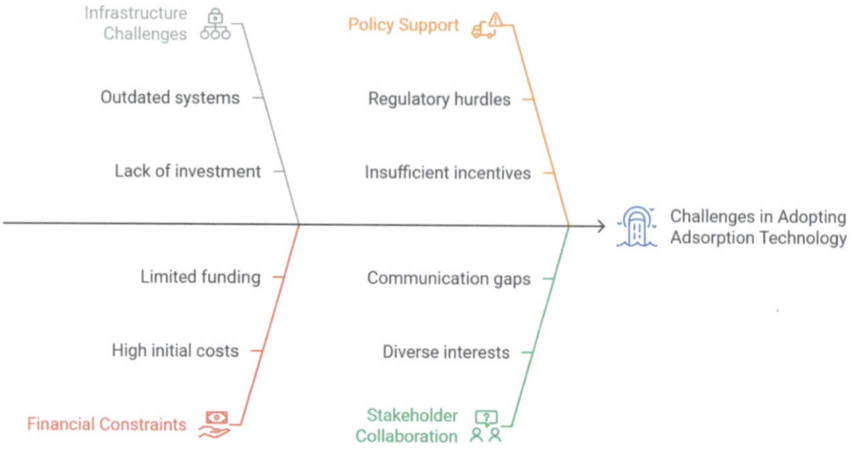

Fig. 6.11 Overcoming challenges in tannery wastewater remediation

However, challenges remain in scaling these methods for broader application. Ensuring consistent performance across varying wastewater compositions and optimizing processes for industrial implementation are critical areas for future research. Addressing these limitations will be essential for realizing the full potential of adsorption-based treatments in Morocco.

6.9.2 Final Remarks on the Potential of Adsorption for Tannery Wastewater Remediation in Morocco

The capacity for adsorption in tannery wastewater remediation in Morocco is sizable. The integration of adsorption with other advanced remedy techniques not only complements pollutant removal but also supports sustainable wastewater control practices. These techniques align with each environmental protection and economic development objectives.

Despite these promising advancements, the tremendous implementation of such technologies faces big challenges. Infrastructure obstacles, high initial costs, and monetary feasibility limit adoption. However, these demanding situations can be mitigated with targeted policy aid, improved studies investment, and powerful stakeholder collaboration.

By addressing those barriers, Morocco can leverage adsorption technology as a cornerstone for sustainable and green tannery wastewater management. This could enhance environmental outcomes and contribute to a circular financial system and lengthy-term aid sustainability (Fig. 6.11).

References

Achmad RT, Budiawan B, Auerkari EI (2017) Effects of chromium on human body. Annu Res Rev Biol 13(2):1–8

Adekeye DK, Asaolu SS, Adefemi SO, Ibigbami OA (2019a) Heavy metal adsorption properties of the basement complex of clay deposit in Ire-Ekiti southwestern Nigeria. IOSR-JESFT 13(2):01–08

Adekeye DK, Asaolu SS, Adefemi SO, Ibigbami OA, Olagboye SA (2019b) Kinetics, isotherm and thermodynamic properties of the basement complex of clay deposit in Ire-Ekiti southwestern Nigeria for heavy metals removal. J Environ Chem Eng 7(3):123–135

Ahmed ElSidig NO, Gasmelseed GA (2020) Cost-effectiveness between methods of treatment of tannery wastewater. Int J Environ Sci Dev 2(4):197–210

Ahmad T, Mustafa S, Naeem A , Anwar F, Mehmood T, Shah KH (2014) Ion Exchange Removal of Chromium (III) from Tannery Wastes by Using a Strong Acid Cation Exchange Resin Amberlite IR-120 H+ and its Hybrids. J Chem Soc Pak 36(5)

Akter S, Naher UHB, Sultana R (2024) Development of low-cost natural adsorbent for the abatement of pollution from tannery effluent—a green technology. https://doi.org/10.1016/j.clwat.2024.100005

Ankareddy S, Matli CS (2023) Experimental Studies on the Removal of Chromium from Tannery Wastewater using Chemical Precipitation and Adsorption Techniques. Curr World Environ 2023 18(1). https://doi.org/10.12944/CWE.18.1.15

Appiah-Brempong M, Essandoh HMK, Asiedu NY, Momade FY (2024) Adsorption du charbon osseux de la DCO et de la couleur des eaux usées des tanneries: analyse de courbe révolutionnaire et modélisation dynamique du lit fixe. Avanc Gén Civ 2024(1):6651094

Application of Artificial Neural Network (ANN) and Adaptive Neuro Fuzzy (ANFIS) Techniques for the Modelling and Optimization of COD Adsorption Process (2022), pp 525–537. https://doi.org/10.1007/978-981-16-2183-3_49

Bhardwaj A, Kumar S, Singh DP (2023) Tannery effluent treatment and its environmental impact: a review of current practices and emerging technologies. Water Qual Res J Can 58(2):128–152. https://doi.org/10.2166/wqrj.2023.002

Bhattacharyya KG, Gupta SS (2008) Adsorption of a few heavy metals on natural and modified kaolinite and montmorillonite: a review. Adv Coll Interface Sci 140(1):114–131

Braghiroli FL, Bouafif H, Neculita CM, Koubaa A (2018) Activated biochar as an effective sorbent for organic and inorganic contaminants in water. Water Air Soil Pollut 229(230):2340–2340

Brauer H (1985) Die Adsorptionstechnik—ein Gebiet mit Zukunft. Chem Ing Tec 57(8):650–663. https://doi.org/10.1002/CITE.330570803

Chand R, Narimura K, Kawakita H, Ohto K, Watari T, Inoue K (2009) Grape waste as a biosorbent for removing Cr (VI) from aqueous solution. J Hazard Mater 163(1):245–250. https://doi.org/10.1016/j.jhazmat.2008.06.084

Chowdhury S, Chowdhury JR, Goswami S (2015) The importance of non high density lipoprotein cholesterol in dyslipidaemia management. J Diabetes Metab 6:623

Cuong DV (2020) Eco-friendly biochar-based materials in removal of contaminants from aqueous solutions: fabrication, characterization and applications. https://doi.org/10.6342/NTU202003877

D'Arcangelo FM, Kruse T, Pisu M (2023) Identifying and tracking climate change mitigation strategies. OECD

Das A, Peu SD, Hossain MS, Nahid MMA, Karim FRB, Chowdhury H, Porag MH, Argha DBP, Saha S, Islam ARMT, Salah MM, Shaker A (2023) Advancements in adsorption-based carbon dioxide capture technologies-A comprehensive review. Heliyon 9(2023):e22341. https://doi.org/10.1016/j.heliyon.2023.e22341

Deepa A, Sonal S, Mishra BK (2022) Application of co-immobilized microbial biochar beads in hybrid biofilter towards effective treatment of chrome tanning wastewater. J Water Process Eng 48:102821–102821. https://doi.org/10.1016/j.jwpe.2022.102821

Dicko M, Guilmont M, Lamari F (2018) Adsorption and biomass: current interconnections and future challenges. Curr Sustain Renew Energy Rep 5(4):247–256. https://doi.org/10.1007/S40518-018-0116-6

Elabbas S, Adjeroud N, Mandi L, Berrekhis F, Pons MN, Leclerc JP, Ouazzani N (2016) Eggshell adsorption process coupled with electrocoagulation for improvement of chromium removal from tanning wastewater. Int J Environ Anal Chem 96(5):345–357

Elabed A, El Abed S, Ibnsouda SK, Erable B (2019) Sustainable approach for tannery wastewater treatment: bioelectricity generation in bioelectrochemical systems. Arab J Sci Eng 44(12):10057–10066. https://doi.org/10.1007/S13369-019-04050-Y

El-Baz AA, Hendy IA, Dohdoh AM, Srour MI (2020) Adsorption technique for pollutants removal; current new trends and future challenges—a review. Egypt Int J Eng Sci Technol 32(1):1–24. https://doi.org/10.21608/EIJEST.2020.45536.1015

Elkarrach K, Merzouki M, Laidi O, Omor A, Biyada S, Benlemlih M (2021) Combination of chemical and biological processes for the treatment of tannery effluent of Fez city in Morocco. Desalin Water Treat 220:109–115. https://doi.org/10.5004/DWT.2021.26989

Elkarrach K, Omor A, Atia F, Laidi O, Benlemlih M, Merzouki M (2023) Traitement des effluents de tannerie par adsorption sur cendres volantes rejetées par les centrales thermiques: Caractérisation, optimisation, cinétique et isothermes. Héliyon 9(1)

Elzinga EJ, Arai Y (2021) Sorption processes in soils and sediments. Soil Syst 5(4):70

FEDIC (2015) Le secteur du cuir au MAROC, 1erAtelier National de l'ENEV du Maroc, 20 et 21 octobre 2015

Fernandez M, Gómez RJ, González PS, Barroso CN, Paisio CE (2023) Sequential application of activated sludge and phytoremediation with aquatic macrophytes on tannery effluents: in search of a complete treatment. Environ Sci Pollut Res 30:74196–74206. https://doi.org/10.1007/s11356-023-27718-3

Food and Agriculture Organization of the United Nations (FAO) (2022) FAO strategy on climate change 2022–2031

Fseha YH, Eniola JO, Sızırıcı B, Samuel S, Yildiz İ, Khaleel A, Adamson AW (2024) Application of natural earth-based materials as adsorbents for the treatment of chromium (VI)-contaminated tannery wastewater: Box-Behnken and fixed-bed column optimization. Sustain Chem Environ 7:100127. https://doi.org/10.1016/j.scenv.2024.100127

Ghazy MF, Gad MA (2014) Performance of Portland cement mixes containing silica fume and mixed with lime-water. HBRC J 10(3):247–257

Gupta VK, Nayak A (2022) Cadmium removal and recovery from aqueous solutions by novel adsorbents prepared from orange peel and Fe_2O_3 nanoparticles. Chem Eng J 180:81–90

Ho YS, McKay G (1999) Pseudo-second order model for sorption processes. Process Biochem 34:401–465. https://doi.org/10.1016/S0032-9592(98)00112-5

Husain A, Ansari MA, Khan AH, Shadan M, Baig MW (2024) Aperçu des méthodes de traitement des eaux usées des tanneries. Dans la série de conférences IOP: Sciences de la Terre et de l'environnement, vol 1326, no 1. Publication IOP, p 012134

Ilou I, Souabi S, Digua K (2014) Quantification of pollution discharges from tannery wastewater and pollution reduction by pre-treatment station

Ines Del Giudice (2024) Environmental issues of tanneries in Fez, Morocco. Prezi

Inyang IO, Efiong EJ, Joshua U (2016) Effectiveness of the mechanisms of fraud prevention and detection in Nigeria. Adv Soc Sci Res J 3(3):206–217

Jahan S, Zahra A, Iftikhar N, Ullah H (2014) Protective effects of different antioxidants against cadmium induced oxidative damage in rat testis and prostate tissues. Syst Biol Reprod Med 60(4):199–205

Karnwal A, Malik T (2024) Nano-revolution in heavy metal removal: engineered nanomaterials for cleaner water. Front Environ Sci 12:1393694

Lagergren S (1898) Zur theorie der sogenannten adsorption geloster stoffe. Kungliga svenska vetenskapsakademiens. Handlingar 24:1–39

Li X, Shen S, Xu Y, Guo T, Dai H, Lu X (2021) Application of membrane separation processes in phosphorus recovery: a review. Sci Total Environ 767:144–346

Lissanddine A, Mandi L, El Achaby M, Mousset E, Rene ER, Ouazzani N, Pons MN, Aziz F (2021) Performance and dynamic modeling of a continuously operated pomace olive packed bed for olive mill wastewater treatment and phenol recovery. Chemosphere 280:130797. https://doi.org/10.1016/j.chemosphere.2021.130797

Lissaneddine A, Aziz K, Ouazzani N, El Achaby M, Haydari I, Mandi L, Aziz F (2023) Continuous treatment of highly concentrated tannery wastewater using novel porous composite beads: central composite design optimization study. J Environ Health Sci Eng 21(2):513–532

Luján-Facundo MJ, Mendoza-Roca JA, Soler-Cabezas JL, Bes-Piá A, Vincent-Vela MC, Pastor-Alcañiz L (2019) Use of the osmotic membrane bioreactor for the management of tannery wastewater using absorption liquid waste as draw solution. Process Saf Environ Prot 131:292–299. https://doi.org/10.1016/j.psep.2019.09.024

Maharana M, Manna M, Sardar M, Sen S (2020a) Heavy metal removal by low-cost adsorbents. In: Environmental chemistry for a sustainable world. Springer, pp 245–276

Maharana SM, Pradhan AK, Pandit MK (2020b) Performance evaluation of mechanical properties of nanofiller reinforced jute-Kevlar hybrid composite. J Nat Fibers 17(5):123–135

Mak J (2019) Determining Economic Feasibility: A Cost Analysis of an Adsorption Based Wastewater Treatment Plant. Lehigh Preserve Institutional Repository

Manna M, Maharana M, Sardar M, Sen S (2023) Application of low-cost adsorbents in environmental cleanup: opportunities and challenges. J Clean Prod 317:128581

Mella B, Benvenuti J, Oliveira RF, Gutterres M (2019) Preparation and characterization of activated carbon produced from tannery solid waste applied for tannery wastewater treatment. Environ Sci Pollut Res 26(7):6811–6817. https://doi.org/10.1007/S11356-019-04161-X

Mia AS (2020) Removal of heavy metals from tannery wastewater by using sawdust and spent tea leaves as an adsorbent. Phys Sci Biophys J 4(1). https://doi.org/10.23880/PSBJ-16000139

Mim S, Hashem MA, Maoya M (2024) Adsorption-oxidation process for dyestuff removal from tannery wastewater. https://doi.org/10.1016/j.enmm.2023.100911

Ministry of Energy Transition and Sustainable Development (2023) Pollution data in Marrakech. Retrieved from Ministère de la Transition Énergétique et du Développement Durable

Misganaw A, Akenaw B, Getu S (2024) Determination of the level of chromium (III) and comparison of chemical precipitating agents to recover and reuse it from tannery waste water. Desal. Water Treat 2024(317):100150. https://doi.org/10.1016/j.dwt.2024.100150

Moktadir MA, Maliha M, Fatematujjohra, Munmun SA, Alam S, Rahman MM (2023) Treatment of tannery wastewater by different membrane bioreactors: a critical review. Environ Adv. https://doi.org/10.1016/j.envadv.2023.100478

Mustapha S, Tijani JO, Ndamitso MM, Abdulkareem S, Shuaib DT, Mohammed AK, Sumaila A (2020) The role of kaolin and kaolin/ZnO nanoadsorbents in adsorption studies for tannery wastewater treatment. Sci Rep 10(1):13068. https://doi.org/10.1038/S41598-020-69808-Z

Nashy E-SHA, Elhadad AMA, El-Khateeb MA (2022) Application of solid hide waste as an adsorbent for the treatment of industrial wastewater. Desalin Water Treatment 274:76–82. https://doi.org/10.5004/dwt.2022.28893

Nayak J, Chakrabortty S, Chakraborty P, Pal P, Pandey S (2021) Clean water reclamation from tannery industrial wastewater in integrated treatment schemes: a substantial review toward a viable solution, pp 199–232. https://doi.org/10.1016/B978-0-12-823031-2.00007-0

Nigam M, Rajoriya S, Singh RS, Kumar P (2019) Adsorption of Cr(VI) ion from tannery wastewater on tea waste: kinetics, equilibrium and thermodynamics studies. J Environ Chem Eng 7(3):103188. https://doi.org/10.1016/J.JECE.2019.103188

Nunes A, Rott O, Hirvonen J, Georgiadis G, Mosanu V, Sochirca A, Laraki K (2014) Morocco environmental performance reviews

Nure JF, Tibebu S, Nkambule T (2023) Optimization of Cr(VI) removal from aqueous solution with activated carbon derived from *Eichhornia crassipes* under response surface methodology. Environ Syst Res 13(1):5427

Nys Y, Gautron J (2007) Structure and Formation of the Eggshell. In: Huopalahti R, López-Fandiño R, Anton M, Schade R (eds) Bioactive Egg Compounds. Springer, Berlin, Heidelberg. https://doi.org/10.1007/978-3-540-37885-3_15

Pal P, Chakrabortty S, Nayak J, Senapati S (2017) A flux-enhancing forward osmosis–nanofiltration integrated treatment system for the tannery wastewater reclamation. Environ Sci Pollut Res 24(18):15768–15780. https://doi.org/10.1007/S11356-017-9206-Z

Pérez Marín D, Rizzolo A, Sortino G (2009) Non-destructive determination of quality parameters in nectarines during on-tree ripening and postharvest storage. Postharvest Biol Technol 52(2):180–188

Qu S, Zhang Q, Wang J (2013) Product-to-parent reversion of trenbolone: unrecognized risks for endocrine disruption. Science 342(6156):347–350

Rais S, Islam A, Ahmad I, Kumar S, Chauhan A, Javed H (2012) Preparation of a new magnetic ion-imprinted polymer and optimization using Box-Behnken design for selective removal and determination of Cu (II) in food and wastewater samples. Food Chem 334:127563. https://doi.org/10.1016/j.foodchem.2020.127563

Raji Z, Karim A, Karam A, Khalloufi S (2023) Adsorption of heavy metals: mechanisms, kinetics, and applications of various adsorbents in wastewater remediation—a review. Waste 1(3):775–805

Ricky R, Shanthakumar S, Ganapathy GP, Chiampo F (2022) Zero liquid discharge system for the tannery industry—an overview of sustainable approaches. Recycling 7(3):31. https://doi.org/10.3390/recycling7030031

Saravana Kumar R, Basu A, Bishayee A, Chatterjee R, Lun Ang W, Pal P, Shah MP, Tripathy SK, Janani VA, Chakrabortty S, Nayak J, Jeon BH (2023) Management of tannery waste effluents towards the reclamation of clean water using an integrated membrane system: a state-of-the-art review. Environ Res 229:115881. https://doi.org/10.1016/j.envres.2023.115881

Sawalha H, Al-Jabari M, El-Hamouz A, Abusafa A, Rene ER (2020) Tannery wastewater treatment and resource recovery options, pp 679–705. https://doi.org/10.1016/B978-0-12-818228-4.00025-3

Sahu JN, Acharya J, Meikap BC (2009) Response surface modeling and optimization of chromium (VI) removal from aqueous solution using Tamarind wood activated carbon in batch process. J Hazard Mater 172(2–3):818–825. https://doi.org/10.1016/j.jhazmat.2009.07.075

Seif M, Nabil R (2019) Modeling and optimization of water treatment plants for tannery effluent management. Environ Model Softw 121:130–140

Sepehr MN, Kazemian H, Amrane A, Ghahramani E (2014) Removal of hardness agents, calcium and magnesium, by natural and alkaline modified pumice stones in single and binary systems. Appl Surf Sci 274:295–305

Singh N, Babu SA (2024) Climate crisis and sustainable solutions: strategies for adaptation, mitigation and sustainable development. Springer

Singh K, Dave H, Prasad B, Kumari M, Dubey D, Rai AK, Ravi R, Manjhi J, Sillanpää M, Prasad KS (2024) nFeO decorated wood biochar as an adsorbent for aqueous Cr(VI) ions: hyphenated, coagulation-column treatment of tannery effluent. J Water Process Eng. https://doi.org/10.1016/j.jwpe.2024.105084

Smith JTH, Warrington A, Linderman SW (2022) Simplified state space layers for sequence modeling. arXiv:2208.04933

Suganthi S, Elango L, Subramanian SK (2013) Groundwater potential zonation by remote sensing and GIS techniques and its relation to the groundwater level in the coastal part of the Arani and Koratalai River Basin, Southern India. Earth Sci Res J 17(2):87–95

Talidi A (2006) Etude de l'élimination du Chrome et du bleu de méthylène en milieux aqueux par adsorption sur la pyrophyllite traitée et non traitée, Thèse de l'université de mohammed v, 2006

Terfie TA, Asfaw SL (2015) Evaluation of selected wetland plants for removal of chromium from tannery wastewater in constructed wetlands, Ethiopia. Afr J Environ Sci Technol 9(5):420–427

Tijani JO, Aminu UA, Bankole MT, Ndamitso MM, Abdulkareem AS (2021) Adsorptive and photocatalytic properties of green synthesized ZnO and $ZnO/NiFe_2O_4$ nanocomposites for tannery wastewater treatment. Niger J Technol Dev 17(4):312–322. https://doi.org/10.4314/NJTD.V17I4.10

Torab-Mostaedi M, Asadollahzadeh M, Hemmati A, Khosravi A (2013) Equilibrium, kinetic, and thermodynamic studies for biosorption of cadmium and nickel on grapefruit peel. J Taiwan Inst Chem Eng 44(2):295–302. https://doi.org/10.1016/j.jtice.2012.11.001

Vijayan JG, Prabhu TN, Singh A (2024) Optimization and automation of adsorption processes in wastewater treatment, pp 74–92. https://doi.org/10.58532/v3bjcs16p1ch7

Wang LK, Dahm DB, Dahm DB, Baier RE, Baier RE, Ziegler RC (2007) Treatment of tannery effluents by surface adsorption. J Appl Chem Biotechnol 25(6):475–490. https://doi.org/10.1002/JCTB.5020250610

Yu H, Tang M, He J, Liu C, Wu P, Jiang W (2024) Intelligent treatment of tannery wastewater via H_2O_2 photocatalytic oxidation coupled adsorption process. J Water Process Eng. https://doi.org/10.1016/j.jwpe.2024.105378

Zaimee MZA, Sarjadi MS, Rahman ML (2021). Heavy Metals Removal from Water by Efficient Adsorbents, Water 13(19):2659. https://doi.org/10.3390/w13192659

Zhao C, Chen W (2019) A review for tannery wastewater treatment: some thoughts under stricter discharge requirements. Environ Sci Pollut Res 26(25):26102–26111. https://doi.org/10.1007/S11356-019-05699-6

Zhao J, Tang Y, Zhou J, Guo H (2022) Tannery wastewater treatment: conventional and promising processes, an updated 20-year review. J Leather Sci Eng 4(1):1–22. https://doi.org/10.1186/s42825-022-00082-7

Zou YH, Huang YB, Si DH, Yin Q, Wu QJ, Weng Z, Cao R (2021) Porous Metal–Organic Framework Liquids for Enhanced CO2 Adsorption and Catalytic Conversion. Angewandte Chemie Novit. https://doi.org/10.1002/ange.202107156

Chapter 7
Electrochemical Treatment of Tannery Effluent: A Sustainable Approach

Sanjeev Kumar Meena, Shiv Om Meena, and Vikas K. Sangal

Abstract The tannery process generates a lot of wastewater in almost every step, and different processes have different wastewater characteristics. Today, the limited available freshwater resources are exposed to a variety of organic pollutants (through industrial wastewaters and chemical fertilizers) that seriously threaten the natural ecosystem and human health. Thus, tannery wastewater must be treated before being released into the environment. Tannery wastewater treatment consists of several stages, including pre-treatment, primary, secondary, and tertiary treatments. While traditional treatment methods are available; however, they are not completely effective for treatment tannery wastewater. This is due to the presence of high concentrations of organic matter, heavy metals, and high levels of chemical oxygen demand (COD) and biological oxygen demand (BOD). Among the common wastewater treatments, advanced oxidation processes (AOPs) are methods used for problematic organic pollutants removal which are difficult to degrade by conventional processes. This process produces high concentration of hydroxyl radical (·OH), which as a strong oxidizer is able to oxidize the majority of organic compounds to CO_2, water, and mineral acids. Combining different processes is a promising strategy for enhancing the efficiency of pollutant removal from water and wastewater. The integration of advanced oxidation processes (AOPs) with other treatment methods has shown improved performance. This chapter provides various treatment processes for tannery wastewater, with a focus on advanced oxidation processes (AOPs) and hybrid technologies. It also explores treatment strategies specifically designed for managing wastewater generated by the tannery industry.

Keywords Advanced oxidation processes · Biological methods · Chemical oxygen demand · Hybrid process · Tannery wastewater

S. K. Meena · S. O. Meena · V. K. Sangal (✉)
Department of Chemical Engineering, Malaviya National Institute of Technology, Jaipur 302017, Rajasthan, India
e-mail: vksangal.chem@mnit.ac.in

M. A. Aboulhassan and S. Souabi (eds.), *Sustainable Practices in the Tannery Industry*,
https://doi.org/10.1007/978-3-031-92037-0_7

7.1 Introduction

Water is essential for sustaining human life, covering approximately 70.8% of the Earth's surface, which amounts to about 361 million km^2. However, only 2.5% of this water is freshwater, and the majority requires purification to ensure it is safe for consumption or other uses.

Less than 1% of the Earth's freshwater is readily available for human use, whereas the vast majority—roughly 97%—is saline water from oceans and seas. According to WHO forecasts, more than three billion people will lack access to potable water by 2025. Nearly one-third of this population currently lives in water-stressed areas, with that percentage expected to rise to two-thirds by 2050. The tremendous expansion in population has resulted in unplanned communities, which dump untreated effluent into rivers and streams. This practice drastically reduces the quality of natural water sources and contributes to water pollution (Bankole et al. 2019).

The rapid growth of population and industrial development has led to a significant increase in untreated wastewater discharge. This trend depletes clean water resources, poses serious environmental challenges, and endangers the health and safety of humans and other living species (Cheng et al. 2016). Due to its toxicity, durability, and carcinogenicity, wastewater including synthetic organic pollutants related to medicines, petrochemicals, dyestuffs, and pesticides has become a global issue (Boczkaj and Fernandes 2017).

The adverse effects of pollutants on water bodies have heightened the focus on wastewater treatment over the years. However, despite conventional treatment processes, the quality of treated effluents has been declining over time due to increasing population growth and industrial development. Water quality analyses from various wastewater treatment facilities reveal the persistent presence of inorganic pollutants and organic micropollutants in treated streams.

The necessity of effectively using water resources, such as treated wastewater, stormwater, rainfall, and industrial effluents, is underscored by the growing demand for clean water driven by human activity and industrialisation. Rapid industrialization has led to the generation of vast quantities of polluted water annually from industries such as pulp and paper, petroleum, leather, textile, paints, pharmaceuticals, tanneries, distilleries, fertilizers, landfill leachate, sugar production, municipal and domestic sectors, hospitals, and agriculture (Edokpayi et al. 2017).

The treatment of pollutants presents a significant challenge for industries due to the absence of highly efficient wastewater treatment technologies, substantial operational costs, and stringent standards imposed by environmental and governmental regulatory bodies. By implementing a number of regulation measures, authorities like India's Central Pollution Control Board are actively work to reduce the amount of industrial effluents released into the environment. Conventional wastewater treatment methods, including physical and chemical flocculation, precipitation, adsorption or ion exchange, and biological processes, have proven inadequate for fully treating tannery effluents due to their complex composition. In recent years, numerous chemical treatment strategies have been explored for their potential effectiveness in addressing this challenge, including advanced oxidation, sonochemical, photochemical, and electrochemical methods (Oturan and Aaron 2014; Sivagami et al. 2018). This study provides a comprehensive overview of the tannery industry and the associated wastewater treatment processes.

7.2 Tannery Wastewater

The trade and production of leather have significantly increased recently. It is a vibrant industry where exports have been rising consistently. The process of treating and turning animal skin into leather is known as tanning.

The tannery industry is widely recognized for its high water consumption and significant discharge of wastewater. Water usage ranges from 25 to 60 m^3 per ton of raw leather, with a yield of 500 kg of finished leather per ton of leather. A ton of leather may require more than 440 kg of chemicals, depending on the method, which is particular to each business. These chemicals dissolve in process water, which produces an effluent that is polluting inorganic substances (chromium, chlorides, ammonium, sulfides, and sulfates) and has a very high organic load.

Wastewater from the tannery industry typically contains a complex mixture of contaminants, including hair, tannins, alkalis, proteins, acids, sulfides, dyes, chlorides, chromium salts, solvents, auxiliary chemicals, and other compounds. These pollutants originate from the incomplete chemical fixation of materials during the multistep tanning process, which transforms raw animal hides and skins into finished leather products.

Tannery effluent is a significant contaminant due to its elevated biochemical oxygen demand (BOD), chemical oxygen demand (COD), high chromium concentration, and rich color content, which can lead to severe environmental issues. Numerous traditional methods were employed to treat wastewater from the tannery industry, including biological techniques (Farabegoli et al. 2004), oxidation process (Schrank et al. 2004), and chemical process (Song et al. 2004), etc. The electrochemical process is one of the most interesting and efficient methods for treating some pollutants in tannery effluent, as it can be implemented through numerous approaches, including electrocoagulation and electrooxidation, etc. Some processes for wastewater treatment has been shown in Fig. 7.1 (Azimi et al. 2017).

Chemical methods are commonly used for treating tannery wastewater but generate large volumes of sludge with high water content, slowing filtration and increasing turbidity and dissolved solids in the effluent. Conventional biological methods are often ineffective due to the low biodegradability of tannery wastewater. Membrane technologies can recover wastewater but are limited by high equipment and monitoring costs. The ion-exchange method faces challenges such as clogging from suspended solids, inefficient operation, costly regenerants, and problematic waste streams (Lefebvre et al. 2005).

Innovative, cost-effective techniques for treating and reusing industrial wastewater are essential for environmental protection. Electrochemical methods, despite being

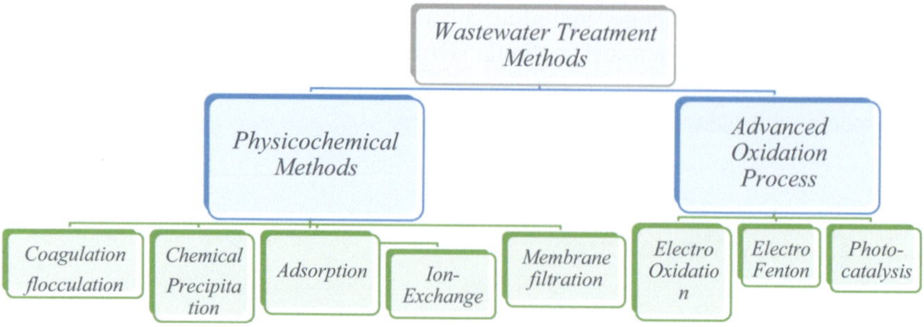

Fig. 7.1 Tree diagram of wastewater treatment methods

over a century old, have recently gained attention for their low operational costs and high treatment efficiency, making them a promising approach for purifying industrial wastewater.

7.3 Physiochemical Characteristics and Effluent Discharge Standards of Tannery Wastewater

Tannery effluent is among the most hazardous industrial pollutants, containing both inorganic and organic materials. Tanneries must therefore treat their wastewater before releasing it into the environment. Based on triplicate tests by researchers for each sample, the physiochemical properties of tannery effluent before and after treatment were determined and has been shown in Table 7.1. The information demonstrated the significant levels of contaminants in raw tannery wastewater. The raw tannery effluent had the greatest levels of COD, total dissolved solids (TDS), ammonium, sulfate, total nitrogen, sulfide, and total chromium.

Tannery wastewater is a significant pollutant with high COD, BOD, chromium levels, and intense coloration, posing serious environmental challenges. Its impact is assessed through parameters such as COD, BOD, TDS, total suspended solids (TSS), total Kjeldahl nitrogen (TKN), and oil and grease content (Kongjao et al. 2008).

Effluent pollutant limits are established primarily through two approaches: adherence to widely recognized standards deemed generally acceptable and the adoption of standards that, while less commonly utilized, are also considered broadly acceptable (IL&FS Ecosmart Limited Hydrabad August 2010). These standards often overlook specific individual circumstances. Along with its capacity to manage contamination from other sources, the receiving water body's dilution capacity is also considered. This evaluation of the recipient water quality after the effluent is discharged and mixed is based on the mass balance of the major contaminants. To regulate and control such pollution, the Central Pollution Control Board (CPCB) of India has established specific discharge standards for tannery effluents (Table 7.2).

Table 7.1 Physiochemical characteristics of tannery wastewater

pH	TDS (mg/L)	Suspended solids (mg/L)	Volatile suspended solids (mg/L)	COD (mg/L)	BOD (mg/L)	Chromium (mg/L)	Sulfide (mg/L)	References
3.7–4.3	17,737	445	–	9900	550	0.11	28.9	Villaseñor-Basulto et al. (2022)
7.0–8.7	13,300–19,700	600–955	–	4100–6700	630–976	11.5–14.3	–	Kongjao et al. (2008)
3.8		–	–	5280	–	29	48	Rezgui et al. (2022)
–	15,152	2004	1660	8000	930	11.2	223	Koteswari and Ramanibai (2003)
7.70	36,800	5300	1300	2200	–	–	–	Lefebvre et al. (2005)
7.1	–	589	–	8800	700	–	–	Azarian et al. (2018)

Table 7.2 Effluent discharged from typical tannery industry and standards (CPCB)

Parameter	Avg. total pollution load	Effluent limits
COD (chemical oxygen demand)	7200	125–250
BOD_5 (biological oxygen demand)	3600	20
SS (suspended solids)	4800	35–100
Total chromium (Cr)	200	1.5–2.0
S^{2-}	200	1–2
TDS (total dissolve solids)	16,000	Locally specific
pH	6–9	6–9

All units in mg/L

7.4 Treatment Technologies of Tannery Wastewater

This section discusses the wastewater treatment processes implemented for managing toxic effluents generated from tannery wastewater. Considering the chemicals involved are very hazardous and carcinogenic, conventional technologies are insufficient for treating tannery effluent streams completely. Therefore, the development of advanced technologies is critical for efficient degradation of toxic effluents in tannery wastewater. Tannery wastewater treatment technologies are aimed at addressing the high levels of organic and inorganic pollutants, heavy metals, and other hazardous compounds found in tannery effluents. These technologies can be broadly categorized into physical, chemical, biological, and advanced treatment methods. Some key technologies include electro-coagulation (EC), electro-Fenton (EF), and electrooxidation (EO) and hybrid techniques for reducing the tannery industrial effluents. Advanced treatment techniques focus on achieving comprehensive removal of both organic and inorganic pollutants while minimizing overall treatment costs. These methods often involve the generation of hydroxyl radicals (OH·) or sulfate radicals (SO_4^-), which are effectively degrade refractory organic compounds and certain inorganic pollutants. Additionally, these techniques can enhance the biodegradability of the wastewater, serving as a pretreatment step to improve the efficiency of subsequent biological treatment processes. These technologies are often tailored to meet site-specific requirements, effluent characteristics, and regulatory standards, making them an essential part of sustainable tannery wastewater management.

7.4.1 Electrocoagulation Process (EC)

EC is a process which utilizes the electro-dissolution of the soluble anodes, resulting in metal cations which destabilize the colloidal particles by surrounding the particles with polyvalent poly-hydroxide complexes and facilitating their coalescence. When hydrogen gas is produced at the cathode, it increases mixing and flocculation. EC produces coagulants in situ and does not require chemical additives (Villaseñor-Basulto et al. 2022).

For example, Eqs. (7.1)–(7.3) present the reactions during the EC process of iron anodes leading to ferric hydroxide ($Fe(OH)_3$) formation:

$$Fe \rightarrow Fe^{2+} + 2e^- \tag{7.1}$$

$$2Fe^{2+} + 5H_2O + 1/2O_2 \rightarrow 2Fe(OH)_3 + 4H^+ \tag{7.2}$$

$$Fe^{2+} + 2OH^- \rightarrow Fe(OH)_2 \tag{7.3}$$

Hydrogen evolution occurs at the cathode according to Eq. (7.4):

$$2H_2O + 2e^- \rightarrow 2HO^- + H_2 \tag{7.4}$$

Finally, in the bulk of the solution, poly-hydroxide production proceeds in agreement with Eq. (7.5):

$$Mn^+ + nOH^- \rightarrow M(OH)_n, \tag{7.5}$$

where M is the anode material, and n is the number of electrons involved in the reaction.

EC has been proposed for water purification and the removal of various pollutants. It is widely used to treat wastewater from different industries, effectively removing dissolved substances and suspended matter. While EC was employed in the twentieth century, its success was limited due to high electricity consumption and operational costs. Recent efforts have focused on optimizing key parameters to reduce energy consumption and maximize effluent treatment efficiency (Jovanović et al. 2021; Meena et al. 2024).

Maha Lakshmi and Sivashanmugam (2013) conducted a study on a laboratory-scale electrocoagulation (EC) system for the removal of chemical oxygen demand (COD) from tannery effluent. The investigation evaluated the effects of key operating parameters, including the type of electrolyte, current density, initial COD concentration, and electrolyte concentration, on the percentage of COD removal and power consumption. Additionally, the quantity of sludge generated was measured after each treatment process. Various combinations of iron (Fe) and aluminum (Al) electrodes were assessed. Under optimal conditions, the study achieved a COD removal efficiency of 89.65%, with an energy consumption of 1.279 kWh/m^3 and an operating cost of \$6.28/m^3. Sengil et al. (2009) conducted an experimental investigation into the removal of chemical oxygen demand (COD), and sulfide from tannery liming drum wastewater using EC. This study examined the influence of key parameters, including current density, initial pH, electrolysis time, and on the treatment efficiency. Mild steel electrodes were employed as anodes, demonstrating effective removal of COD, sulfide, and oil–grease from the aqueous phase. Optimal operating ranges for the investigated parameters were determined through experimentation. The results confirmed the efficacy of the EC process in achieving significant removal of the targeted contaminants. The summary of some electrocoagulation processes has been shown in Table 7.3.

7.4.2 Electro-Fenton Process

Fenton's reagent, discovered by Henry J. H. Fenton in the 1890s, is a mixture of hydrogen peroxide (H_2O_2) and an iron (II) salt, typically ferrous sulfate ($FeSO_4$). It is used to oxidize and break down organic compounds. In the presence of Fe^{2+} as a catalyst, hydrogen peroxide decomposes to produce hydroxyl (OH·) radicals. The Fenton process achieves optimal performance at a pH of 3. A key advantage of the electro-Fenton process is the in situ production of at least one reagent required for the Fenton reaction within an electrochemical reactor. This approach is regarded as one of the most promising advanced oxidation processes (AOPs) for the remediation of highly contaminated water. However, a notable limitation of this method is the generation of ferrous sludge, which requires proper disposal.

Organic compounds oxidize more quickly in solutions containing H_2O_2 and iron (II) salts, which react to produce hydroxyl radicals:

$$H_2O_2 + Fe^{2+} \rightarrow Fe^{3+} + OH^- + HO\cdot \tag{7.6}$$

Recently, Fenton's reagent has been electrochemically generated in situ using two distinct methods:

(a) by addition of a catalytic amount of Fe^{2+} ions, reduction of oxygen and regeneration of Fe^{2+} on a suitable cathode in acidic media:

$$O_2 + 2H^+ + 2e^- \rightarrow H_2O_2 \qquad E_o = 0.67\,V \tag{7.7}$$

$$Fe^{3+} + e^- \rightarrow Fe^{2+} \qquad E_o = 0.771\,V \tag{7.8}$$

(b) by addition of H_2O_2 and electrochemical production of Fe^{2+} ions with a sacrificial Fe anode according to the anodic reaction:

$$Fe \rightarrow Fe^{2+} + 2e^- \tag{7.9}$$

In this process, a portion of Fe^{3+} ions is generated through Fenton's reaction (7.8) between the supplied H_2O_2 and electro-regenerated Fe^{2+}, which subsequently precipitates as $Fe(OH)_3$.

The mechanism of the electro Fenton process is shown in Fig. 7.2 (Nidheesh and Gandhimathi 2012).

Table 7.3 Summary of electrocoagulation processes on the treatment of tannery wastewater

Process	Type of tannery wastewater	Optimum conditions	COD (%) removal	BOD (%) removal	Remark	References
Electro-coagulation	Real TWW Initial COD—25,300 mg/L BOD—10,850 mg/L	Current density 35 mA/cm^2 at pH—3, time 10 min and energy consumptions were 5.768 kWh/m^3	82	–	Mild steel electrode is superior	Sengil et al. (2009)
Electro-coagulation	Real TWW Initial COD—7680 mg/L	Aluminum cathode, current density—75 A/m^2 Time—45 min	90 <	90 <	Cr, total carbon and sulfide create sludge problems	Benhadji et al. (2011)
Electro-coagulation	Real TWW	pH—7.8, electrode—Al and Fe Time range—30–45 min	95	96		Jovanović et al. (2021)
Electro-coagulation	Real TWW Initial COD—21,000 mg/L	Fe/Fe electrodes and 20 mA/cm^2 current density	89.65%, COD removal, 1.279 kWh/m^3 energy consumption			Maha Lakshmi and Sivashanmugam (2013)
Electro-coagulation	Real TWW Initial COD—3300 mg/L Cr—16.8	Coagulant—aluminum sulfate and ferric chloride pH—7.8	30–37% removal of total COD from settled tannery wastewaters and 74–99% removal of chromium at an initial concentration of 12 mg/L Ferric chloride produced better results than aluminum sulfate			Song et al. (2004)
Electro-coagulation	Synthetic TWW	Electrode—Fe/Fe	83.54% COD removal			Meena et al. (2024)

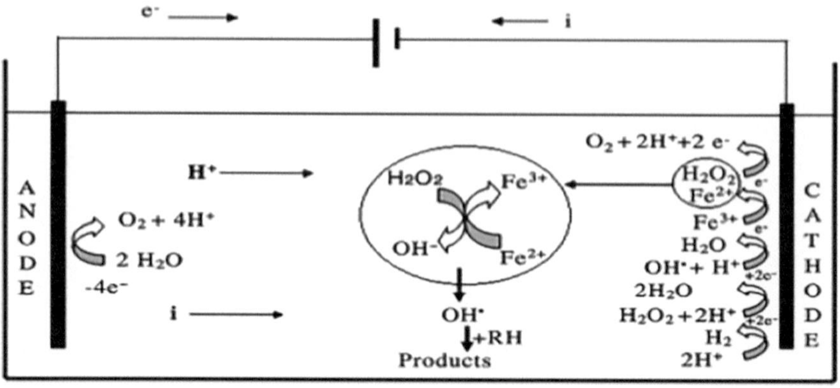

Fig. 7.2 Reaction mechanism of E-Fenton process

The treatment of tannery effluent using EF processes with cost-effective graphite electrodes was investigated. The degradation performance was assessed based on chemical oxygen demand (COD) removal. The effects of key parameters, including current density, electrolysis time, and electrolyte concentration, were analyzed to identify the optimal conditions for maximum efficiency and minimal energy consumption. Under the optimized electro-Fenton process, a maximum COD removal efficiency of 67.25% was achieved (Doumbi et al. 2022). Chromium (Cr), recognized as both mutagenic and carcinogenic, is a significant environmental pollutant widely utilized in various industrial applications. Due to its harmful health effects, it is crucial to remove Cr from effluents before discharging them into aquatic environments. The aim of this study was to evaluate the effectiveness of the EF process using Fe electrodes in the presence of sodium sulfate as a supporting electrolyte for the removal of Cr(VI) from aqueous solutions (Rahmani et al. 2015). The summary of some electro Fenton processes has been shown in Table 7.4.

7.4.3 Electrochemical Oxidation Process

Electrochemical oxidation (EO) is an environmentally friendly technology capable of completely mineralizing nonbiodegradable organic matter and removing nitrogen species. The electrochemical oxidation of pollutants takes place through the following two major pathways:

(1) Direct anodic oxidation, which in general allows to destroy pollutants at the anode surface.
(2) Indirect oxidation: An electrochemically generated oxidizer ($H_2S_2O_8$, HClO, O_3) acts as the mediator for oxidizing contaminants. Both oxidation processes may appear at the same time in the electrooxidation of aqueous effluents. The mechanism of the reaction is illustrated in Fig. 7.3 (Särkkä et al. 2015).

Table 7.4 Summary of electro Fenton processes on the treatment of tannery wastewater

Process	Type of tannery wastewater	Optimum conditions	COD (%)	BOD (%)	Remark	References
Electro Fenton	Real TWW Initial COD—2700–4000 mg/L BOD—900–1100 mg/L	At optimum concentration of H_2O_2 is 44.40 g/L and $FeSO_4$ is 6 g/L	69	72	Low COD removal	Mandal et al. (2010)
		At H_2O_2 is 220 g/L and $FeSO_4$ is 6 g/L	95		Cost is high	
Electro Fenton	Real TWW Initial COD—3320 mg/L	Electrode—graphite pH—3 Current intensity—200 mA Time—6 h $[Fe^{2+}] = 0.1$ mM	67.25		Low COD removal	Doumbi et al. (2022)
Electro Fenton	Real TWW Initial TOC—1950 mg/L	Electrode—BDD Fe ion conc.—2.0 mM/L Current density—pH—3, 70 mA/cm^2 Time—30 min			TOC removal—25% Energy consumption—74.5 kWh/m^3	Villaseñor-Basulto et al. (2022)
Electro Fenton	Real TWW Initial COD—5380 mg/L Cr—29 mg/L	Current intensity—(0.1–2 A) Electrode—BDD/carbon Fe^{2+}—1.2 mM Time—2 h pH—3	COD removal—51%		Cr removal—100% ECC—0.024 kWh/g COD	Rezgui et al. (2022)

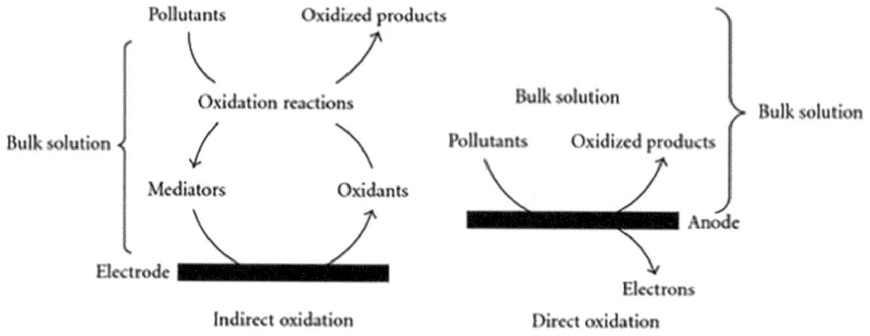

Fig. 7.3 Reaction mechanism of electrooxidation process

Tannery wastewater was treated using an electrochemical oxidation method, employing Ti/Pt, Ti/PbO_2, and Ti/MnO_2 anodes, along with a Ti cathode. The variations in colour concentration, sulfide, COD, ammonia (NH_4^+), and total chromium were assessed as a function of treatment applied current density and time (Rao et al. 2001).

Dimensionally stable anodes (DSA) made of titanium, tin, iridium, and ruthenium were used in electrochemical treatment of tannery wastewater. The current efficiency was computed as a function of the electrolysis time, and the impacts of the current density, oxide composition, and effluent conductivity were examined. The findings showed that the concentrations of total phenolic compounds and total organic carbon decreased with each of the electrodes under investigation (Costa et al. 2008). The summary of some electrooxidation processes has been shown in Table 7.5.

7.4.4 Studies of Hybrid Processes on the Treatment of Tannery Wastewater

When the establishment of complete water treatment plants is not cost-effective or required, the electrocoagulation (EC) method is an effective treatment method for both drinking water for small to medium-sized municipalities, as well as, water utilized in industrial processes. And its effectiveness has been demonstrated for water decolorization, turbidity reduction, water hardness removal, and purification of fluoride, boron, phosphate, iron, mercury, cadmium, lead, chromium, nickel, strontium, copper, cobalt, and cesium ion contaminants. Because the EC process eliminates over 85% of pollutants, it is an efficient alternative for waste treatment. Yet industrial wastewater treatment is still challenging due to its high volume and types of hazardous pollutants. The EC process removed COD to a limited extent, therefore not bringing the water to a potential state of reuse in the industrial and domestic sector. To improve pollutant removal efficiency, shorten electrolysis time, and lower

Table 7.5 Summary of electrooxidation processes on the treatment of tannery wastewater

Process	Type of tannery wastewater	Optimum conditions	Performance	References
Electrooxidation	Real TWW Initial COD—4400 mg/L	Ti/RuO_2, Ti/IrO_2 and Ti/BDD electrodes were used	COD removal—82%	Tien and Luu (2020)
Electrooxidation	Real TWW Initial TOC—1005 mg/L	DSA anodes	TOC removal—40.5%	Costa et al. (2008)
Electrooxidation	Real TWW Initial COD—2389 mg/L Cr—29 mg/L Sulfide—176 mg/L	Ti/Pt–Ir, Ti/PbO_2, $Ti/RhO_x–TiO_2$, and $Ti/PdO–Co_3O_4$ electrodes were used	COD, sulfides and Cr was monitored	Szpyrkowicz et al. (2005)
Electrooxidation	Real TWW	Anode—$Ti/SnO_2/PdO_2/RuO_2$ Current density—1.5 A/dm^2	COD removal 59%	Naumczyk et al. (2019)
Electrooxidation	Real TWW Initial COD—330 mg/L	Current density—7.5 mA/cm^2 Anode—RuO_2/SnO_2 Cathode—stainless steel Time—150 min	COD removal 58.9%	Huang et al. (2013)
Electrooxidation	Real TWW	Electrode—$TiO_2/RuO_2/IrO_2$	COD removal—87.5%	Sundaramoorthy et al. (2022)

costs, the EC process is often combined with other treatments, such as electrochemical oxidation or the Fenton process. Asaithambi et al. (2022) achieved an 85% COD removal using the EC-peroxidation (ECP) process. By adding 243 mg/L of hydrogen peroxide with Fe electrodes and an initial pH of 5, they increased COD removal by 13%. However, the use of strong oxidants like H_2O_2 reduces the environmental friendliness of the ECP process. Sharma et al. (2020) discovered that the EC-EO method uses less energy and removes 95% of organic contaminants more effectively than the ECF method. The electrocoagulation-membrane method, which combines electrocoagulation with microfiltration and nanofiltration membranes, is one exciting new hybrid technology. Harsh chemicals and undesirable byproducts are avoided with this procedure. The EC procedure can be used as a pretreatment to lessen membrane fouling, which is still a problem. The summary of hybrid processes is given in Table 7.6.

Table 7.6 Summary of hybrid processes studies on the treatment of tannery wastewater

Processes		Process parameters	Performance	References
EF with EC	EF	Current intensity—(0.1–2 A) Electrode—BDD/carbon Fe^{2+}—1.2 mM Time—2 h pH—3	COD removal—51% Cr removal—100% ECC—0.024 kWh/g COD	Rezgui et al. (2022)
	EC	Current intensity—(1–4 A) pH—3.8 Time—3 h	COD removal—66% Cr removal—98% ECC—0.007 kWh/g COD	
	EF/EC	Current intensity—0.7 A pH—natural Fenton catalyst (2 mmol/L of Fe^{2+}) Time—5 h	The final COD (628 mg/L) and Cr (< 0.5 mg/L)	
EC with EO	EO	Current density—42 mA/cm^2 Time—180 min Electrode—stainless steel/PbO_2 pH—7.2	COD removal 99% Color removal 98% Energy consumption—60 kWh/m^3	Azarian et al. (2018)
	EC	Current density—17 mA/cm^2 Time—10 min Electrode—Al/Al pH—7.8	COD removal 82% Color removal 68% Energy consumption—4.9 kWh/m^3	
	EC/EO	EC: current density—17 mA/cm^2 Time—10 min Electrode—Al/Al pH—6 EO: current density—42 mA/cm^2 Time—40 min Electrode—stainless steel/PbO_2 pH—8.1	COD $\leq$ 5 mg/L Total energy consumption—18 kWh/m^3	
EC with EF	EC	Electrode—iron Current density—9.7 mA/cm^2 Fe ion conc.—0.66 mM/L Time—60 min	TOC removal—32%	Villaseñor-Basulto et al. (2022)

(continued)

Table 7.6 (continued)

Processes		Process parameters	Performance	References
	EF	Electrode—BDD Fe ion conc.—2.0 mM/L Time—30 min Current density—70 mA/cm^2 pH—3	TOC removal—25% Energy consumption—74.5 kWh/m^3	
	EC/EF	Current density of 9.7 mA/cm^2 at 217 min	TOC removal—64% Energy consumption as low as 22 kWh/m^3	
EC with EO	EC	Electrode—Al/Fe Current density—2.5–25 $Ti/IrO_2/SnO_2$	TOC removal—68.6% Energy consumption—6.36 kWh/m^3	Sayin et al. (2024)
	EC/EO	EC—Al/Fe Current density—2.5–25 mA/cm^2 EO—$Ti/IrO_2/SnO_2$ Current density—100 mA/cm^2	TOC removal—84.39%	

7.4.5 Membrane Technology with Electrochemical Advanced Oxidation Processes (EAOPs)

Emerging technologies known as EAOPs have the ability to mineralise organic compounds into CO_2, H_2O, and other inorganic species, or they can break them down into less refractory products. Their positive attributes—such as their environmental friendliness, steady performance, ease of process control, and lack of reliance on chemical reagents—have drawn more and more attention (Moreira et al. 2017). Each EAOP is less effective for commercial use due to its limitations for large-scale industrial applications, such as weak mass transfer of pollutants in the reactor and relatively high energy consumption (Ganiyu et al. 2015). Furthermore, EAOPs are not the best option for treating vast amounts of wastewater that contain low quantities of contaminants. Nonetheless, membrane fouling can be addressed and separation performance enhanced by combining membrane technology with EAOPs. Energy consumption can be decreased and process efficiency increased using this synergistic strategy (Pan et al. 2019).

Because of its high separation selectivity, low energy consumption, lack of chemical requirements, ease of scalability, and continuous operation, membrane technology is thought to have promise for decontaminating water (Goh and Ismail 2018). In recent years, membrane-based processes have been developed and applied for various applications, including nanofiltration (NF), particle filtration (PF), ultrafiltration (UF), microfiltration (MF), reverse osmosis (RO), membrane distillation

(MD), forward osmosis (FO), and membrane bioreactors (MBR) (Lee et al. 2016). Fouling is a common problem for membrane operations, as pollutants build up on the membrane's surface or inside its pores as a result of rejection, impairing performance (Jhaveri and Murthy 2016). Furthermore, membrane separation is a physical procedure that concentrates impurities instead of totally purging the effluent. For wastewater with complex compositions, where complete retention of pollutants is challenging or impossible, this problem is especially important.

7.5 Health and Safety Aspects in Tannery Industry

The toxicity of industrial effluent discharge is a highly complex subject due to the diverse range of aquatic organisms and the multitude of components present in the effluent. These components vary in their chemical nature and degradation rates, adding to the complexity. Consequently, specific and comprehensive data regarding their impact on aquatic ecosystems are often unavailable. Moreover, it is challenging to isolate and quantify the effects of individual components or complex mixtures on specific groups such as fish, invertebrates, algae, and daphnia. Leather processing is associated with numerous significant environmental challenges. The tanning process, in particular, generates substantial volumes of wastewater at various stages. Tanneries are infamous for their unpleasant odors and the presence of unused harmful chemicals in their effluents, underscoring their polluting nature. The waste produced during tanning includes liquid effluents containing high levels of organic and inorganic suspended particles, elevated chemical oxygen demand (COD), and potentially hazardous metals. These factors make tannery discharges a major concern for environmentalists (Kanagaraj et al. 2008). LD50 (lethal dose 50%) is a commonly used toxicity metric representing the dose required to cause mortality in 50% of a test species population (Asia 2000). The response to a given exposure can vary significantly among species, with different reactions to the same toxin dosage. When reporting LD50 values, it is essential to specify the test species and the evaluation period, which is typically 24, 96 h, or 14 days. Aquatic ecosystems are evaluated and protected using environmental quality criteria that are based on factors including chemical destiny, behavior, and aquatic toxicity.

These criteria are based on highly sensitive data collected from diverse studies and species. In cases where sufficient data are unavailable, expert judgment may be employed to guide decision-making. The highest permitted concentration of a chemical that must not be surpassed in order to protect certain aquatic uses and ecological integrity is known as an environmental quality standard.

7.6 Conclusion

This review provides a comprehensive summary of tannery wastewater treatment, encompassing the characteristics of tannery effluent, applicable effluent standards, wastewater treatment technologies, and associated health and safety considerations. The tannery industry is a significant source of environmental pollution, discharging substantial quantities of hazardous chemicals due to its reliance on outdated and traditional processing technologies. The study also evaluates the feasibility of advanced oxidation processes (AOPs) and hybrid processes as a potential solution for the effective treatment of effluents generated by the tannery industry. The tannery industry faces several challenges, including economic constraints, social obstacles and a lack of effective monitoring facilities. To address these issues, leather manufacturers have begun adopting modern technologies both in production processes and effluent management to reduce environmental pollution and minimize waste generation. Despite these efforts, the effluents from the leather industry continue to contain high concentrations of hazardous compounds and heavy metals, posing significant threats to environmental quality and sustainability.

References

Asaithambi P, Yesuf MB, Govindarajan R, Hariharan NM, Thangavelu P, Alemayehu E (2022) A review of hybrid process development based on electrochemical and advanced oxidation processes for the treatment of industrial wastewater. Int J Chem Eng 2022:e1105376. https://doi.org/10.1155/2022/1105376

Asia S (2000) Regional programme for pollution control in the tanning industry

Azarian G, Miri M, Nematollahi D (2018) Combined electrocoagulation/electrooxidation process for the COD removal and recovery of tannery industry wastewater. Environ Prog Sustain Energy 37:637–644. https://doi.org/10.1002/ep.12711

Azimi A, Azari A, Rezakazemi M, Ansarpour M (2017) Removal of heavy metals from industrial wastewaters: a review. ChemBioEng Rev 4:37–59. https://doi.org/10.1002/cben.201600010

Bankole MT, Abdulkareem AS, Mohammed IA, Ochigbo SS, Tijani JO, Abubakre OK, Roos WD (2019) Selected heavy metals removal from electroplating wastewater by purified and polyhydroxylbutyrate functionalized carbon nanotubes adsorbents. Sci Rep 9:4475. https://doi.org/10.1038/s41598-018-37899-4

Benhadji A, Taleb Ahmed M, Maachi R (2011) Electrocoagulation and effect of cathode materials on the removal of pollutants from tannery wastewater of Rouïba. Desalination 277:128–134. https://doi.org/10.1016/j.desal.2011.04.014

Boczkaj G, Fernandes A (2017) Wastewater treatment by means of advanced oxidation processes at basic pH conditions: a review. Chem Eng J 320:608–633. https://doi.org/10.1016/j.cej.2017.03.084

Cheng M, Zeng G, Huang D, Lai C, Xu P, Zhang C, Liu Y (2016) Hydroxyl radicals based advanced oxidation processes (AOPs) for remediation of soils contaminated with organic compounds: a review. Chem Eng J 284:582–598. https://doi.org/10.1016/j.cej.2015.09.001

Costa CR, Botta CMR, Espindola ELG, Olivi P (2008) Electrochemical treatment of tannery wastewater using DSA electrodes. J Hazard Mater 153:616–627. https://doi.org/10.1016/j.jhazmat.2007.09.005

Doumbi RT, Bertrand Noumi G, Ngobtchok B, Domga (2022) Tannery wastewater treatment by electro-Fenton and electro-persulfate processes using graphite from used batteries as free-cost electrode materials. Case Stud Chem Environ Eng 5:100190. https://doi.org/10.1016/j.cscee.2022.100190

Edokpayi JN, Odiyo JO, Durowoju OS, Edokpayi JN, Odiyo JO, Durowoju OS (2017) Impact of wastewater on surface water quality in developing countries: a case study of South Africa. Water Qual. IntechOpen. https://doi.org/10.5772/66561

Farabegoli G, Carucci A, Majone M, Rolle E (2004) Biological treatment of tannery wastewater in the presence of chromium. J Environ Manage 71:345–349. https://doi.org/10.1016/j.jenvman.2004.03.011

Ganiyu SO, van Hullebusch ED, Cretin M, Esposito G, Oturan MA (2015) Coupling of membrane filtration and advanced oxidation processes for removal of pharmaceutical residues: a critical review. Sep Purif Technol. Environmental nanotechnology and sustainability in water treatment in honor of Professor Chin-Pao Huang, Ph.D., Harvard University, Donald C. Philips, Professor Civil and Environmental Engineering, University of Delaware, USA, vol 156, pp 891–914. https://doi.org/10.1016/j.seppur.2015.09.059

Goh PS, Ismail AF (2018) A review on inorganic membranes for desalination and wastewater treatment. Desalin Rev Res Dev Desalin 434:60–80. https://doi.org/10.1016/j.desal.2017.07.023

Huang LL, Liu JF, Zhang N, Sun B, Feng YJ (2013) Treatment of tannery wastewater by electrochemical oxidation process. Appl Mech Mater 361–363:697–700. https://doi.org/10.4028/www.scientific.net/AMM.361-363.697

Jhaveri JH, Murthy ZVP (2016) A comprehensive review on anti-fouling nanocomposite membranes for pressure driven membrane separation processes. Desalination 379:137–154. https://doi.org/10.1016/j.desal.2015.11.009

Jovanović T, Velinov N, Petrović M, Najdanovic S, Bojić D, Radović Vučić M, Bojić A (2021) Mechanism of the electrocoagulation process and its application for treatment of wastewater: a review. Adv Technol 10:63–72. https://doi.org/10.5937/savteh2101063J

Kanagaraj J, Babu N, Mandal A (2008) Recovery and reuse of chromium from chrome tanning waste water aiming towards zero discharge of pollution. J Clean Prod 16:1807–1813. https://doi.org/10.1016/j.jclepro.2007.12.005

Kongjao S, Damronglerd S, Hunsom M (2008) Simultaneous removal of organic and inorganic pollutants in tannery wastewater using electrocoagulation technique. Korean J Chem Eng 25:703–709. https://doi.org/10.1007/s11814-008-0115-1

Koteswari Y, Ramanibai R (2003) The effect of tannery effluent on the colonization rate of plankters: a microcosm study. Turk J Biol 27:163–170

Lee A, Elam JW, Darling SB (2016) Membrane materials for water purification: design, development, and application. Environ Sci Water Res Technol 2:17–42. https://doi.org/10.1039/c5ew00159e

Lefebvre O, Vasudevan N, Torrijos M, Thanasekaran K, Moletta R (2005) Halophilic biological treatment of tannery soak liquor in a sequencing batch reactor. Water Res 39:1471–1480. https://doi.org/10.1016/j.watres.2004.12.038

Maha Lakshmi P, Sivashanmugam P (2013) Treatment of oil tanning effluent by electrocoagulation: influence of ultrasound and hybrid electrode on COD removal. Sep Purif Technol 116:378–384. https://doi.org/10.1016/j.seppur.2013.05.026

Mandal T, Maity S, Dasgupta D, Datta S (2010) Advanced oxidation process and biotreatment: their roles in combined industrial wastewater treatment. Desalination 250:87–94. https://doi.org/10.1016/j.desal.2009.04.012

Meena SK, Meena SO, Sangal VK (2024) Once through continuous mode electrocoagulation for tannery wastewater remediation: process optimization and performance evaluation. J Indian Chem Soc 101:101212. https://doi.org/10.1016/j.jics.2024.101212

Moreira FC, Boaventura RAR, Brillas E, Vilar VJP (2017) Electrochemical advanced oxidation processes: a review on their application to synthetic and real wastewaters. Appl Catal B Environ 202:217–261. https://doi.org/10.1016/j.apcatb.2016.08.037

Naumczyk JH, Kucharska MA, Ładyńska JA, Wojewódka D (2019) Electrochemical oxidation process in application to raw and biologically pre-treated tannery wastewater. Desalin Water Treat 162:166–175. https://doi.org/10.5004/dwt.2019.24349

Nidheesh PV, Gandhimathi R (2012) Trends in electro-Fenton process for water and wastewater treatment: an overview. Desalination 299:1–15. https://doi.org/10.1016/j.desal.2012.05.011

Oturan MA, Aaron J-J (2014) Advanced oxidation processes in water/wastewater treatment: principles and applications. A review. Crit Rev Environ Sci Technol 44:2577–2641. https://doi.org/10.1080/10643389.2013.829765

Pan Z, Song C, Li L, Wang H, Pan Y, Wang C, Li J, Wang T, Feng X (2019) Membrane technology coupled with electrochemical advanced oxidation processes for organic wastewater treatment: recent advances and future prospects. Chem Eng J 376:120909. Emerging advanced oxidation technologies and developing perspectives for water and wastewater treatment. https://doi.org/10.1016/j.cej.2019.01.188

Rahmani AR, Hossieni E, Poormohammadi A (2015) Removal of chromium (VI) from aqueous solution using electro-Fenton process. Environ Process 2:419–428. https://doi.org/10.1007/s40710-015-0068-4

Rao NN, Somasekhar KM, Kaul SN, Szpyrkowicz L (2001) Electrochemical oxidation of tannery wastewater. J Chem Technol Biotechnol 76:1124–1131. https://doi.org/10.1002/jctb.493

Rezgui S, Ghazouani M, Bousselmi L, Akrout H (2022) Efficient treatment for tannery wastewater through sequential electro-Fenton and electrocoagulation processes. J Environ Chem Eng 10:107424. https://doi.org/10.1016/j.jece.2022.107424

Särkkä H, Bhatnagar A, Sillanpää M (2015) Recent developments of electro-oxidation in water treatment—a review. J Electroanal Chem 754:46–56. https://doi.org/10.1016/j.jelechem.2015.06.016

Sayin FE, Gengeç E, Özbay B, Özbay İ (2024) Combination of electrocoagulation and electrooxidation processes for treatment of real wood processing effluents. Int J Environ Sci Technol. https://doi.org/10.1007/s13762-024-06255-7

Schrank SG, José HJ, Moreira RFPM, Schröder HF (2004) Elucidation of the behavior of tannery wastewater under advanced oxidation conditions. Chemosphere 56:411–423. https://doi.org/10.1016/j.chemosphere.2004.04.012

Sengil IA, Kulaç S, Ozacar M (2009) Treatment of tannery liming drum wastewater by electrocoagulation. J Hazard Mater 167:940–946. https://doi.org/10.1016/j.jhazmat.2009.01.099

Sharma S, Aygun A, Simsek H (2020) Electrochemical treatment of sunflower oil refinery wastewater and optimization of the parameters using response surface methodology. Chemosphere 249:126511. https://doi.org/10.1016/j.chemosphere.2020.126511

Sivagami K, Sakthivel KP, Nambi IM (2018) Advanced oxidation processes for the treatment of tannery wastewater. J Environ Chem Eng 6:3656–3663. https://doi.org/10.1016/j.jece.2017.06.004

Song Z, Williams CJ, Edyvean RGJ (2004) Treatment of tannery wastewater by chemical coagulation. Desalination 164:249–259. https://doi.org/10.1016/S0011-9164(04)00193-6

Sundaramoorthy S, Singh N, Taube C, Katiyar R, Muralidharan V, Saravanan P (2022) Electro-oxidation of tannery wastewater to achieve zero discharge—a step towards sustainability. Environ Technol 44:1–29. https://doi.org/10.1080/09593330.2022.2049887

Szpyrkowicz L, Kaul SN, Neti RN, Satyanarayan S (2005) Influence of anode material on electrochemical oxidation for the treatment of tannery wastewater. Water Res 39:1601–1613. https://doi.org/10.1016/j.watres.2005.01.016

Technical EIA guidance manual for leather/skin/hide processing industry prepared for the Ministry of Environment and Forests Government of India by IL&FS Ecosmart Limited Hyderabad August 2010

Tien TT, Luu TL (2020) Electrooxidation of tannery wastewater with continuous flow system: role of electrode materials. Environ Eng Res 25:324–334. https://doi.org/10.4491/eer.2018.349

Villaseñor-Basulto DL, Picos-Benítez A, Pacheco-Alvarez M, Pérez T, Bandala ER, Peralta-Hernández JM (2022) Tannery wastewater treatment using combined electrocoagulation and electro-Fenton processes. J Environ Chem Eng 10:107290. https://doi.org/10.1016/j.jece.2022.107290

Part III
Sustainable Practices and Effluent Management in the Tannery Sector

In Part 3, the focus shifts to sustainable practices that can be implemented within the tannery sector to reduce environmental impact. This section discusses the importance of integrating sustainable effluent management strategies and resource recovery techniques into the production process.

Chapter 8 introduces sustainable treatment technologies that aim to reduce pollution and improve the biodegradability of tannery effluents, making them safer for the environment. These treatment methods not only help meet regulatory standards but also enhance the overall sustainability of the tanning process. Chapter 9 explores the innovative practice of sustainable salt recovery from tannery effluent residues, which plays a vital role in reducing waste, conserving resources, and cutting operational costs. The chapter outlines successful case studies and the potential for scaling these practices in tannery operations worldwide. In Chap. 10, the book turns to the concept of the circular economy, discussing how tannery operations can integrate circular economy principles to promote recycling, reduce waste, and increase resource efficiency. This chapter emphasizes the importance of rethinking production processes and designing products with their end-of-life value in mind.

Chapter 8
Sustainable Treatment of Tannery Effluents: Pollution Reduction and Biodegradability Improvements

Rachid Faouzi, Hassan Oumesaoud, Soukaina Harif, and Moulay Abdelazize Aboulhassan

Abstract The improper handling of various wastes, particularly wastewater generated by tanning processes, has led to significant environmental issues and adversely affected human health. Despite ongoing efforts to mitigate tannery wastewater pollution, end-of-pipe treatment methods remain the most commonly used approach. This study investigates the treatment of tannery wastewater using coagulation-flocculation processes with three coagulants: $Al_2(SO_4)_3{\cdot}18H_2O$, $FeCl_3$, and $Ca(OH)_2$. The research focuses on optimizing pH and coagulant dosages to achieve maximum pollutant removal efficiency, minimal sludge production, and enhanced biodegradability. Results demonstrate that up to 71% COD, 92% color, and 97% chromium removal can be achieved under optimal conditions. $Al_2(SO_4)_3{\cdot}18H_2O$ and $Ca(OH)_2$ proved more effective than $FeCl_3$ in pollutant removal and biodegradability improvement. However, $Ca(OH)_2$ produced higher sludge volumes, affecting economic feasibility. The study concludes that $Al_2(SO_4)_3{\cdot}18H_2O$ is the most suitable coagulant, offering high removal efficiency, low sludge production, and improved biodegradability, making it viable for integration into biological treatment processes. These findings contribute to sustainable tannery wastewater management strategies.

Keywords Tannery wastewater · Coagulation · Pollutant removal · Sludge production · Biodegradability index

---

R. Faouzi (✉) · H. Oumesaoud · S. Harif · M. A. Aboulhassan (✉)
Laboratory of Mechanic Process Energy and Environment, National School of Applied Sciences, Ibn Zohr University, Agadir, Morocco
e-mail: rachid.faouzi@edu.uiz.ac.ma

M. A. Aboulhassan
e-mail: a.aboulhassan@uiz.ac.ma

M. A. Aboulhassan
Laboratory of Engineering Sciences, Engineering Sciences Team, Management and Educational Innovation—ESTEM, Casablanca, Morocco

M. A. Aboulhassan and S. Souabi (eds.), *Sustainable Practices in the Tannery Industry*,
https://doi.org/10.1007/978-3-031-92037-0_8

8.1 Introduction

The tanning of hides and skins into leather has been a vital industrial activity for centuries, providing essential materials for numerous products. However, this process is highly water-intensive, requiring approximately 30–40 m^3 of water per ton of hide processed (Suthanthararajan et al. 2004). Recent estimates indicate that global leather production has surpassed **13 x 10^9 kg annually**, generating approximately **39–52 x 10^{10} L of liquid effluent**, which significantly contributes to global water pollution and places immense pressure on water resources (Hasan et al. 2021). The continuous growth in leather production further exacerbates challenges in ensuring water availability and managing effluent discharge sustainably.

Tannery wastewater is an extremely complex and highly polluted mixture, containing biogenic matter, organic and inorganic chemicals, high salinity, and toxic pollutants such as sulfides, chromium, and other heavy metals (Al-Jabari et al. 2021; Humayra et al. 2023). Additionally, the effluent includes dissolved and suspended solids, ammonia, organic nitrogen, and dyes, making it hazardous to ecosystems and difficult to treat (Roš and Gantar 1998). Chromium (Cr), a critical component in tanning, presents severe environmental risks as it oxidizes into Cr(VI), a highly toxic and soluble form that contaminates groundwater and poses serious health hazards (Avudainayagam et al. 2003; Sumathi et al. 2005). The discharge of untreated or poorly treated tannery effluents disrupts aquatic ecosystems, impedes light penetration, and introduces long-lasting toxicity to food chains and biological treatment systems (Talarposhti et al. 2001). The growing scale of leather production amplifies the urgency of addressing these environmental and health concerns.

Efforts to mitigate tannery wastewater pollution have led to advancements in physicochemical and biological treatment technologies. Among these, coagulation-flocculation has emerged as a widely employed method due to its simplicity, cost-effectiveness, and ability to remove suspended solids, organic pollutants, and heavy metals. Recent studies have explored the integration of coagulation-flocculation with advanced treatment processes. For instance, Hasan et al. (2021) demonstrated that electrocoagulation achieves over 95% chromium and organic pollutant removal, while Souza et al. (2020), Babu Ponnusami et al. (2023) and Kaswan and Kaur (2023) emphasized the potential of advanced oxidation processes (AOPs), such as Fenton oxidation and photocatalysis, for breaking down recalcitrant dyes. Furthermore, membrane technologies like nanofiltration and reverse osmosis have been shown to recover water and remove dissolved salts, achieving separation efficiencies of up to 99% (Egea-Corbacho et al. 2021; Wang et al. 2024).

Biological treatments have gained attention due to their environmental sustainability. Processes like anaerobic digestion and constructed wetlands have demonstrated biodegradability indices (BOD_5/COD) exceeding 0.8, indicating their effectiveness in degrading organic matter (Barros et al. 2008; Cossu et al. 2017). Hybrid systems, combining coagulation-flocculation with biological treatments, have shown improved pollutant removal efficiency and reduced sludge generation (Collivignarelli et al. 2019; Badawi and Zaher 2021; Mehrkhah et al. 2024). However, significant gaps remain in the optimization of coagulation-flocculation processes, particularly for the removal of color and chromium and the minimization of sludge production.

This study aims to address these challenges by evaluating the efficiency of coagulation-flocculation processes in treating tannery wastewater. Specifically, it focuses on optimizing coagulant dosages and pH to maximize the removal of organic matter, color, and chromium while minimizing sludge production. Additionally, the impact of these processes on improving biodegradability indices is investigated to support their integration into subsequent biological treatments. By addressing these objectives, this research contributes to the development of more sustainable and effective strategies for tannery wastewater management.

8.2 Materials and Methods

8.2.1 Methodology

The treatment of tannery wastewater in this study follows a systematic approach, as illustrated in Fig. 8.1. The process begins at the tannery plant, where wastewater is generated as part of the leather tanning process. Wastewater sampling is conducted to collect representative effluent samples from various stages of production.

These samples undergo Jar Test experiments to evaluate the efficiency of different coagulants. In particular, $Al_2(SO_4)_3 \cdot 18H_2O$ (Aluminum Sulfate) is used in this study as the coagulant for coagulation-flocculation processes. This stage determines the optimal coagulant dosage and pH for maximum pollutant removal.

Following coagulation, analytical methods are employed to assess key water quality parameters such as turbidity, pH, and chromium content, using specialized instruments such as turbidity meters and ion-specific analyzers. These analyses help evaluate the effectiveness of the coagulation process and assess the potential need for further biological treatment based on the treated water quality.

This methodology provides a comprehensive framework for managing tannery effluent, optimizing coagulation processes, and ensuring the sustainability of the treatment approach for reducing pollutants in tannery wastewater.

8.2.2 Sample Collection

Samples were collected from a tannery located in Mohammedia, Morocco. The tannery generates wastewater from various streams at different times, resulting in significant variability in effluent characteristics, as illustrated in Table 8.1. Since the facility lacks an equalization tank, hourly samples were collected over an 8-h production cycle. These samples were thoroughly mixed in a drum to create a

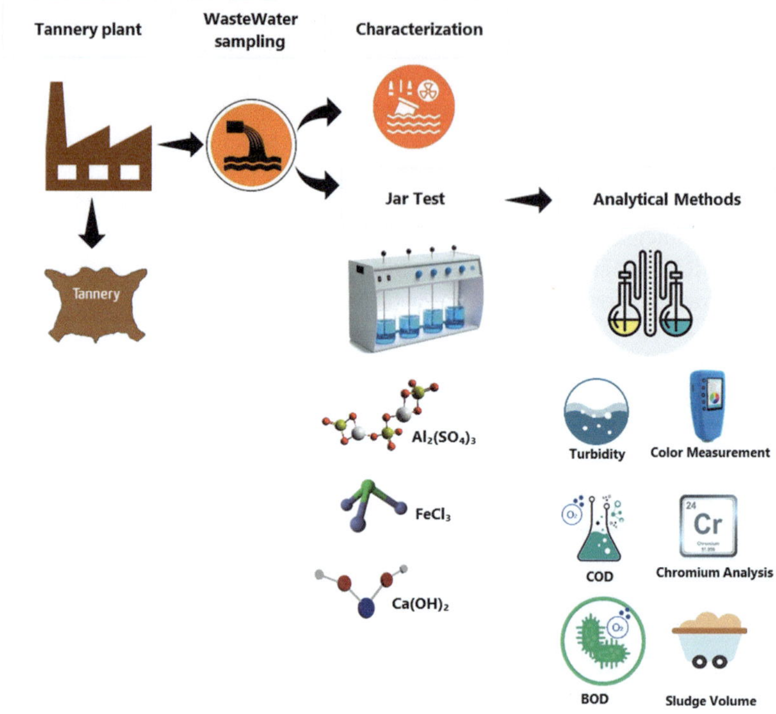

Fig. 8.1 Schematic diagram of the methodology for tannery wastewater treatment using coagulation-flocculation

representative composite sample for subsequent analysis. The coagulants used in this study were $Al_2(SO_4)_3 \cdot 18H_2O$, $FeCl_3$, and $Ca(OH)_2$.

8.2.3 *Jar Test Procedure*

Jar test experiments were conducted under controlled laboratory conditions using a standard jar test apparatus. Four equal-volume polyethylene beakers were prepared to evaluate the effects of varying coagulant dosages or initial pH values in each experimental run. To ensure uniformity, sample bottles were shaken thoroughly to resuspend any settled solids, and the required volume of sample was transferred into each jar test beaker.

The jar test process consisted of three sequential stages:

Rapid Mixing: Samples were mixed at 160 rpm for 5 min to ensure uniform dispersion of the coagulant.

Slow Mixing: The samples were stirred at 30 rpm for 20 min to facilitate floc formation.

Table 8.1 Characterization of the main drain tannery wastewaters[a]

Parameters	Min.	Max.	Average
pH	5.6	6.6	6.2
Conductivity (ms/cm)	9.5	41.2	35.6
Turbidity (FTU)	177	383	296
SM (mg/L)	550	1160	1152
Sulfate (mg/L)	1060	6801	4662
Chloride (mg/L)	1065	2485	1860
Total phosphorus (mg/L)	7	14	12
TKN (mg/L)	168	284	252
BOD (mg/L)	102	492	371
COD (mg/L)	1056	4723	3226
Chromium (mg/L)	10	130	65
H_2S (mg/L)	65	160	118

[a] Number of samples = 8

Settling: The samples were allowed to settle undisturbed for 1 h.

8.2.4 *Analytical Methods*

The efficiency of the coagulants was evaluated by measuring the following parameters: **turbidity, chemical oxygen demand (COD), biological oxygen demand (BOD), color, chromium content, and sludge volume**. The methods used for each parameter are detailed below:

(a) **Color Measurement**:

- Prior to measurement, the samples were filtered using a 0.45 μm Millipore membrane filter to eliminate turbidity interference.
- A UV–visible spectrophotometer (Model 7800 UV/VIS) was used to determine color by measuring absorbance at three wavelengths (436, 525, and 620 nm). The sum of these absorbance values was used to quantify color, as described by Olthof and Eckenfelder (1975) and Pala and Tokat (2002). This approach was adopted to account for the variability in dye composition in the wastewater.

(b) **Chromium Analysis**:

- The concentration of chromium in the liquid phase was measured using graphite furnace atomic absorption spectrometry (Shimadzu AA-6800).

(c) **Turbidity**:

- Turbidity was determined using a HI 93703 microprocessor turbidity meter.

(d) **COD, BOD, and Physicochemical Parameters**:

- These parameters were analyzed following **standard methods** outlined by AFNOR (1999).

(e) **Sludge Volume**:

- After the slow mixing stage, the contents of each beaker were transferred to Imhoff cones and allowed to settle for 1 h. The volume of the settled sludge was recorded using the volumetric method described by Rice et al. (2012).

8.3 Results and Discussion

8.3.1 Effect of pH on Turbidity Removal

The coagulation efficiency of tannery wastewater treatment depends heavily on the solution pH, which influences the formation of coagulant hydrolysis products. As shown in Fig. 8.2, the optimal turbidity removal occurs within specific pH ranges for each coagulant. For $Al_2(SO_4)_3 \cdot 18H_2O$, the optimal pH range is 7–8, while for $FeCl_3$, it is 5–7. At these ranges, the hydrolyzed metal hydroxides form insoluble precipitates that effectively remove colloidal particles, resulting in maximum turbidity reduction. The low solubility of amorphous metal hydroxides $Me(OH)_3$ at intermediate pH values plays a key role in the coagulation process, as it minimizes soluble metal species and facilitates the removal of suspended solids.

The results align with previous studies (Duan and Gregory 2003), which demonstrate that optimizing pH during coagulation significantly improves pollutant removal. Additionally, maintaining the pH within these optimal ranges ensures that

Fig. 8.2 Effect of coagulation pH on turbidity removal

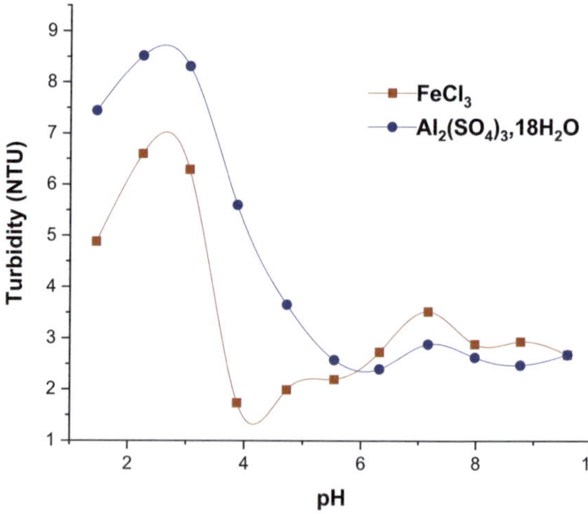

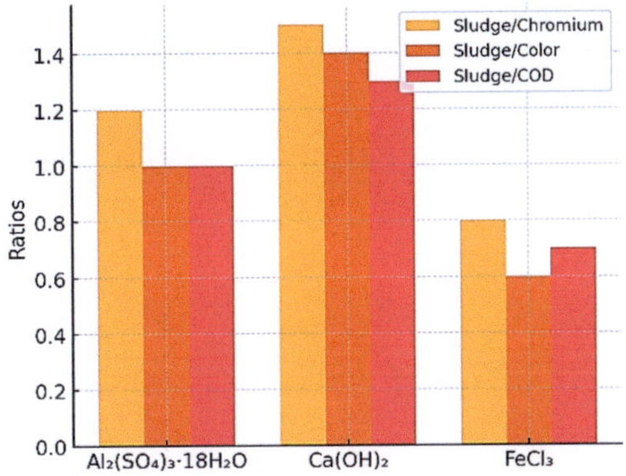

Fig. 8.3 Ratios between the volumes of sludges produced (mL) and the COD. Cr and color reduction (%) for different coagulants at optimal doses

the residual concentrations of aluminum and iron stay below the regulatory limits of 0.5 mg/L and 2 mg/L, respectively (Letterman and Driscoll 1988; Sang et al. 2008) (Fig. 8.3).

8.3.2 COD Removal Efficiency

Figure 8.4 illustrates the reduction in chemical oxygen demand (COD) at the optimal pH for each coagulant. At an initial COD concentration of 3442 mg/L, COD removal efficiencies reached 71% for both $Al_2(SO_4)_3 \cdot 18H_2O$ and $Ca(OH)_2$ at coagulant doses of 600 mg/L and 1000 mg/L, respectively. In contrast, $FeCl_3$ achieved a slightly lower maximum removal efficiency of 67% at a dose of 400 mg/L.

The residual COD concentrations were 996 mg/L for $Al_2(SO_4)_3 \cdot 18H_2O$ and $Ca(OH)_2$ and 1122 mg/L for $FeCl_3$. This difference is likely attributed to the partial

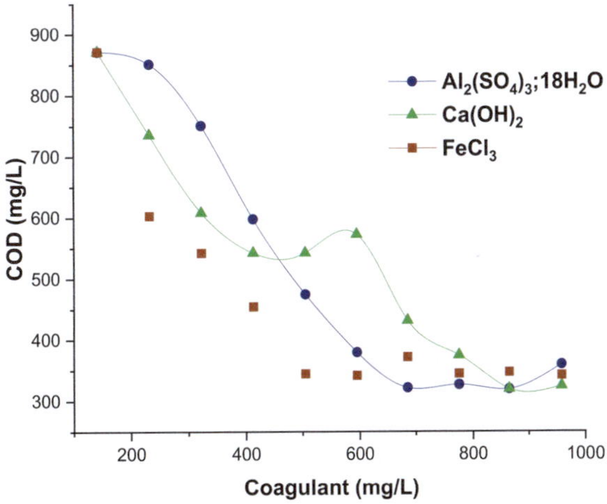

Fig. 8.4 Effect of coagulant doses on COD removal

Fig. 8.5 Color removal using different coagulants

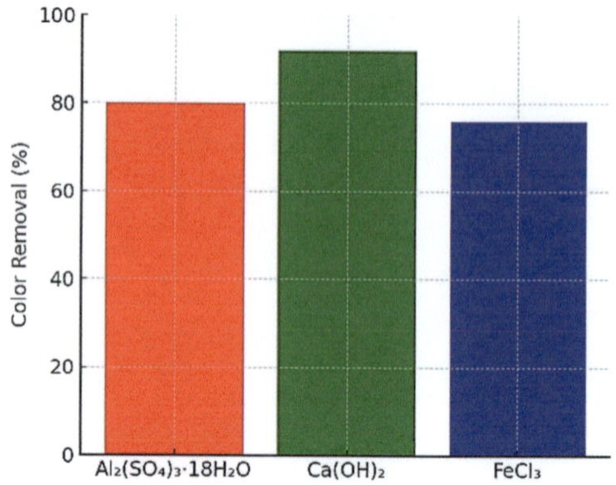

solubility of organic compounds that remain unremovable by coagulation and sedimentation. These findings align with previous studies by Song et al. (2004), which identified pH ranges of 6.5–7.5 for $Al_2(SO_4)_3 \cdot 18H_2O$ and 6.5–8.5 for $FeCl_3$ as optimal for COD removal in tannery wastewater.

8.3.3 Color Removal Efficiency

The presence of dyes in tannery wastewater poses challenges for biological treatment due to their resistance to biodegradation. Figure 8.5 shows that the coagulation-flocculation process significantly reduces color, achieving maximum removal rates of 92%, 81%, and 77% using $Ca(OH)_2$, $Al_2(SO_4)_3 \cdot 18H_2O$, and $FeCl_3$, respectively. The superior performance of $Ca(OH)_2$ can be attributed to its ability to raise pH, which enhances the agglomeration of dye molecules into settleable flocs.

However, $FeCl_3$ exhibited the lowest color removal efficiency, likely due to its reaction with sulfide or vegetable tannins, which can produce color-rich precipitates (Bousher et al. 1997). The high sulfide content in tannery wastewater, coupled with the use of vegetable tanning processes, exacerbates this issue.

8.3.4 Chromium Removal

As shown in Fig. 8.6, chromium removal efficiencies reached 97%, 89%, and 79% for $Ca(OH)_2$, $Al_2(SO_4)_3 \cdot 18H_2O$, and $FeCl_3$, respectively. The results suggest that $Ca(OH)_2$ is the most effective coagulant for chromium removal, as its addition increases wastewater pH to an alkaline range (~ pH 9.1), which promotes the precipitation of insoluble chromium hydroxides. In contrast, $FeCl_3$ operates in an acidic pH range, resulting in lower chromium removal efficiency due to limited hydroxide formation.

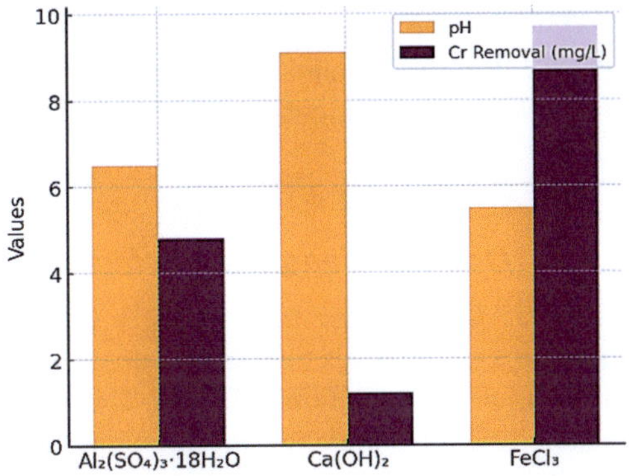

Fig. 8.6 Effect of coagulants at optimal doses on pH and Cr

The findings corroborate those of Roš and Gantar (1998), who reported that chromium removal is maximized under alkaline conditions. Residual chromium concentrations were 1.2 mg/L, 4.8 mg/L, and 9.7 mg/L for $Ca(OH)_2$, $Al_2(SO_4)_3{\cdot}18H_2O$, and $FeCl_3$, respectively, highlighting the effectiveness of $Ca(OH)_2$ in achieving compliance with regulatory limits.

8.3.5 Biodegradability Index Improvement

The biodegradability index (BOD_5/COD ratio) serves as an indicator of wastewater suitability for biological treatment. According to Fig. 8.7, the treated wastewater exhibits the highest biodegradability index when $FeCl_3$ is used, exceeding the threshold value of 0.8, which classifies it as readily biodegradable. This is likely due to the lower chromium and color content in wastewater treated with $Al_2(SO_4)_3{\cdot}18H_2O$ and $Ca(OH)_2$, which enhances the biodegradability index.

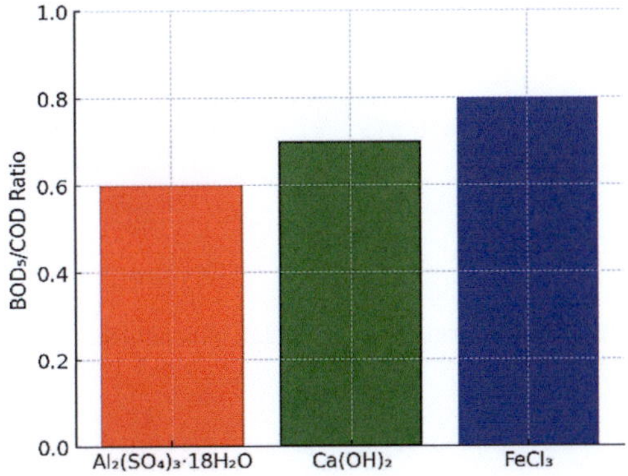

Fig. 8.7 Biodegradability index of wastewaters at optimal

However, $FeCl_3$-treated wastewater may require further biological treatment to achieve discharge standards, as residual chromium and color levels are higher compared to the other coagulants.

8.3.6 *Optimal Coagulant Selection and Cost Analysis*

Figure 8.8 summarizes the optimal doses and costs of the coagulants. While $FeCl_3$ offers a low-cost solution (€0.30/m^3), its relatively lower removal efficiencies and higher residual pollutant concentrations limit its application. $Al_2(SO_4)_3 \cdot 18H_2O$, despite its higher cost (€0.88/m^3), provides an excellent balance between pollutant removal efficiency, sludge production, and biodegradability improvement, making it the most suitable coagulant for tannery wastewater treatment.

Coagulants	$FeCl_3$	$Al_2(SO_4)_3 \cdot 18H_2O$	$Ca(OH)_2$
Doses (mg/L)	400	600	1000
COD removal (%)	67	71	71
Color removal (%)	76	80	92
Cr removal (%)	79	89	97
Sludge (mg/L)	88	103	140
Cost (€ /m³)	0.30	0.88	0.32

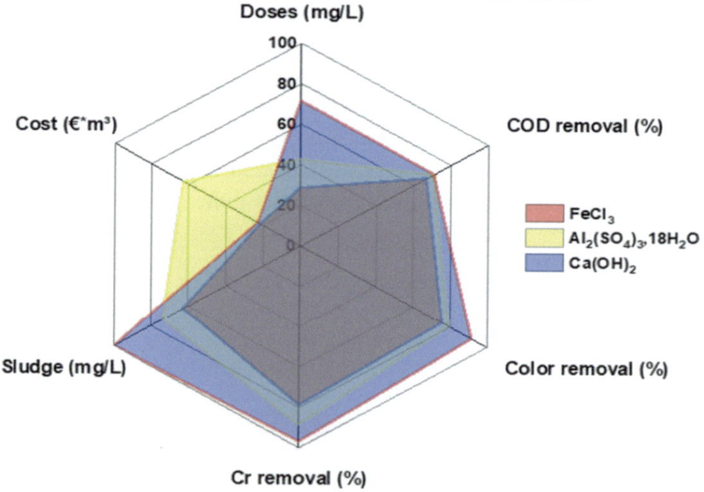

Fig. 8.8 Performance comparison of coagulants (normalized)

8.4 Conclusion

This research provides a valuable contribution to the field of tannery wastewater treatment by demonstrating the effectiveness of coagulation-flocculation with $Al_2(SO_4)_3 \cdot 18H_2O$, $FeCl_3$, and $Ca(OH)_2$ as coagulants. The study effectively showcases the impact of these coagulants on reducing pollutants such as COD, Color, and chromium while improving biodegradability. Among the coagulants tested, $Al_2(SO_4)_3 \cdot 18H_2O$ stands out as the most efficient in pollutant removal with minimal sludge production, making it a viable option for integration into biological treatment processes.

The research offers important insights into optimizing coagulation parameters for tannery wastewater treatment, contributing to more sustainable practices in the leather industry. The findings suggest that while $FeCl_3$ is a cost-effective solution, it is less efficient in terms of pollutant removal and may require additional treatment steps. On the other hand, $Al_2(SO_4)_3 \cdot 18H_2O$ strikes an optimal balance between cost, Effectiveness, and environmental impact, making it the most suitable choice for large-scale applications.

In conclusion, this study presents a comprehensive and sustainable approach to managing tannery effluents, and its recommendations for the selection of coagulants based on pollutant removal efficiency and sludge production will be valuable for industries aiming to meet environmental regulations while ensuring cost-effective operations.

References

Al-Jabari M, Sawalha H, Pugazhendhi A, Rene ER (2021) Cleaner production and resource recovery opportunities in leather tanneries: technological applications and perspectives. Bioresour Technol Rep

Avudainayagam S, Megharaj M, Owens G, Kookana RS, Chittleborough D, Naidu R (2003) Chemistry of chromium in soils with emphasis on tannery waste sites. Rev Environ Contam Toxicol. https://doi.org/10.1007/0-387-21728-2_3

Babu Ponnusami A, Sinha S, Ashokan H, Paul MV, Hariharan SP, Arun J, Gopinath KP, Hoang Le Q, Pugazhendhi A (2023) Advanced oxidation process (AOP) combined biological process for wastewater treatment: a review on advancements, feasibility and practicability of combined techniques. Environ Res

Badawi AK, Zaher K (2021) Hybrid treatment system for real textile wastewater remediation based on coagulation/flocculation, adsorption and filtration processes: performance and economic evaluation. J Water Process Eng. https://doi.org/10.1016/j.jwpe.2021.101963

Barros P, Ruiz I, Soto M (2008) Performance of an anaerobic digester-constructed wetland system for a small community. Ecol Eng. https://doi.org/10.1016/j.ecoleng.2008.02.015

Bousher A, Shen X, Edyvean RGJ (1997) Removal of coloured organic matter by adsorption onto low-cost waste materials. Water Res. https://doi.org/10.1016/S0043-1354(97)00037-7

Collivignarelli MC, Abbà A, Carnevale Miino M, Damiani S (2019) Treatments for color removal from wastewater: state of the art. J Environ Manage

Cossu R, Fantinato G, Pivato A, Sandon A (2017) Further steps in the standardization of BOD_5/COD ratio as a biological stability index for MSW. Waste Manag. https://doi.org/10.1016/j.wasman.2017.06.035

Duan J, Gregory J (2003) Coagulation by hydrolysing metal salts. Adv Colloid Interface Sci. https://doi.org/10.1016/S0001-8686(02)00067-2

Egea-Corbacho A, Gutiérrez S, Coello D, Quiroga JM (2021) Comparison in the removal of stimulants and antibiotics from wastewater for its subsequent reuse with different technologies. Chem Eng Res Des. https://doi.org/10.1016/j.cherd.2020.12.006

Hasan MA, Hashem MA, Arman MN, Momen MA (2021) Batch electrocoagulation process for removal of chromium from tannery wastewater. J Eng Sci. https://doi.org/10.3329/jes.v12i1.53098

Humayra S, Hossain L, Hasan SR, Khan MS (2023) Water footprint calculation, effluent characteristics and pollution impact assessment of leather industry in Bangladesh. Water (Switzerland). https://doi.org/10.3390/w15030378

Kaswan V, Kaur H (2023) A comparative study of advanced oxidation processes for wastewater treatment. Water Pract Technol. https://doi.org/10.2166/wpt.2023.061

Letterman RD, Driscoll CT (1988) Survey of residual aluminum in filtered water. J Am Water Work Assoc. https://doi.org/10.1002/j.1551-8833.1988.tb03020.x

Mehrkhah R, Park SY, Lee JH, Kim SY, Lee BH (2024) A comparative study of advanced oxidation-based hybrid technologies for industrial wastewater treatment: an engineering perspective. Chem Eng Sci. https://doi.org/10.1016/j.ces.2023.119675

Olthof MG, Eckenfelder WW (1975) Color removal from pulp and paper wastewaters by coagulation. Water Res. https://doi.org/10.1016/0043-1354(75)90030-5

Pala A, Tokat E (2002) Color removal from cotton textile industry wastewater in an activated sludge system with various additives. Water Res. https://doi.org/10.1016/S0043-1354(01)00529-2

Rice E, Baird R, Eaton A, Clesceri L (2012) Standard methods for the examination of water and wastewater

Roš M, Gantar A (1998) Possibilities of reduction of recipient loading of tannery wastewater in Slovenia. Water Sci Technol

Sang Y, Sun T, Chen J, Liang C (2008) Flocculation process optimization of a novel coagulant for landfill leachate pretreatment. In: 2nd international conference on bioinformatics and biomedical engineering, iCBBE 2008

Song Z, Williams CJ, Edyvean RGJ (2004) Treatment of tannery wastewater by chemical coagulation. Desalination. https://doi.org/10.1016/S0011-9164(04)00193-6

Souza EA, Cerqueira UMFM, Silva LA (2020) Comparative study of various advanced oxidation processes for the treatment of tannery wastewater. Desalin Water Treat. https://doi.org/10.5004/dwt.2020.25183

Sumathi KMS, Mahimairaja S, Naidu R (2005) Use of low-cost biological wastes and vermiculite for removal of chromium from tannery effluent. Bioresour Technol. https://doi.org/10.1016/j.biortech.2004.04.015

Suthanthararajan R, Ravindranath E, Chitra K, Umamaheswari B, Ramesh T, Rajamani S (2004) Membrane application for recovery and reuse of water from treated tannery wastewater. Desalination. https://doi.org/10.1016/S0011-9164(04)00174-2

Talarposhti AM, Donnelly T, Anderson GK (2001) Colour removal from a simulated dye wastewater using a two-phase anaerobic packed bed reactor. Water Res. https://doi.org/10.1016/S0043-1354(00)00280-3

Wang H, Yang J, Zhang H, Zhao J, Liu H, Wang J, Li G, Liang H (2024) Membrane-based technology in water and resources recovery from the perspective of water social circulation: a review. Sci Total Environ

Chapter 9
Sustainable Salt Recovery from the Solid Residue of Tannery Common Effluent Treatment Plants

Parul Sahu, Rahul J. Sanghavi, Sumesh C. Upadhyay, and Arvind Kumar

Abstract India, a leading leather producer with over 3000 tanneries—70% of which are in Tamil Nadu—treats tannery effluent through individual or common effluent treatment plants (CETPs). Secondary treated wastewater is further processed using membrane systems for water recovery and reuse. Following the adoption of Zero Liquid Discharge (ZLD) technology, many tanneries have implemented reverse osmosis (RO) for water recovery followed by multiple-effect evaporation (MEE), resulting in a residue of salts and other materials stored in CETPs. Disposal of this solid residue poses a significant threat to the environment. As the solid residue mainly contains inorganic salts and organic matter, resource recovery can help manage this waste and make the process sustainable and cost-effective. This study addresses the treatment of solid residue from tanneries' CETPs by presenting sustainable processes for minimizing waste and recovering sodium chloride (NaCl), sodium sulphate (Na_2SO_4) and gypsum ($CaSO_4 \cdot 2H_2O$). Two process configurations have been designed to valorize the solid waste to recover added-value products. An alternative process is also developed to modify ZLD process with enhanced separation and recovery of salts with a better composition for potential commercial use. Utilizing solid residues can contribute to the circular economy by minimizing waste and promoting resource recovery. As tanneries explore these options, they can also improve their public image and meet growing consumer demand for environmentally responsible practices. Overall, the combined efforts of the industry and regulatory bodies can lead to significant advancements in sustainable practices within the leather sector.

Keywords Tannery industry · Common effluent treatment plants (CETPs) · Solid waste residue · Resource recovery · Waste minimization · Sustainability

---

P. Sahu (✉) · R. J. Sanghavi · S. C. Upadhyay · A. Kumar
Salt and Marine Chemicals Division, CSIR—Central Salt & Marine Chemicals Research Institute (CSMCRI), Bhavnagar, Gujarat, India
e-mail: paruls@csmcri.res.in

P. Sahu · S. C. Upadhyay · A. Kumar
Academy of Scientific and Innovative Research (AcSIR), Ghaziabad, India

M. A. Aboulhassan and S. Souabi (eds.), *Sustainable Practices in the Tannery Industry*,
https://doi.org/10.1007/978-3-031-92037-0_9

9.1 Introduction

The leather industry is a significant global market focused on converting animal skin and hides into valuable leather products. India, with over 3000 tanneries, is a leading producer, especially in Tamil Nadu, where about 70% of hides are tanned. Hides need preservation before tanning, typically using common salt (NaCl). Various processes like soaking, liming, pickling, and tanning are employed, involving numerous chemicals such as sodium chloride and chromium salts. However, these methods also result in high levels of salt and pollutants in wastewater, generating approximately 30–50 m^3 of effluent for each ton of hide processed. Therefore, the leather industry has faced criticism for its environmental impact, with the tanning sector identified as a significant source of pollution (Suthanthararajan et al. 2004).

The treatment processes typically used in effluent treatment plants of tannery follow a conventional system that combines physio-chemical treatment with biological treatment. This approach effectively removes pollutants like BOD, COD, suspended solids, and chromium. However, the treated tannery wastewater still contains a high level of total dissolved solids (TDS), which cannot be removed by conventional treatment methods. The components contributing to TDS include sodium, chloride, sulphate, calcium, magnesium, and residual organic impurities. As a result, the treated effluent is not suitable for reuse in the tanning process due to its high TDS content (Suthanthararajan et al. 2004). The growing concerns over water and soil pollution in areas with tannery clusters near water bodies led pollution control authorities to enforce stricter adherence to TDS discharge limits.

The Zero Liquid Discharge (ZLD) system has been introduced, previously unused in the tanning industry. ZLD works by concentrating dissolved solids through Reverse Osmosis (RO) and Multi-Effect Evaporation (MEE) until only damp solid waste remains. This solid waste is then disposed of (landfill disposal), while nearly all the water is reclaimed and reused. A schematic of tannery waste management in India is shown in Fig. 9.1.

The effluent treatment plants require significant land, capital, power, and specialized personnel, making them unaffordable for small-scale tanneries. Consequently, the effluent from all tanneries is collected at a centralized Common Effluent Treatment Plant (CETP) in India, which is managed and maintained by small-scale tanners through a cooperative society (Pathe et al. 2004). Several existing CETPs have been upgraded with RO and MEE technologies, along with additional processes like tertiary treatment and water softening. The disposal of solid residue in secured landfill sites is prohibited by pollution control agencies due to concerns that its constituent ions may leach into groundwater, and the leachate's high salinity would complicate treatment. As a result, it is typically collected and stored in storage yards in many leather industries, with no viable reuse options available. The high concentration of both inorganic salts and organic matter in residue further restricts its disposal (Boopathy et al. 2013). This significant volume of solid residue produced by the CETPs requires proper treatment and reclamation instead of being disposed as solid waste in the environment (Boopathy et al. 2012).

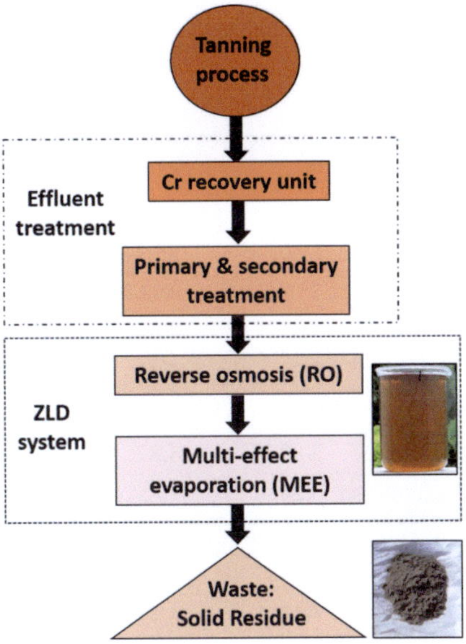

Fig. 9.1 Process flow schematic of tannery waste management in India

9.2 Characterization of Tannery CETP Residue

Solid residue and RO reject liquid samples for the study were collected from different tannery CETPs of Tamil Nadu (TN), India. The ionic composition of inorganics presents in the solid residue samples of tannery CETPs of Tamil Nadu, India, was examined using standard analysis processes and presented in Table 9.1. Sulphate ions were detected gravimetrically as barium sulphate, while chloride ions were identified using Mohr's method. Calcium and magnesium ions were determined through complexometric titrations with EDTA (Vogel and Mendham 2000). Sodium and potassium ions were measured using the flame photometer method. The tabulated data indicate that the solid residue produced during tanning primarily consists of sodium chloride, sodium sulphate, and calcium sulphate, along with trace amounts of magnesium chloride and other soluble and insoluble impurities.

The solid residue comprised of mixture of inorganic salts (sodium chloride, sodium sulphate, and calcium sulphate) and organic impurities is processed to recover the available salts in pure form. This study presents the methods developed to extract secondary products from the solid waste of tannery CETPs.

Table 9.1 Major inorganic compositions of solid residue samples of tannery CETPs

Sample ID →	S1	S2	S3	S4	S5	S6
Location	Ranitec, TN	Ranitec, TN	Ranitec, TN	Ranitec, TN	Amburtec, TN	Ranitec, TN
	Ionic composition (wt%)					
Cl^-	54.08	55.94	55.95	57.46	47.30	50.24
Na^+	35.09	36.17	36.17	37.15	30.76	35.43
SO_4^{2-}	5.61	4.02	5.42	3.75	14.71	11.89
Ca^{2+}	0.34	0.45	0.01	0.34	0.60	2.38
Mg^{2+}	0.04	0.02	0.04	0.04	0.04	0.03
Moisture	3.80	4.87	NA	NA	0.18	2.87
Insolubles	0.24	1.85	0.03	0.06	0.03	NA
	Probable salt composition (wt%)					
NaCl	91.32	92.01	92.00	94.49	78.24	82.70
Na_2SO_4	7.02	4.35	8.00	4.33	19.49	9.01
$CaSO_4$	1.18	1.54	0.01	1.17	2.40	8.22
Other	0.22	0.13	0.16	0.17	0.19	0.11

NA data not available

9.3 Processes for Salt Recovery from Tannery CETP Residue

The composition of major ions in the solid residue varies depending on the tannery location. This variation influences the choice of process for recovering salts from the residue. Boopathy and co-workers have extensively worked on recovering value added chemicals from solid residue after evaporation. According to their study, chloride and sulphate ions could be successfully separated by a selective ion precipitation techniques from saturated solution of solid residue (Boopathy et al. 2012; Boopathy and Sekaran 2013). In a follow-up study, they investigated the electrochemical oxidation using different electrode systems to treat organic compounds (Boopathy et al. 2014). The developed process involves the use of mineral acids, such as HCl, to recover NaCl, and energy-intensive methods like calcination to recover calcium sulphate. Additionally, electrochemical oxidation processes for removing organic components can substantially raise both the cost and complexity of the process.

We have developed alternative processes using well-established and low-cost unit operations and processes to recover inorganic salts (sodium chloride, Glauber's salt and gypsum), while eliminating organic compounds using flocculation and clarification. This section presents two process variants developed to recover salts for potential commercial use from solid residue as raw materials. These processes are suitable for reclaiming the existing solid waste residue stored at CETPs. An alternative process is also depicted to valorize RO concentrate with the modified ZLD process to enhance separation and recovery of salts and water.

9.3.1 Process-I to Recover Sodium Chloride and Sodium Sulphate

A cyclic process is developed to recover pure NaCl and Glauber's salt $Na_2SO_4{\cdot}10H_2O$ from solid residue having lower $CaSO_4$ concentrations (e.g. Sample S1–S5 in Table 9.1) (Kumar et al. 2020). This process shown in Fig. 9.2 aims to recover two valuable salts—sodium chloride (NaCl) and sodium sulphate (as Glauber's salt $Na_2SO_4{\cdot}10H_2O$)-from the tannery CETP residue through a series of physical and chemical steps. The CETP residue, which contains a mix of salts, is first washed with a saturated brine solution. This high-concentration salt solution helps to displace soluble salts and organic contents, from the solid residue, facilitating their separation. The washing step can be performed using a screw type washery system or Hydrocyclone. The washed salt is primarily NaCl devoid of soluble salt (Na_2SO_4) and organic impurities. The salt is further washed with fresh water to remove any remaining impurities, adhered brine or soluble by-products, ensuring high purity of the recovered salt. The final product is > 99% NaCl, which can be dried before storage. The photographs of a process demonstration to recover NaCl from solid residue are presenter in Fig. 9.3.

The washing liquor, which is a mixture of dissolved salts (Na_2SO_4, $CaSO_4$, and NaCl) and other impurities, undergoes clarification. This process involves settling or using a coagulant to remove suspended solids, ensuring a clear liquid that contains the dissolved salts. The next stage of sodium sulphate recovery takes leverage of

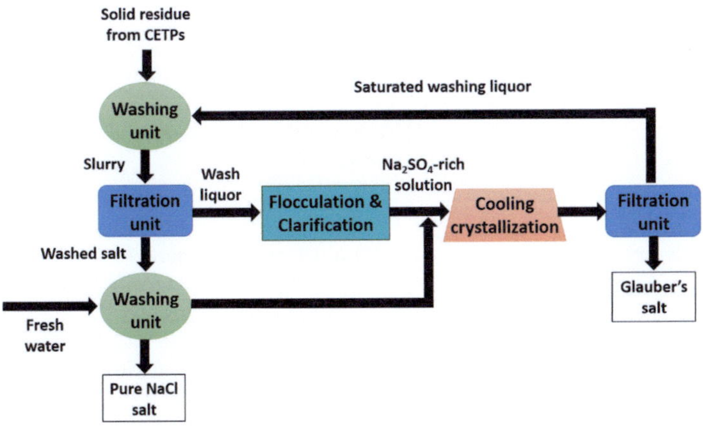

Fig. 9.2 Process-I depicting recovery of NaCl and Glauber's salt

Fig. 9.3 Photographs demonstrating the washing process of solid residue to recover NaCl

the facts that (i) sodium chloride solubility is practically independent of temperature while (ii) sodium sulphate exhibits significant temperature dependent solubility. Hence, the wash liquor, upon chilling, yields nearly pure sodium sulphate (as sodium sulphate decahydrate/Glauber's salt). Glauber's salt ($Na_2SO_4{\cdot}10H_2O$) yield can be maximized by chilling the liquor to sub-zero temperature. The remaining liquor devoid of Na_2SO_4 is recycled back to washing of the solid residue. A heat recovery system can be effectively implemented by utilizing the cold, Na_2SO_4-depleted liquor to pre-cool the clarified Na_2SO_4-rich solution. In this system, the temperature of the Na_2SO_4-depleted liquor-having already undergone a separation process and cooled down during the treatment is harnessed to reduce the temperature of the clarified Na_2SO_4-rich solution before it enters the next stage of processing. An indicative material flow for processing 1 metric ton of solid residue (Sample S4 in Table 9.1) is presented in Table 9.2.

Table 9.2 Indicative material flow for Process-I

Solid residue (raw material)	Mass = 1 ton Composition (wt%) NaCl = 94.49% Na_2SO_4 = 4.33% $CaSO_4$ = 1.17% Other = 0.17%
Utility	Fresh water = 0.1 ton
Product recovery	NaCl = 0.8 ton (> 98% purity) Glauber's salt = 0.11 ton (> 98% purity)
Sludge	Insolubles/organics = 0.05 ton

9.3.2 Process-II to Recover Sodium Chloride

Solid residue samples from some CETPs, such as Sample S6 in Table 9.1, contain a significant amount of $CaSO_4$. Since Process-I relies on the dissolution of undesired sulphate salts during the washing step, samples with high $CaSO_4$ content may not yield high-purity NaCl, as $CaSO_4$ has low solubility in saturated brine. To address this issue, an alternate process has been developed to recover high-purity NaCl. In this method, calcium, silica, and sulphate impurities are removed as insoluble solids by dissolving the solid residue in brine or freshwater. The resulting solution is then filtered to prepare a saturated solution, free from these impurities. The details of the lab-scale experiments, parametric optimization, and techno-economic analysis of the process are provided in our previous publication (Sahu et al. 2024). The process schematic is shown in Fig. 9.4. The CETP residue is first dissolved in water or brine to dissolve the soluble salts, including sodium chloride. The solution contains NaCl and Na_2SO_4 from the residue. The dissolved mixture is then clarified, typically through sedimentation or filtration, to remove insoluble impurities such as organic matter, calcium sulphate, and suspended particles. The sodium sulphate (Na_2SO_4) present in the solution is converted into insoluble gypsum by using calcium chloride ($CaCl_2$) as a desulphatation agent. The desulphated solution is filtered to ensure that any remaining fine particles or precipitates are removed, leaving behind a clear brine solution rich in sodium chloride. The clear brine solution is then evaporated under controlled conditions, typically using a multi-effect evaporator or a vacuum evaporator. Subject to availability of low-cost land around CETP and suitable arid climatic conditions, solar evaporation in open ponds can be explored. As the water evaporates, the concentration of sodium chloride increases, and it eventually reaches saturation, leading to the crystallization of pure sodium chloride crystals. Table 9.3 presents a representative material flow for processing 1 metric ton of solid residue (Sample S6 in Table 9.1).

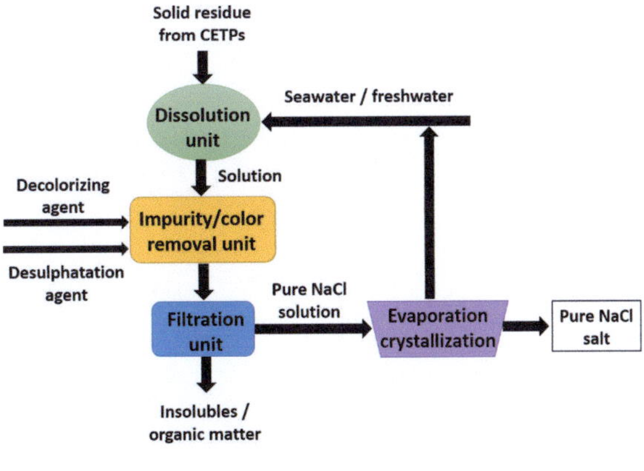

Fig. 9.4 Process-II illustrating the recovery of NaCl

Table 9.3 Indicative material flow for Process-II

Solid residue (raw material)	Mass = 1 ton Composition (wt%) NaCl = 82.70% Na_2SO_4 = 9.01% $CaSO_4$ = 8.22% Other = 0.11%
Chemicals for treatment	Activated charcoal = 10 kg Desulfating agent = 6.7 kg
Product recovery	NaCl = 0.78 ton (> 98% purity)
Sludge	Insolubles/organics = 0.2 ton

9.3.3 *Process-III to Recover Sodium Chloride and Gypsum*

An alternative approach to recovering valuable products from the RO reject of CETP's ZLD system, prior to multi-effect evaporation, is outlined in Fig. 9.5 (Sanghavi 2023). This RO reject is brown coloured liquid having obnoxious odour and TDS of ~ 2 to 6 wt%.

The characteristics of RO reject streams generated at tannery CETPs reported in literature are presented in Table 9.4 (Ramasamy 2020). The clarified liquor primarily contains total dissolved solids (sodium, sulphate, and chloride ions), and organic matter.

The first step involves treating the coloured liquid with a decolorizing agent (such as activated carbon or bleaching agents) to remove organic contaminants and colour. After decolorization, the solution undergoes clarification with alum, resulting in a clarified solution with low turbidity (< 1 NTU). This clarification process helps remove colour, odour, and prevents the entrapment of unwanted salts in the final

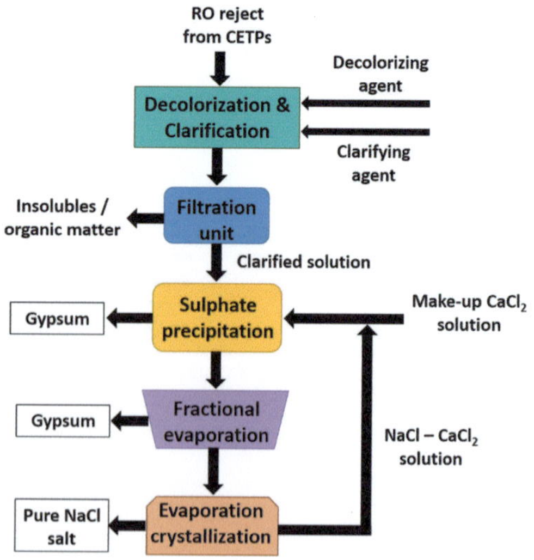

Fig. 9.5 Schematic of Process-III for the recovery of NaCl and gypsum

Table 9.4 Characteristics of RO reject stream generated in tannery CETPs

Parameters	pH[a]	TDS	BOD	COD	Sodium	Chloride	Sulphate
Range (g/l)	6.85–7.55	20 ± 6.14	0.40 ± 0.32	1.53 ± 0.85	3.96 ± 1.67	4.99 ± 2.20	5.55 ± 1.95

[a] All values in g/l except pH

NaCl product, ensuring its high purity for specific applications. The clarified tannery liquor was treated with a precipitating agent (e.g. calcium chloride) to precipitate sulphates, such as calcium sulphate (gypsum). Cost-effective desulphating agents, such as distiller waste from soda ash plants, can be effectively utilized for the desulphation process, provided they are readily available in the local area. As shown in Fig. 9.5, gypsum is separated at two stages, first at the low brine concentration (TDS ~ 5 wt%) and later at higher concentration (TDS ~ 25 wt%) achieved through fractional evaporation. The $CaCl_2$ solution was added in a 20% excess over the stoichiometric amount to ensure complete desulphatation of the solution and to maintain a surplus of Ca for the separation of residual gypsum (at saturation) dissolved in the liquid. This approach is based on the principle of the solubility product (K_{sp}), where the K_{sp} of $CaSO_4$ in pure water is 2.4×10^{-5} and the K_{sp} of gypsum in a brine system is 2.2×10^{-4}. When the ionic product exceeds the solubility product, the salt precipitates out of the solution. The precipitate is separated through filtration or settling, leaving behind a solution rich in sodium chloride and unreacted $CaCl_2$. The sulphate-free, concentrated liquid is further evaporated until solution concentration of ~ 27% to crystallize NaCl salt. The solid fraction removed at this stage, containing > 99% NaCl on a dry basis, has minimal impurities of Ca^{2+}, Mg^{2+}, and SO_4^{2-}. This NaCl meets grade I industrial specifications for NaCl content. After NaCl recovery, the remaining concentrated liquor, rich in NaCl (~ 15 wt%) and $CaCl_2$ (~ 22 wt%), is recycled for further desulphatation of the clarified liquor. This developed process demonstrates an innovative method for recovering pure NaCl and gypsum from the RO reject liquor, highlighting the efficient recovery of valuable products through this system. The surface morphology of crystallized gypsum and sodium chloride was analysed using FE-SEM (JFM 7100 F: Oxford Inc.). Figure 9.6a, b shows the SEM images of needle-shaped gypsum crystals. Figure 9.7a, b shows the SEM image of cubic crystals obtained for salt fractions during brine evaporation crystallization. Table 9.5 gives a representative material flow for processing 1 m^3 of RO concentrated liquor.

Fig. 9.6 **a**, **b** SEM images of needle-shaped gypsum crystals

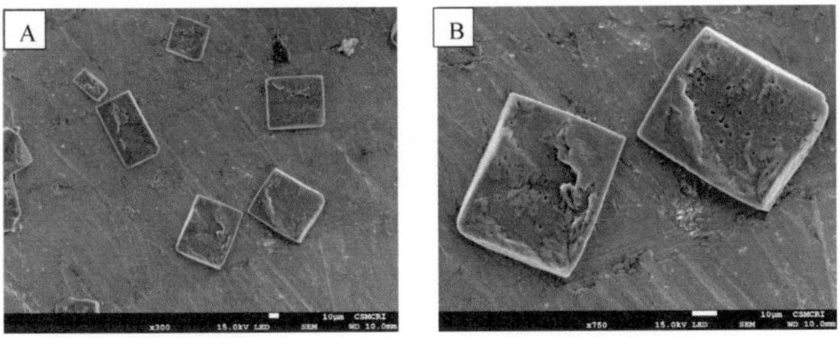

Fig. 9.7 **a**, **b** SEM images of crystallized salt (NaCl)

Table 9.5 Indicative material flow for Process-III

Decolorized and clarified liquor (raw material)	Volume = 1 m^3 Composition NaCl = 40.44 g/l Na_2SO_4 = 9.30 g/l $CaSO_4$ = 3.20 g/l Others = 1.40 g/l
Chemicals for treatment	Desulfating agent = 10 kg
Product recovery	NaCl = 31 kg (> 98% purity) Gypsum = 10.55 kg

9.4 Conclusion

It can be established that Process-I and Process-II reported in Sect. 9.3 provide effective methods for valorization of industrial waste by recovering valuable salts from tannery residues lying in CETPs. Process-I enables the simultaneous recovery of both sodium chloride and sodium sulphate, while Process-II is focused on recovering pure sodium chloride through evaporation techniques. The concentration of $CaSO_4$ in the CETP residue determines the selection of the appropriate process. Process-III presents process intensification by modifying the existing ZLD system, incorporating additional unit operations such as decolorization, clarification, and gypsum precipitation before the conventional multi-effect evaporation process.

The developed processes (Fig. 9.8) can meet stringent environmental regulations while also generating valuable salts of high purity. The indicative material flow for each process serves as a crucial foundation for assessing the techno-economic feasibility and sustainability of the developed processes. The potential for commercializing these recovered salts further enhances the economic feasibility of the system, making it a viable option for tanneries seeking to minimize their environmental. These sustainable and viable solutions for reusing accumulated salt residue can be actively pursued by tannery industries, especially with the backing of Pollution Control Boards. By implementing such initiatives, tanneries can not only mitigate their environmental impact but also transform waste into valuable resources.

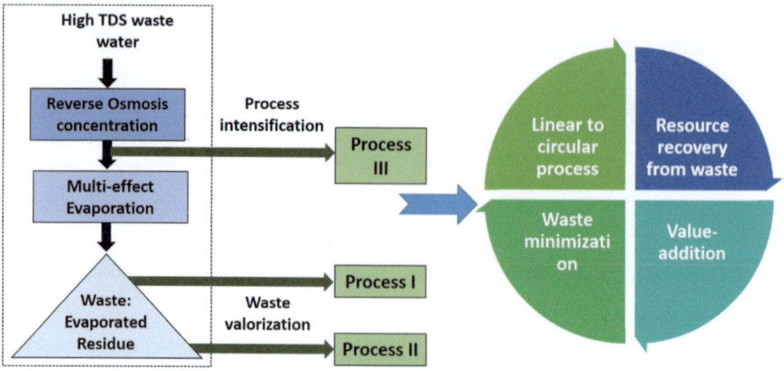

Fig. 9.8 Sustainable approach for repurposing the waste from tannery CETPs

The involvement of Pollution Control Boards can facilitate this process by providing guidance on best practices, regulatory compliance, and financial incentives for adopting sustainable waste management strategies. This collaborative approach can help tanneries effectively manage their salt residues, potentially reducing disposal costs and enhancing overall operational efficiency.

References

Boopathy R, Sekaran G (2013) First report on separation of sulfate ions from evaporated residue of RO reject stream generated from the leather industry. Ind Eng Chem Res 52:9239–9246. https://doi.org/10.1021/ie400731w

Boopathy R, Gnanamani A, Mandal AB, Sekaran G (2012) A first report on the selective precipitation of sodium chloride from the evaporated residue of reverse osmosis reject salt generated from the leather industry. Ind Eng Chem Res 51:5527–5534. https://doi.org/10.1021/ie201735s

Boopathy R, Karthikeyan S, Mandal AB, Sekaran G (2013) Characterisation and recovery of sodium chloride from salt-laden solid waste generated from leather industry. Clean Technol Environ Policy 15:117–124. https://doi.org/10.1007/s10098-012-0489-y

Boopathy R, Mandal AB, Sekaran G (2014) Electrochemical treatment of evaporated residue of reverse osmosis concentrate generated from the leather industry. RSC Adv 4:54614–54624. https://doi.org/10.1039/C4RA09278C

Kumar A, Upadhyay SC, Sanghavi RJ, Hamidani AHU, Vyas BG, Makwana MH, Dobariya RV, Bhatti SJ, Gauswami MV (2020) A process for the recovery of high purity salts from crudes containing sodium chloride and sodium sulphate, 202011009669

Pathe PP, Suresh Kumar M, Kharwade, Kaul SN (2004) Common effluent treatment plant (CEPT) for wastewater management from a cluster of small scale tanneries. Environ Technol 25:555–563. https://doi.org/10.1080/09593332608618562c

Ramasamy B (2020) Short review of salt recovery from reverse osmosis rejects, chap 4. In: Çinku MC, Karabulut S (eds) Salt in the earth. IntechOpen, Rijeka. https://doi.org/10.5772/intechopen.88716

Sahu P, Upadhyay SC, Liya K, Gohil D, Srinivasan SV, Kumar A (2024) Waste management and salt recovery from solid residue of tannery common effluent treatment plant. Indian Chem Eng 1–13. https://doi.org/10.1080/00194506.2023.2300148

Sanghavi RJ (2023) Studies on recovery of value-added chemicals from brines using additives/effluent streams of different chemical industries. Charotar University of Science and Technology

Suthanthararajan R, Ravindranath E, Chits K, Umamaheswari B, Ramesh T, Rajamam S (2004) Membrane application for recovery and reuse of water from treated tannery wastewater. Desalination 164:151–156. https://doi.org/10.1016/S0011-9164(04)00174-2

Vogel AI, Mendham J (2000) Vogel's textbook of quantitative chemical analysis. Prentice Hall, Harlow

Chapter 10
Promoting Circular Economy: Implementation Principles in the Tannery Sector

Sachin C. Hallad, V. Kavan Kumar, R. Mahendiran, and N. L. Panwar

Abstract The concept of the circular economy (CE) centers on the optimization of resource productivity and eco-efficiency, with a particular emphasis on the industrial structure, new technology development, application, equipment renewal, and management renovation. While the leather business indirectly contributes to the growth of the local economy, it also significantly pollutes the environment. Tannery fleshings is a solid waste generated during the processing of hides in terms of trimmings and lime fleshing these are a rich source of protein and also have benefits in the production of renewable energy which benefits the management of disposal problems which causing the environmental pollution. These fleshings can be utilized in the biochemical and thermos-chemical conversion process to produce energy in terms of biogas, ethanol, biodiesel, carbon, bio-oil and syngas. These reduce the utilization of raw materials, water, power in leather industries and also the leather processing expenses. The conversion of leather fleshings into the renewable energy sources reduces the emission of greenhouse gases to the environment. The utilization of tannery waste materials is much more needed to fulfill the sustainable development goals (SDGs) for better future.

Keywords Tannery fleshings · Trimmings · Circular economy · Renewable energy sources · Greenhouse gases

---

S. C. Hallad · V. Kavan Kumar (✉) · N. L. Panwar
Department of Renewable Energy Engineering, CTAE, MPUAT, Udaipur, Rajasthan, India
e-mail: kavankumarreddy07@gmail.com

R. Mahendiran
Department of Renewable Energy Engineering, AEC&RI, TNAU, Coimbatore, Tamil Nadu, India

M. A. Aboulhassan and S. Souabi (eds.), *Sustainable Practices in the Tannery Industry*,
https://doi.org/10.1007/978-3-031-92037-0_10

10.1 Introduction

Energy has long been regarded as one of the most important aspects of daily existence. Humans rely mostly on coal, petroleum, and natural gas, all of which are non-renewable energy sources that will eventually run out. Petroleum fuel combustion causes environmental problems such as air pollution and global warming (Periyasamy et al. 2017). Solid residues are by-products of industrial and human operations that have been linked to harmful consequences for humans and the environment when improperly disposed off. As a result, reducing the waste generation of this trash and properly managing it have become an urgent concerns for government authorities (Vinicius et al. 2020). In developing countries, waste disposal or usage from diverse sectors, such as large and small enterprises and families, is a big concern. Because of numerous differences in data collection methods, definitions, and seasonal variations, determining the precise amount of garbage generated in poor nations is challenging. 700 kg of biomass wastes are produced for every tonne of raw hide/skin processed. The percentage of solid trash produced is as follows: The weight of raw hide is 7–12% for trimmings, 5.5–6% for chrome trimmings, 10% for chrome shaving, 4–6% for chrome splits, 0.2–0.4% for buffing dust, and 1.2% for skin trimming, and hair 10–12% of the animal's weight (Fathima et al. 2012). Out of this 250 kg of leather fleshings are produced for every ton of raw hides (Sundar et al. 2011). Hide is a natural substance made up of collagen fibers joined together in a 3D structure, and hide components are tanned for conservation and to improve their physical, chemical, and biological qualities (Amalraj and Pius 2014). Along from that, various other mechanical operations are required to standardize the thickness of the skin/leather, resulting in enormous amounts of solid waste, such as trimmings from hides and skins, fats, buffing dust, and effluent sludge (Yilmaz et al. 2007). Solid wastes created mostly during the shaving and cutting stages are classified as Class I harmful due to their toxicity as determined by the ABNT NBR 10.005 leaching test (ABNT 2004). However, between 2003 and 2013, the leftovers obtained during the stage of leather thickness adjusting, as well as lime and wet blue shavings, declined significantly, probably due to their usage in fertilizer production and the making of gelatin and animal feed (Priebe et al. 2016). In order to harmonize science, technology, and sustainability applications other than landfill disposal have been investigated, with the goal of finding a new use for tannery wastes while also lowering the impact of the manufacturing process (Dettmer et al. 2010). The solid wastes generated from the tannery industries also been used for the production of high efficiency of biodiesel (Booramurthy et al. 2020, 2022), anaerobic digestion to produce biogas (Agustini et al. 2020; Bayrakdar 2020), biopolymers (Ammasi et al. 2020), adsorbent materials through thermo-chemical processes (Gomes et al. 2016; Palani et al. 2017), agricultural applications by composting of wastes (Nabavinia et al. 2015; Malafaia et al. 2016). Thus, the objective of this book chapter is to addressing the various renewable conversion technologies to convert the leather wastes into useful end products and their importance in the real-life situations. The above-mentioned concepts were discussed elaborately with the process in this chapter with respect to tannery waste treatment to produce the value-added products from the conversion processes and their importance toward the circular economy of tannery sectors.

10.2 Materials and Methods

Utilizing these materials and techniques, leather waste is converted into renewable energy sources to support waste recovery into energy according to commercial and industrial needs. The promising results of each technology are mentioned and still to overcome in the area of waste management. Through a combinatorial keyword search using terms like "leather solid wastes," "tannery wastes," "circular economy," "solid waste management," and "valorization of leather waste," the literature for this review was methodically chosen from a large database. The latest papers from peer-reviewed journals as well as the Google Scholars database were used as sources for the literature.

10.2.1 Agriculture

The availability of nutrients in the soil throughout a crop's growth is correlated with its productivity. Fertilizers must be added to the soil to meet the needs of the plant and guarantee that it has all the macro and micronutrients needed for growth. Because of the abundance of organic matter and nitrogen they contain, tanning wastes have been employed extensively as fertilizers to increase agricultural output. The plan in Fig. 10.1 illustrates the several applications for this trash that have been suggested by studies.

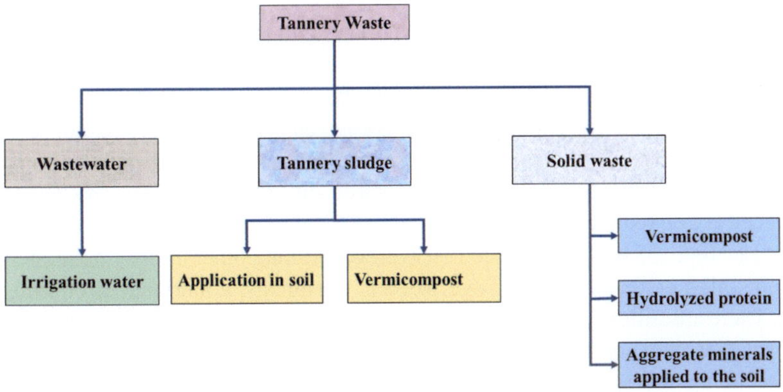

Fig. 10.1 Origin and destination of tannery wastes for application in agriculture

Tannery wastes can be applied as organic manure to soil. Nabavinia et al. (2015) examined into how biochar and tannery wastes affected the physicochemical and chemical characteristics of soil as well as the yield of radish. Three levels of tannery waste (0, 2.5, and 5%) and two levels of biochar (0 and 2.5 t ha^{-1}) made up the treatments in a fully randomized design (factorial) with three replications. After three months of field capacity conditions in the treated pots, five radish seeds were put in each pot's soil at a depth of one centimeter. The findings demonstrated that while tannery wastes and biochar treatments increased the amounts of nitrogen and phosphorus in the soil as well as their concentrations in radish roots and above-ground biomass, they had no effect on the potassium content of the soil and reduced the electrical conductivity of the soil. Post plant harvest, the tannery waste treatment raised the pH of the soil. Although the biochar treatment produced a little rise, it also greatly boosted the fresh and dry matter weight of radish. It is determined that applying biochar and tannery wastes, which have high nutrient contents, to soils deficient in organic carbon in dry and semi-arid locations may be beneficial for enhancing plant growth and yield. Sludge from a tannery wastewater treatment facility was used by Silva et al. (2010) to grow the ornamental plant Capsicum sp. Tannery sludge, cow dung, and cane straw (T1) and tannery sludge, cow dung, and carnauba straw (T2) were the two treatments that were used. It was discovered that the plant produced more leaves and fruits for both treatments. The T1 treatment had a higher chlorophyll content, demonstrating that a species' increased pigment content is proportionate to its high nutritional content. A fertilizer derived from chromium-tanned leather waste was studied by Nogueira et al. (2010) using phosphate (P) and potassium (K). To extract the chromium from the garbage, the authors employed acid hydrolysis at 50 °C. After being hydrolyzed, the residue was dried and crushed into various particle sizes after being submerged in P and K solutions. Commercial fertilizer and residue were used for treatments in order to assess rice plant development. Although the residue-containing formulations produced reduced dry mass, they enhanced plant grain output, indicating that they were a source of nutrients for the crop.

The beamhouse (pre-tanning) accounts for more than 80% of the organic pollution load in BOD terms; a large portion of this is caused by deteriorated hide, skin, and hair debris. For every ton of hides, at least 300 kg of chemicals (such as salt and lime) are added throughout the tanning process. Wastewater will have an excess of salts that have not been used. These substances have the potential to precipitate and increase the amount of suspended solids or solid waste due to pH changes. Wastewater is produced at every stage of the tanning process, save for finishing processes. For every ton of raw hide, 35 m^3 are produced on average. Among other things, the wastewater contains high concentrations of dye, solvent chemicals, ammonia, chromium, and salts. There are still some concerns regarding the utilization of tannery wastes, despite the fact that they show promise for use as fertilizers. Ali et al. (2018) claim that the use of wastewater from the tannery effluent treatment plant limited the growth and development of wheat biomass. The accumulation of metals in the soil, such as chromium, which stresses plants and suppresses the activity of antioxidant molecules, is connected to this effect.

10.2.2 *Utilization of Wastewater from Tannery Fleshing in Renewable Energy Sources*

Tannery fleshing produces effluent that is high in organic stuff, such as proteins, lipids, and other biodegradable substances. Although this wastewater's high pollutant load and possible toxicity make management difficult, there is a lot of promise for producing renewable energy from it. Technologies such as algal bioenergy systems, microbial fuel cells (MFCs), and anaerobic digestion can transform this waste into valuable energy resources. This promotes sustainable waste management and energy generation. Through anaerobic digestion, microbes break down the organic materials to create biogas, which is mainly made up of carbon dioxide and methane and can be utilized as fuel for cars, heating, or electricity production. Wastewater's chemical and biological oxygen demand is also decreased by this treatment. Similar to this, dark fermentation—a process in which some bacteria convert organic substances into biohydrogen can be used to manufacture hydrogen. To address inhibiting substances like chromium, pre-treatment is required. Using electrochemically active microorganisms, microbial fuel cells (MFCs) immediately transform organic matter in wastewater into electrical power, cleaning the wastewater and offering a low-maintenance, sustainable energy source. Furthermore, algal bioenergy systems use wastewater as a source of nutrients to grow algae, which can then be converted into biofuels like biogas, bioethanol, or biodiesel. In addition to removing organic contaminants, nitrogen, and phosphorus, this process also captures carbon dioxide, so promoting carbon neutrality. Waste reduction, resource recovery, increased energy sustainability, and economic feasibility are just a few advantages of using tannery fleshing effluent for electricity. It offers an environmentally beneficial way to manage industrial waste while producing renewable energy.

10.2.3 *Renewable Energy Sources*

The tannery fleshings were utilized in a variety of conversion processes, including pyrolysis, hydrothermal, biogas, biodiesel, biopolymer, and biohydrogen briquettes. These are the various biomass conversion methods used to produce value-added products for further practical uses (Fig. 10.2; Moktadir et al. 2023).

10.2.3.1 Biogas from Tannery Fleshing

The amount of CO_2 in the atmosphere has been steadily rising since the industrial revolution and is predicted to reach 750 ppm by the end of the century (Barzagli and Mani 2019). The greenhouse effect is caused by CO_2, which blocks the sun's infrared radiation and raises the earth's temperature as a result. The quest for new

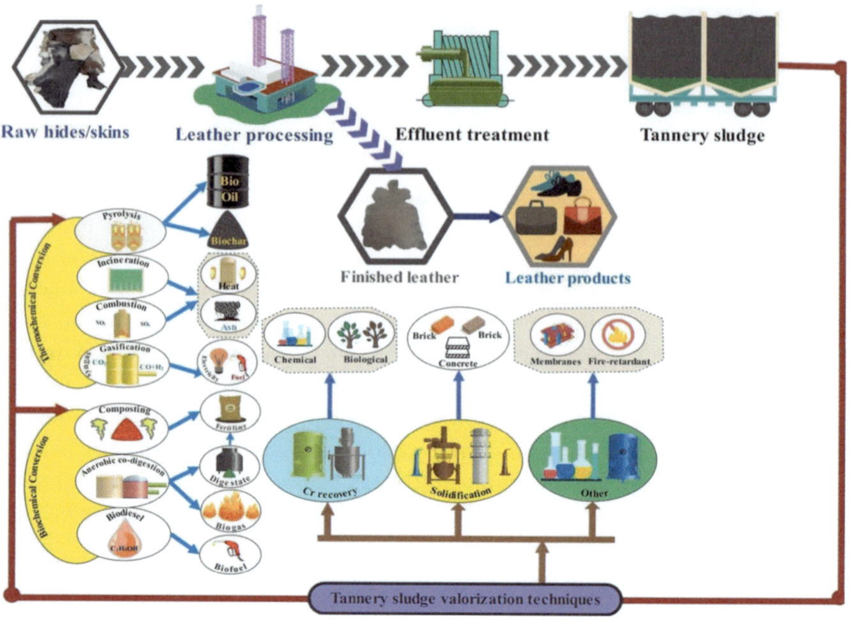

Fig. 10.2 Tannery sludge utilization

alternative energy sources is necessitated by this effect's wide-ranging effects, and the generation of biogas stands out as a potent renewable energy source.

The biological breakdown of several organic wastes, including chicken, swine, and bovine dung, as well as the utilization of enzymes and microbes, can result in biogas, a type of sustainable energy source. The high organic content of tannery fleshing makes it a potential substrate for the anaerobic digestion process that produces biogas. Tannery waste can be converted into biogas, however it is usually prepared using certain procedures before anaerobic digestion. Animal fat is the most noteworthy waste product of the leather industry because of its high thermal value. Because of their high protein content, these residues can be treated using an anaerobic digester. Each stage of digestion hydrolysis, acidogenesis, acetogenesis, and methanogenesis—is contingent upon the various microorganisms' metabolic conditions. Biogas, which is mostly composed of carbon dioxide (CO_2) and methane (CH_4), is produced via anaerobic digestion, in which microbes break down organic material without oxygen. A suitable feedstock for the production of biogas, which addresses waste management concerns and provides a renewable energy source that encourages resource recovery and environmental sustainability, is tanning fleshing, which is rich in biodegradable components. Tannery fleshing is first collected and pre-treated to get rid of impurities like salts, lime, and leftover chemicals. Size reduction, dehairing, and homogeneity are examples of pre-treatment procedures that improve the material's biodegradability. The fleshing that has been pre-treated is then put into an anaerobic digester, where microorganisms break down the organic matter in phases. For example, hydrolysis breaks down complex macromolecules into simpler compounds like fatty acids and amino acids; acidogenesis turns these into volatile fatty acids, alcohols, and gases; acetogenesis turns intermediates into hydrogen, CO_2, and acetic acid; and methanogenesis produces methane and CO_2 to create biogas. The biogas is collected and filtered to remove impurities including hydrogen sulfide (H_2S) and water vapor in order to produce methane-enriched biogas, which may be utilized as a renewable energy source. The digestate, sometimes referred to as residual slurry, is a nutrient-rich by-product that can be used as biofertilizer in agriculture.

The primary research on using industrial tannery waste for biogas production is given in Table 10.1, along with the primary parameter values. The variables pH, TS, VC, COD, and TKN directly affect the amount of biogas obtained and can also affect how much of it is burned, according to research by Booramurthy et al. (2020). Even though the pH of 7.8 is beyond the appropriate range of tannery waste and trimmings in the mixtures to be biodegraded, these pH values approached the anaerobic digestion (AD) range, suggesting that the pH of the residues was not the limiting issue. Due to the longer retention time needed for sludge hydrolysis, Sri Bala Kameswari et al. (2011) employed ozonation as a substitute to raise the chemical demand for soluble oxygen (SCOD). When ozone is used for cellular solubilization and hydrolysis of the sludge, the cell wall is broken down and SCOD is released, enhancing the production of biogas. The generation of biogas was raised to 310 mL/g by using 0.18 g of O_3/g TS.

To lower the pH of the leather fleshing to an appropriate level, deliming was done using ammonium salts such as ammonium sulfate and ammonium chloride. This procedure guaranteed that the fleshing pH would drop below 8.5. By lowering the pH as previously indicated, the waste water from deliming can be used to prepare a slurry with the inoculum, the anaerobic digestion can be fastened by this addition. Various inoculum to substrate ratios can be used in the batch scale trials. So that the waste water generated throughout the process can be utilized in the slurry preparation of biogas production.

The AD of solid waste from chromium-containing tanneries has been investigated by researchers. Investigations were conducted into the thermophilic AD of wastewater sludge and leather shavings (Zupančič and Jemec 2010). In both sludge and shavings, pH measurements began on average at 7.6 and rose to 8.2 (Li et al. 2011). The pH in the AD process starts the acid and neutralizes it throughout the process. Trivalent chromium salts, the most often employed compounds in tanneries, provide the basis for the most common way of tanning leather. Between 60 and 80% of the total chromium administered during the tanning process is consumed; the remaining chromium is left in the tanning float and eventually ends up as wastewater. Before wastewater may proceed to the biological treatment phase, where flocculated/sedimented contaminants are disposed of as sludge, it must first undergo physical chemical sedimentation. Tanning sludge cannot be directly disposed of in a landfill because of the high levels of organic and trivalent chromium, which make it toxic. As a result, anaerobic digestion presents itself as a viable option for recycling this chromium (III)-containing trash and generating biogas (Booramurthy et al. 2020).

Table 10.1 Main parameter analyzed to obtain biogas

Feedstock	Parameters						References
	Biogas generation (mL/g)	pH	TS (mg/g)[a] or (mg/L)[b]	VS (mg/g)[a] or (mg/L)[b]	COD (mg/g)[a] or (mg/L)[b]	TKN (mg/g)[a] or (mg/L)[b]	
Sludge	19.3	7.8	–	9810[b]	–	–	Booramurthy et al. (2020)
Raw tannery waste effluent	68.57	7.5	18,850[b]	4650[b]	–	–	Saxena et al. (2019)
Leather fleshings	460.0	12.5	2240[b]	1670[b]	221[a]	14[a]	Bayrakdar (2020)
Treatment sludge		7.9	2240[b]	1490[b]	252[a]	11.1[a]	
Shavings	19.3	4.03	–	9400[b]	–	–	Agustini et al. (2018)
Primary sludge	385.0	8.1	35,000[b]	18,900[b]	24,950[b]	29[a]	Sri Bala Kameswari et al. (2014)
Second sludge		7.3	15,220[b]	7610[b]	9890[b]	42[a]	
Leather fleshings	20.22	–	5452	5055	–	–	Bavaresco et al. (2015)
Fleshings	–	11.1	100[a]	650[a]	900[a]	76[a]	Sri Bala Kameswari et al. (2011)

[a]TS (mg/g); [b](mg/L)

Tannery fleshing biogas can be used for a variety of purposes, including as generating heat and electricity through combined heat and power (CHP) units, replacing traditional fuels in cooking and heating, and serving as a clean automobile fuel once biomethane is converted. Among the many benefits of the process are waste valorization through the conversion of waste into valuable energy, a sustainable substitute for fossil fuels, economic benefits through the creation of new revenue streams for tannery industries, and resource efficiency through the production of both energy and nutrient-rich biofertilizer.

10.2.3.2 Biodiesel from Tannery Fleshing

Biodiesel is a renewable and biodegradable fuel derived from natural sources such as vegetable oils, animal fats, or waste cooking oils. Oils or fats combine with an alcohol (often methanol or ethanol) in the presence of a catalyst to produce fatty acid methyl esters (FAME), also referred to as biodiesel, through a chemical process called transesterification. The outcome of this process is glycerol. The two types of biodiesel that can be used in diesel engines are pure (B100) and mixed with petroleum diesel in various ratios (e.g., B20, which is 20% biodiesel and 80% diesel). Because of its various advantages, including lower sulfur content, biodegradability, and fewer greenhouse gas emissions, it is more environmentally friendly than conventional diesel fuel. In terms of trade and the potential for raw material use, biodiesel is a viable substitute for the fossil fuel like diesel. Soybeans, corn, cellulose, algal oil, and sugar cane are the most widely used sources; nevertheless, these food supplies are taken out of the global food supply, and their oils are costly.

The main sources of biodiesel in India are Mahua, Pongamia, and Jatropha, which account for 33–42% of the oil content. Additionally, using leftover tallow, animal fat, and cooking oil was discovered to be sustainable and shown a greener method of producing energy. The creation of energy and the management of solid waste are two advantages of using organic waste energy. When the fleshing waste is hydrolyzed using a proteolytic enzyme, its high protein and fat content is released, together with NH_3, H_2, CO_2, and volatile fatty acids (VFA) (Shanmugam and Horan 2009). When exposed for longer than a minute, this VFA causes the production of H_2S, which damages the human neurological system and can result in coma and death. A photochemical process in the atmosphere causes the generated H_2S to impact the ozone layer (Hashem et al. 2015). A novel method for producing biodiesel from the waste of the leather industry has been explored as a way to circumvent the negative environmental effects of fatty waste materials. Between 14 and 20% of the fleshing waste is fat. Because the Soxhlet extraction method recovers 90–98% of the oil from the sample, it is the most widely used method for extracting fat from fleshing waste. However, the use of hazardous solvents and a high energy need are this technique's main drawbacks (Shivani et al. 2011). Another method for removing the fat and oil component from the samples is alkaline hydrolysis. In contrast to the Soxhlet method, it is less effective (70–75%). While solvent extraction is not suitable for laboratory-scale analysis, the alkaline hydrolysis method can be used to extract oil from a significant amount of waste materials with little energy input. By using less resources, the use of leather fleshing (LF) oil improves sustainability and lowers the cost of feedstock for the manufacture of biodiesel.

The production of biodiesel from tannery fleshing is a creative way to convert waste materials into renewable energy. The process of manufacturing leather involves removing a fatty, meaty waste from animal skins called "tannery fleshing." This garbage typically contains significant amounts of lipids, which can be utilized as feedstock for the production of biodiesel. Large-scale leather production worldwide generates a lot of waste, which offers a substitute for producing biodiesel because it contains a lot of fat between 40 and 60% in the waste from fleshing cows. Therefore, it is possible to create a product that has both added value and an environmental bias by lowering these residues that are released into the environment. Fatty acid esters, which are created by transesterifying vegetable or animal fats with low molecular weight alcohol, are combined to create biodiesel. One result of this procedure is glycerol. A homogeneous alkaline catalyst is usually used in the transesterification reaction between methanol and triglycerides, which are produced when three fatty acids combine. This catalyst requires raw materials with a low fatty acid concentration, up to 1 mg KOH/g. One can categorize transesterification as either catalytic or non-catalytic. Using basic or acid catalysts is frequent in catalysis; the majority of the time, basic catalysts are utilized. However, non-catalytic enzyme utilization has a high production value, but it also presents challenges with enzyme reuse and separation, among other issues. There are several crucial phases involved in turning tannery fleshing into biodiesel. In order to eliminate water, proteins, and non-lipid contaminants, the fleshing is first gathered from tannery facilities and pre-treated. Methods like chemical treatment, mechanical pressing, and drying are used to accomplish this. After that, lipids are removed from the fleshing by mechanical means or chemical solvents like hexane; enzymatic techniques may also be used to increase extraction effectiveness and lessen environmental impact. Transesterification, a chemical reaction involving triglycerides and methanol in the presence of a catalyst such sodium or potassium hydroxide, is the next step in processing the extracted lipids. Glycerol is a by-product of this process, which transforms the lipids into fatty acid methyl esters (FAME), the main ingredients of biodiesel. Glycerol is then isolated for possible industrial application once the biodiesel has been purified to exclude contaminants, unreacted alcohol, and catalyst residues. In order to confirm that the biodiesel is suitable as a renewable fuel, it is lastly tested to make sure it satisfies quality requirements like ASTM D6751 or EN 14214. Figure 10.3 illustrates the processes involved in the processing and manufacture of biodiesel.

Animal fats often include around 5 and 30% fatty acids, which allows soap to develop and complicates separation. In order to confirm that they contain more fatty acids than vegetable oil, Booramurthy et al. investigated the makeup of the fatty acids found in the fat from the tannery wastes (Table 10.2) (Booramurthy et al. 2022). This issue can be resolved by using a pre-treatment that involves the esterification of methanol and sulfuric acid as a catalyst. The fat can then be transesterified using

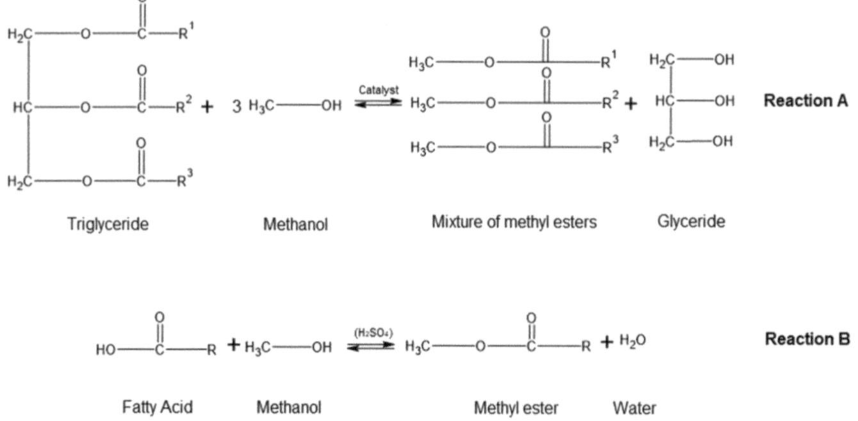

Fig. 10.3 Reactions of biodiesel production and pre-treatment

Table 10.2 Main fatty acids present in the fat extracted from tannery wastes

Type of fatty acid	Carbon chain	%
Miristic acid	C14:0	4.18
Palmitic acid	C16:0	24.22
Steric acid	C18:0	30.23
Oleic acid	C18:1	37.76
Linoleic acid	C18:2	2.78
Linolenic acid	C18:3	0.17
Araquic acid	C20:0	0.20
Undetectable	–	0.46

the standard procedure to produce methyl esters. The two primary processes in the pre-treatment process are deacidification, or the removal of fatty acids, and refining. It is feasible to reduce the amount of water, ash, acids, or perhaps even nitrogen at the refining step. A physical extraction procedure, typically employing methanol, takes place in the event of deacidification (Ramos et al. 2008).

The National Agency of Petroleum, Natural Gas, and Biofuels (ANP) of Brazil (Ramos et al. 2008) states that the amount of water in biodiesel should not exceed 500 mg/kg because any more than this could cause the fuel to corrode and shorten engine life. However, according to ABNT NBR 14448, the acid value stays at a maximum of 0.5 mg KOH/g, which also affects corrosion and storage stability. When some basic characteristics, as those mentioned in European (EN/ISO) and American (ASTM) regulations, are examined and adjusted to the current restrictions, using waste from the leather sector generally turns out to be a highly promising alternative to manufacture biodiesel. The development of soap during transesterification, which necessitates a pre-treatment, viscosity, and cloud point, is one of the primary obstacles in the process of producing biodiesel from leather waste. Research is therefore required to find ways to improve these parameters because using this plentiful waste as a substitute raw material can assist meet a portion of the market demand for biodiesel while lowering the consumption of food sources. There are various benefits to producing biodiesel from tannery fleshing. Using alkaline methanolic solutions (methylammonium tetrahydroxide) in place of pure methanol, Nogueira et al. (2010) examined the pre-treatment of fat and found that fatty acid levels were reduced by 98.4%. As a result, biodiesel that mostly conforms to the European standard EN14214 of 2012 was produced through transesterification. Table 10.3 lists various research that used tannery waste to make biodiesel.

By efficiently using trash, it turns a troublesome waste into a useful resource and lessens environmental impacts. By providing an alternative to fossil fuels, this renewable energy source lowers greenhouse gas emissions. By giving waste products new value, this technique also boosts the leather industry's adoption of circular economy principles. But there are obstacles to overcome. Fleshing requires extensive pre-treatment and extraction procedures due to its high moisture content and impurities. Achieving cost-effective and scalable operations is still a major obstacle to

Table 10.3 Comparison of research on B100, petro-diesel, and biodiesel produced using tannery waste

	mg KOH/g	Petro-diesel[c]	B100[d]	TW[e]			
		Kubendran et al. (2017)		Kubendran et al. (2017)	Šánek et al. (2015)	Booramurthy et al. (2022)	Booramurthy et al. (2020)
Density at 15 °C[a] or 30 °C[b]	kg/l	0.834[a]	0.86–0.9[b]	0.855[b]	0.878[a]	–	0.87[a]
Viscosity at 40 °C	mm^2/s	1.9–4.1	1.9–6	4.2	4.7	4.4	4.6
Flash point	°C	60–80	100–170	160	161.5	153	150
Cloud point	°C	– 15 to 5	– 3 to 12	4	–	16	8
Cetane number	CN	48–52	48–65	–	67.4	62	58
Water	%	0.05	0.05	< 0.01	–	0.05	–
Acidity	mg KOH/g	0.5	0.5	0.07	0.16	0.4	0.3

[a] Describing the temperature 15 °C, referring to the different temperatures in which density was determined
[b] Describing the temperature 30 °C, referring to the different temperatures in which density was determined
[c] Petro-diesel: diesel from a petrochemical source only
[d] B100: biodiesel made from exclusively non-petrochemical sources
[e] Biodiesel obtained from tannery waste

commercial viability, and the use of chemical solvents for lipid extraction can provide environmental risks if improperly managed. Energy output and trash management can both be enhanced by sustainability improvements. Biogas recovery from non-lipid organic components combined with the production of biodiesel is one example. Advanced catalysts, such as heterogeneous or enzyme-based varieties, can lessen their negative effects on the environment and increase process efficiency. Industries may address waste management issues and produce sustainable biofuels by using tannery fleshing, which is in line with international goals for environmental preservation and renewable energy.

10.2.3.3 Biopolymers

Biopolymers are synthetic or natural polymers that are biodegradable or generated from biological sources. Because of their many uses and environmentally favorable qualities, these materials are becoming more and more well-known as sustainable substitutes for traditional, petroleum-based polymers. Biopolymers are renewable biological materials derived from plants, algae, or waste products, which ensures sustainable production; they are biodegradable, meaning they can break down naturally into water, carbon dioxide, and biomass under the right conditions, minimizing environmental impact; they have a variety of properties, from flexible to rigid structures, which make them adaptable for a variety of applications while significantly reducing reliance on fossil fuels and associated pollution; and they find applications in a variety of industries, including packaging, where they are used for biodegradable food containers, films, and bags; in the medical and pharmaceutical industries, they are used in drug delivery systems, wound dressings, and sutures. They are utilized in building as bio-based adhesives, composites, and insulating materials; in textiles as sustainable fibers for apparel and upholstery; and in agriculture as slow-release fertilizers and biodegradable mulch films. Numerous advantages of biopolymers include their sustainability due to their use of renewable resources, which lessens dependency on non-renewable raw materials; their environmental friendliness due to their biodegradability, which lowers long-term pollution; and their versatility, which allows them to be used in a wide range of industrial applications, lowering carbon emissions during production and disposal. The benefit of biodegradable films is their faster rate of decomposition, which is directly related to their solubility and microbial activity since the solubilization makes the film's constituents accessible to the soil's microbiota, which consists of bacteria, fungi, and protozoa.

Petroleum-derived polymer pollution is a significant environmental issue today, with an estimated 8 million of the 300 million tons produced globally each year ending up in the oceans, and this number is expected to treble by 2034 (Plastic Oceans 2020). The rate at which the material degrades is a significant factor that influences the spread of organic pollutants to humans and biota. Businesses and scientists are therefore giving the investigation and control of environmental pollution top priority. In this context, converting industrial waste into useful material promotes economic flow. When pollution from fossil polymers occurs, a sustainable alternative is the extraction of collagen and/or gelatin from waste products from the leather processing sector for use in biodegradable films (Muralidharan et al. 2020). Tannery fleshing, a by-product of leather production, comprises a considerable amount of proteins, lipids, and other organic components. This trash can be effectively exploited as a raw material for creating biopolymers, aligning with sustainable waste management and the circular economy. In addition to lowering environmental pollution, turning tannery fleshing into biopolymers gives something that would otherwise be regarded as waste value. A number of crucial stages are involved in the manufacture of biopolymers from tannery fleshing. In order to eliminate undesirable ingredients like salts and lime used in the leather-making process, tannery fleshing is first gathered and pre-treated. In order to get the material ready for additional processing, this pre-treatment entails washing, dehairing, and size reduction. Following enzymatic or chemical hydrolysis, collagen is separated from the fleshing to produce hydrolyzed proteins or gelatin. These can be combined with other materials or chemically altered to produce biopolymer gels or films. Microbial fermentation can also occur in the organic fractions, such as proteins and lipids. In this stage, the recovered components are processed under controlled

Fig. 10.4 Biodegradable films made from tannery solid waste for use in agriculture, **a** trimming leather waste, **b** recovered gelatin, **c**, **d** simulated soil application test: time 0 and 60 days, respectively (Rosseto 2020)

conditions using certain bacteria to create polyhydroxyalkanoates (PHAs) or other biopolymers. Depending on the desired use, the extracted or manufactured biopolymers are next polymerized and processed into useable forms as coatings, sheets, or films.

The procedure of acid hydrolysis to obtain gelatin from raw bovine skin from tannery wastes was studied by Masilamani et al. (2017). After that, the gelatin was crushed and dried for use in movies. To achieve a performance comparable to the low-density polyethylene (LDPE) film, which was employed as a reference, the scientists optimized the film's formulation during the investigation by adding a polymer formed of polyvinyl alcohol (PVA), glycerol, and glutaraldehyde. Tests for printing, thermal sealability, and mechanical properties have produced positive outcomes for package applications. The standardized film samples were submerged in water for one hour after being scribed and stamped with various inks for the printing test. The potential uses of gelatin derived from chromium-tanned leather waste are examined in the study by Rosseto (2020), with characteristics improved by enzymatic and antibacterial treatment. Since the remaining chromium bonded to the protein molecules makes their uses in the food business unfeasible, the use of these films in agriculture is intriguing. The authors highlight the potential of using land cover as one of the application possibilities (Fig. 10.4).

10.2.3.4 Biohydrogen from Tannery Fleshing

Biohydrogen can be produced from tannery fleshings through dark fermentation or photo fermentation processes. A rich blend of organic molecules, including proteins, lipids, and other biodegradable components, can be found in tanning fleshing, a by-product of the leather industry. It is a possible feedstock for the synthesis of biohydrogen because of these organic fractions. Hydrogen, a sustainable and clean energy source that combines the benefits of waste management and renewable energy generation, may be produced biologically through anaerobic digestion and fermentation processes. Utilized in fuel cells and industrial settings, hydrogen is a clean fuel with a high energy content. The high calorific value of hydrogen (122 kJ g^{-1}) and its lack of adverse environmental effects make it one of the most viable energy vectors of the future. Anaerobic fermentation techniques utilize the carbon content of organic waste molecules and generate H_2 more effectively than photosynthetic processes.

Aathika et al. (2018) evaluated the efficient conversion of leather solid waste into bio-hydrogen using anaerobic hydrogenic activity in a batch reactor. They also looked at key performance metrics such as pH, total solids, volatile solids, chemical oxygen demand, specific hydrogen yield, and degradation efficiency of the crude organic matter to see if the bio-hydrogen potential could be maximized. The optimal waste mixing ratio was found to be LF:S:HS (2:1:1), with a removal rate of 92.3% volatile solids and a pH of 5.5 at 37 °C. The cumulative bio-hydrogen production rose from 41 mL in the control to 323 mL, an eightfold increase. This study demonstrated improved material and energy efficiency and is unquestionably more cost-effective in terms of trash utilization. The goal of research and development is to increase the tannery fleshing's capacity to produce biohydrogen while also making it more efficient and scalable. Advancements include genetically engineering microorganisms, co-digesting with other organic wastes, and sophisticated pre-treatment methods have the potential to increase hydrogen production.

10.2.3.5 Pyrolysis of Tannery Fleshing

Pyrolysis thermally decomposes tannery fleshing in the absence of oxygen to produce biochar, bio-oil, and syngas. These products can serve as soil enhancers, liquid fuels, and energy carriers, respectively. Pyrolysis is the thermal breakdown of organic wastes, including food and agricultural wastes, sewage sludge, plastic wastes, scrap tires, and others, in an inert atmosphere to create carbonaceous residue (biochar), synthetic gas (syngas), and pyrolytic oil (bio-oil). Yilmaz et al. looked into the energy conversion of several kinds of tanned wastes using pyrolysis at temperatures between 450 and 600 °C. They found that the highest solid residue yield is around 48.5%, while the largest oil production is approximately 23%. Activated carbon derived from tanned leather waste was the focus of multiple studies, which revealed that pyrolyzed leather produced good surface area activated carbon (Yilmaz et al. 2007). Using a fixed-bed pyrolyzer, Kluska et al. (2019) examined the pyrolysis of leather waste at 300–500 °C and came to the conclusion that thermal cracking of this waste for energetic use makes sense above 400–450 °C. These results led to an emphasis on recovering carbonaceous residues from leather waste using the pyrolysis technique, with little consideration given to liquid and gaseous chemicals as important by-products. Kantarli and Yanik (2009) investigated the process of turning tannery waste sludge into valuable material. This was accomplished by analyzing the thermal behavior of sewage sludge using thermogravimetry (TG) and a vertical fixed-bed reactor. Thermogravimetric analysis was used to calculate the temperature of 750 degrees Celsius using different heating rates. The major pyrolytic breakdown phase was seen at temperatures between 248 and 500 °C. In the fixed-bed reactor, waste sludge was pyrolyzed in a nitrogen atmosphere between 450 and 600 °C. The pyrolysis products were sorted into char, gas, and liquid. It was found that the product's dispersion was slightly impacted by temperature. The pyrolysis gas has a noteworthy gross calorific value of 23.22 MJ N m^{-3}. Despite having a respectable calorific value, the pyrolytic oils were rich in oxygenated and nitrogenated chemicals.

Tanners and leather manufacturers can use Tannery Fleshing Waste (TFW) pyrolysis as a creative solution because it can lower TFW management expenses (transport and landfilling), create revenue for industry by creating creative and affordable products (biofuels, biofertilizers, and bioadsorbents), and guarantee environmentally responsible and sustainable business practices. Both liquid and gaseous chemicals as valuable products received little consideration based on the surface area of carbon.

Using digestate or tannery sludge as an adsorbent for wastewater treatment offers a simple, cost-effective, and eco-friendly solution that is suitable for poor nations. Tannin digestate contains a sizable amount of residual carbon that can be physically or chemically activated. Geethakarthi and Phanikumar (2012) developed an effective activated carbon adsorbent with a greater BET surface area of 188 m^2/g and a micropore size distribution of less than 20 Ao using tannery sludge. Anionic azo reactive dye RR31 supplied at a concentration of 40 mg/L was removed with up to 95% efficiency by the produced activated carbon (1.5 g/L). Mella et al. (2019) were able to remove Acid Orange 142 and Acid Brown 414 azo dyes from TWW with removal efficiencies of up to 71% and 73%, respectively.

10.2.3.6 Gasification of Tannery Wastes

Gasification has the potential to improve sustainability in the leather industry and offer substantial cost savings in terms of waste management and electricity generation. The process of gasification turns any carbon-containing material into a combustible gas that may be used as fuel to generate heat and electricity. This gas is mostly composed of carbon monoxide, hydrogen, and methane. Many tannery wastes can be macerated, flash dried, densified, and gasified to produce clean syngas that can be utilized in boilers or other Combined Heat and Power systems. Consequently, up to 70% of the waste's intrinsic energy value can be recovered as syngas, of which up to 60% is surplus to process drying requirements and can be utilized for on-site boiler applications or thermal energy recovery.

Tanneries can account for as much as 70% of the processed hide weight and produce large amounts of sludge, shavings, trimmings, hair, buffing dust, and other miscellaneous wastes. Because it produces ash that contains chrome (III), the gasification method offers a feasible alternative thermal treatment for materials containing chromium due to chemically reducing circumstances. Due to its reconstructed nature, this ash has substantial commercial potential. With pre-treatment techniques like maceration, drying, and subsequent densification or briquetting, all of the tannery's wastes can be gasified. The tannery might use the combustible gas produced by a combination drying and gasification process as a direct source of tax-exempt renewable energy in addition to removing solid waste. According to gasification tests, it is possible to recover up to 70% of the intrinsic energy value of the wastes that are typically disposed of as "synthesis gas" energy.

A tanyard on Norway's west coast has been using a proprietary technology called PyroArc for commercial purposes since the middle of 2001. Both gasification and plasma cracking are used in the process. In terms of excess energy and ferrochrome, a safe alloy that is frequently used by the metallurgical industry, the PyroArc method offers the potential to transform the tannery waste issue into a valuable source that could benefit the plant owner. The procedure produces a non-leaching slag that can be used for civil engineering projects instead of ashes, so there are no leftovers for landfill disposal.

10.2.3.7 Circular Economy from Tannery Sector

The biogas produced from the tannery fleshings contains the methane gas up to 70%. This helps in the production of heat and electricity for the tannery industry to reduce the electrical and fuel charges. Also, the same can helpful in the production of biohydrogen. The fat content present in the tannery fleshings (green and limed) was utilized for the production of bioethanol and biodiesel. These fuels can be blending with the conventional fuels to run the engines/machineries within the site. The other value-added products like biochar, bio-oil, syngas and biopolymers from the dried tannery fleshings have more commercial value in the market. Thus, the chapter explains various renewable technologies and their valuable products are playing a crucial role in the circular economy of tannery sector and also the on-site treatment of tannery fleshings reduces the landfilling, incarnation, and leaching of salts into the agricultural fields and water bodies. The overall concept of the utilization of waste released from tannery sector is much needed to highlight the UN sustainable Development Goals (SDGs) for better life and the pollution control to bring green environment.

10.3 Conclusion

The efficient use of tannery waste, such as wastewater and solid residues, has great potential to promote sustainable waste management and renewable energy. Anaerobic digestion, dark fermentation, microbial fuel cells, and algal bioenergy systems are some of the technologies that can be used to transform these waste materials into resources such as biogas, biodiesel, biohydrogen, and biofuels. These practices address the environmental problems associated with garbage disposal while also reducing greenhouse gas emissions and dependence on non-renewable energy sources. Utilizing organic matter recovered from tannery effluents for industrial and agricultural use further demonstrates the possibilities of a circular economy approach. Resource efficiency, economic sustainability, and environmental conservation can all be greatly enhanced by tannery waste, provided that it is managed properly and waste conversion methods are improved. This chapter illustrates the possibility to minimize ecological effect while converting industrial by-products into high-value solutions for practical energy demands.

References

Aathika ARS, Kubendran D, Yuvarani M, Thiruselvi D, Amudha T, Karthik P et al (2018) Enhanced biohydrogen production from leather fleshing waste co-digested with tannery treatment plant sludge using anaerobic hydrogenic batch reactor. Energy Sources Part A Recov Util Environ Eff 40:586–593. https://doi.org/10.1080/15567036.2018.1435754

ABNT NBR (2004) Procedure for obtention leaching extract of solid wastes. Ass Bras Norm Técn, Rio Janeiro

Agustini CB, da Fontoura JT, Mella B, Gutterres M (2018) Evaluating co-substrates to supplement biogas production from tannery solid waste treatment—cattle hair, microalgae biomass, and silicone. Biofuels Bioprod Bioref 12:1095–1102. https://doi.org/10.1002/bbb.1929

Agustini CB, da Costa M, Gutterres M (2020) Tannery wastewater as nutrient supply in production of biogas from solid tannery wastes mixed through anaerobic co-digestion. Process Saf Environ Prot 135:38–45. https://doi.org/10.1016/j.psep.2019.11.037

Ali S, Rizwan M, Waqas A, Hussain MB, Hussain A, Liu S et al (2018) Fulvic acid prevents chromium-induced morphological, photosynthetic, and oxidative alterations in wheat irrigated with tannery waste water. J Plant Growth Regul 37:1357–1367. https://doi.org/10.1007/s00344-018-9843-6

Amalraj A, Pius A (2014) Removal of selected basic dyes using activated carbon from tannery wastes. Sep Sci Technol 49:90–100. https://doi.org/10.1080/01496395.2013.819018

Ammasi R, Victor JS, Chellan R, Chellappa M (2020) Amino acid enriched proteinous wastes: recovery and reuse in leather making. Waste Biomass Valorization 11:5793–5807. https://doi.org/10.1007/s12649-019-00912-6

Barzagli F, Mani F (2019) The increased anthropogenic gas emissions in the atmosphere and the rising of the earth's temperature: are there actions to mitigate the global warming? Substantia 3:101–111. https://doi.org/10.13128/Substantia-69

Bavaresco L, Perondi D, Dettmer A (2015) Steam explosion utilization as pre-treatment of chrome leather waste in order to produce biogas. In: XXXIII IULTCS congress 2015

Bayrakdar A (2020) Anaerobic co-digestion of tannery solid wastes: a comparison of single and two-phase anaerobic digestion. Waste Biomass Valorization 11:1727–1735. https://doi.org/10.1007/s12649-019-00902-8

Booramurthy VK, Kasimani R, Subramanian D, Pandian S (2020) Production of biodiesel from tannery waste using a stable and recyclable nano-catalyst: an optimization and kinetic study. Fuel 260. https://doi.org/10.1016/j.fuel.2019.116373

Booramurthy VK, Kasimani R, Pandian S (2022) Biodiesel production from tannery waste using a nano catalyst (ferric-manganese doped sulphated zirconia). Energy Sources Part A Recov Util Environ Eff 44:1092–1104. https://doi.org/10.1080/15567036.2019.1639849

Dettmer A, Nunes KGP, Gutterres M, Marcílio NR (2010) Production of basic chromium sulfate by using recovered chromium from ashes of thermally treated leather. J Hazard Mater 176:710–714. https://doi.org/10.1016/j.jhazmat.2009.11.090

Fathima N, Rao R, Nair BU (2012) Tannery solid waste to treat toxic liquid wastes: a new holistic paradigm. Environ Eng Sci 29:363–372. https://doi.org/10.1089/ees.2010.0445

Geethakarthi A, Phanikumar BR (2012) Characterization of tannery sludge activated carbon and its utilization in the removal of azo reactive dye. Environ Sci Pollut Res 19:656–665. https://doi.org/10.1007/s11356-011-0608-z

Gomes CS, Piccin JS, Gutterres M (2016) Optimizing adsorption parameters in tannery-dye-containing effluent treatment with leather shaving waste. Process Saf Environ Prot 99:98–106. https://doi.org/10.1016/j.psep.2015.10.013

Hashem M, Tomal MNA, Mondal B (2015) Generation of fleshings at beamhouse in tannery and its environmental impact assessment: Bangladesh perspective. Bangladesh J Sci Ind Res 50:227–232. https://doi.org/10.3329/bjsir.v50i3.25590

Kantarli IC, Yanik J (2009) Use of waste sludge from the tannery industry. Energy Fuels 23:3126–3133. https://doi.org/10.1021/ef8011068

Kluska J, Ochnio M, Kardaś D, Heda Ł (2019) The influence of temperature on the physicochemical properties of products of pyrolysis of leather-tannery waste. Waste Manag 88:248–256. https://doi.org/10.1016/j.wasman.2019.03.046

Kubendran D, Salma Aathika AR, Amudha T, Thiruselvi D, Yuvarani M, Sivanesan S (2017) Utilization of leather fleshing waste as a feedstock for sustainable biodiesel production. Energy Sources Part A Recov Util Environ Eff 39:1587–1593. https://doi.org/10.1080/15567036.2017.1349218

Li Y, Park SY, Zhu J (2011) Solid-state anaerobic digestion for methane production from organic waste. Renew Sustain Energy Rev 15:821–826. https://doi.org/10.1016/j.rser.2010.07.042

Malafaia G, de Araúj FG, da Costa ED, Guimarães ATB, Leandro WM, de Lima Rodrigues AS (2016) Corn production in soil containing in natura tannery sludge and irrigated with domestic wastewater. Agric Water Manag 163:212–218. https://doi.org/10.1016/j.agwat.2015.09.018

Masilamani D, Srinivasan V, Ramachandran RK, Gopinath A, Madhan B, Saravanan P (2017) Sustainable packaging materials from tannery trimming solid waste: a new paradigm in wealth from waste approaches. J Clean Prod 164:885–891. https://doi.org/10.1016/j.jclepro.2017.06.200

Mella B, Benvenuti J, Oliveira RF, Gutterres M (2019) Preparation and characterization of activated carbon produced from tannery solid waste applied for tannery wastewater treatment. Environ Sci Pollut Res 26:6811–6817. https://doi.org/10.1007/s11356-019-04161-x

Moktadir MA, Ren J, Zhou J (2023) A systematic review on tannery sludge to energy route: current practices, impacts, strategies, and future directions. Sci Total Environ 901:166244

Muralidharan V, Arokianathan MS, Balaraman M, Palanivel S (2020) Tannery trimming waste based biodegradable bioplastic: facile synthesis and characterization of properties. Polym Test 81. https://doi.org/10.1016/j.polymertesting.2019.106250

Nabavinia F, Emami H, Astaraee A, Lakzian A (2015) Effect of tannery wastes and biochar on soil chemical and physicochemical properties and growth traits of radish. Int Agrophys 29:333–339. https://doi.org/10.1515/intag-2015-0040

Nogueira FGE, do Prado NT, Oliveira LCA, Bastos ARR, Lopes JH, de Carvalho JG (2010) Incorporation of mineral phosphorus and potassium on leather waste (collagen): a new NcollagenPK-fertilizer with slow liberation. J Hazard Mater 176:374–380. https://doi.org/10.1016/j.jhazmat.2009.11.040

Palani Y, Rao Jonnalagadda R, Fathima NN (2017) Adsorption on activated carbon derived from tannery fleshing waste: adsorption isotherms, thermodynamics, and kinetics. Environ Prog Sustain Energy 36:1725–1733. https://doi.org/10.1002/ep.12637

Periyasamy K, Santhalembi L, Mortha G, Aurousseau M, Guillet A, Dallerac D et al (2017) Production, partial purification and characterization of enzyme cocktail from *Trichoderma citrinoviride* AUKAR04 through solid-state fermentation. Arab J Sci Eng 42:53–63. https://doi.org/10.1007/s13369-016-2110-x

Plastic Oceans (2020) The plastic oceans initiative

Priebe GPS, Kipper E, Gusmão AL, Marcilio NR, Gutterres M (2016) Anaerobic digestion of chrome-tanned leather waste for biogas production. J Clean Prod 129:410–416. https://doi.org/10.1016/j.jclepro.2016.04.038

Ramos L, Fernandes R, Crispim A, Ramalho E, Caetano NSP (2008) Biodiesel production from leather industry wastes. In: 10th international chemical and biological engineering conference

Rosseto M (2020) Enzymatic and antimicrobial treatment in corn starch and gelatin films recovered from chromium (III)-tanned wastes. Dissertation, University of Passo Fundo

Šánek L, Pecha J, Kolomazník K, Bařinová M (2015) Biodiesel production from tannery fleshings: feedstock pretreatment and process modeling. Fuel 148:16–24. https://doi.org/10.1016/j.fuel.2015.01.084

Saxena S, Saharan VK, George S (2019) Modeling & simulation studies on batch anaerobic digestion of hydrodynamically cavitated tannery waste effluent for higher biogas yield. Ultrason Sonochem 58. https://doi.org/10.1016/j.ultsonch.2019.104692

Shanmugam P, Horan NJ (2009) Optimising the biogas production from leather fleshing waste by co-digestion with MSW. Bioresour Technol 100:4117–4120. https://doi.org/10.1016/j.biortech.2009.03.052

Shivani P, Khushbu P, Faldu N, Thakkar V, Shubramanian RB (2011) Extraction and analysis of *Jatropha curcas* L. seed oil. Afr J Biotechnol 10:18210–18213. https://doi.org/10.5897/AJB11.776

Silva JDC, Leal TTB, Araújo ASF, Araujo RM, Gomes RLF, Melo WJ et al (2010) Effect of different tannery sludge compost amendment rates on growth, biomass accumulation and yield responses of Capsicum plants. Waste Manag 30:1976–1980. https://doi.org/10.1016/j.wasman.2010.03.011

Sri Bala Kameswari K, Kalyanaraman C, Thanasekaran K (2011) Effect of ozonation and ultrasonication pretreatment processes on co-digestion of tannery solid wastes. Clean Technol Environ Policy 13:517–525. https://doi.org/10.1007/s10098-010-0334-0

Sri Bala Kameswari K, Kalyanaraman C, Porselvam S, Thanasekaran K (2011) Enhancement of biogas generation by addition of lipase in the co-digestion of tannery solid wastes. Clean Soil Air Water 39:781–786. https://doi.org/10.1002/clen.201000408

Sri Bala Kameswari K, Kalyanaraman C, Umamaheswari B, Thanasekaran K (2014) Enhancement of biogas generation during co-digestion of tannery solid wastes through optimization of mix proportions of substrates. Clean Technol Environ Policy 16:1067–1080. https://doi.org/10.1007/s10098-013-0706-3

Sundar VJ, Gnanamani A, Muralidharan C, Chandrababu NK, Mandal AB (2011) Recovery and utilization of proteinous wastes of leather making: a review. Rev Environ Sci Biotechnol 10:151–163. https://doi.org/10.1007/s11157-010-9223-6

Vinicius C, Rigueto T, Rosseto M, Dal D, Krein C, Elisangela B et al (2020) Alternative uses for tannery wastes: a review of environmental, sustainability, and science. J Leather Sci Eng

Yilmaz O, Cem Kantarli I, Yuksel M, Saglam M, Yanik J (2007) Conversion of leather wastes to useful products. Resour Conserv Recycl 49:436–448. https://doi.org/10.1016/j.resconrec.2006.05.006

Zupančič GD, Jemec A (2010) Anaerobic digestion of tannery waste: semi-continuous and anaerobic sequencing batch reactor processes. Bioresour Technol 101:26–33. https://doi.org/10.1016/j.biortech.2009.07.028

Part IV
Advancing Circular Economy and Overcoming Challenges in the Tannery Sector

The final part of the book addresses the future of the tannery industry through the lens of the circular economy, which seeks to redefine traditional production and consumption patterns. This section offers strategies for advancing circular economy principles within the tannery sector and overcoming the barriers to adopting sustainable practices.

Chapter 11 discusses various strategies for advancing circular economy initiatives in the tannery industry, from waste reduction to the recovery of valuable materials. It emphasizes how businesses can benefit from adopting circular economy principles and showcases successful examples of circular practices already in place within the industry. Chapter 12 identifies the key challenges faced by the industry, particularly in technological development, and explores the barriers to implementing sustainable solutions. This chapter presents potential pathways for overcoming these obstacles and ensuring that the industry remains competitive while embracing sustainability. Chapter 13 focuses on the utilization of tannery solid waste, showcasing how waste byproducts can be repurposed into valuable materials, such as biogas, compost, or alternative raw materials for other industries. Chapter 14 concludes the book by presenting strategies for fostering the principles of circular economy across the entire tannery sector, highlighting the importance of collaboration, policy reform, and industry-wide commitment to sustainability.

Through this comprehensive exploration, *Sustainable Practices in the Tannery Industry* not only addresses the urgent need for change but also offers a roadmap for transforming the leather industry into a more sustainable and eco-friendlier sector. By focusing on technological innovation, waste management, and the adoption of circular economy principles, the book lays the groundwork for an industry that is not only more responsible in its practices but also better positioned to thrive in a future that increasingly values sustainability and environmental stewardship.

Chapter 11
Strategies for Advancing Circular Economy in the Tannery Sector

Juan F. Saldarriaga and Julián E. López

Abstract The tanning industry is one of the primary sources of water and soil pollution in developing countries, largely due to the generation of waste that combines organic materials from animal skins with inorganic chemicals used in tanning processes. This mixture results in hazardous waste, the management and valorization of which present significant challenges due to its complexity and high potential for pollution. However, within the framework of circular economy and sustainability strategies, certain types of waste, such as animal hair, have demonstrated their viability as raw materials for the production of innovative biomaterials. One such product is biochar, a carbonaceous material with numerous environmental applications, particularly in agriculture and the removal of pollutants from various soil and water matrices. Research has confirmed that biochar derived from tannery waste can efficiently remove emerging pollutants, including pharmaceutical compounds, from water. Studies have shown that biochar can achieve removal rates of up to 90% for drugs such as amoxicillin, diclofenac, azithromycin, and erythromycin. Furthermore, activating biochar has significantly enhanced its ability to adsorb these compounds, thereby expanding its potential as a tool for decontaminating water resources. This approach not only helps to mitigate the environmental impact associated with the tanning industry but also promotes the valorization of waste by transforming it into high-value products. In this way, it encourages a transition toward more sustainable industrial practices that align with the principles of the circular economy and contribute to the reduction of adverse effects on the environment.

Keywords Circular economy · Leather · Biochar · Hazardous waste · Pollutant removal · Sustainability

---

J. F. Saldarriaga (✉)
Department of Civil and Environmental Engineering, Universidad de los Andes, Bogotá, Colombia
e-mail: jf.saldarriaga@uniandes.edu.co; juanfelorza@gmail.com

J. E. López
Faculty of Architecture and Engineering, Environmental Engineering Program, Institución Universitaria Colegio Mayor de Antioquia, Medellín, Colombia
e-mail: julian.lopez@colmayor.edu.co

M. A. Aboulhassan and S. Souabi (eds.), *Sustainable Practices in the Tannery Industry*,
https://doi.org/10.1007/978-3-031-92037-0_11

11.1 Introduction to the Circular Economy in the Leather Sector

The Circular Economy in the leather sector is presented as a viable alternative to mitigate the environmental impact of a historically polluting industry (Takacs et al. 2022). Leather production, which generates a global market valued at approximately USD 400 billion annually, is based on the processing of animal hides and skins, mainly as a by-product of the meat industry (Prev 2024; Amobonye et al. 2023; Muralidharan et al. 2022). However, this activity has been criticized for its contribution to deforestation, excessive use of water resources and greenhouse gas emissions (Omoloso et al. 2021).

Historically, the leather industry has faced significant challenges in terms of sustainability. Extensive cattle ranching, necessary to obtain the skins, has led to the destruction of vital ecosystems such as the Amazon, exacerbating problems such as climate change and biodiversity loss (Gerber et al. 2015; FAO 2017). In addition, the traditional tanning process uses chemicals that are harmful to both workers and the environment. In this context, the circular economy emerges as an approach that seeks not only to reduce the waste generated by this industry, but also to maximize the efficient use of available resources (Wrzesińska-Jędrusiak et al. 2023; Singh et al. 2023).

The circular economy in the leather sector implies a shift toward more sustainable practices, where waste generated by the meat industry is valued as a raw material. This approach promotes the reuse and recycling of leather, allowing up to 10 million tons per year to be recycled and transformed into new products (Ahmed et al. 2022; Dragomir and Dumitru 2022). Initiatives such as the Leather Working Group are working to establish stricter environmental standards and encourage the use of less polluting methods in tanning, such as vegetable tanning, which uses natural extracts instead of toxic chemicals (LWG 2024).

The transition to a circular economy in the leather sector is not only fundamental to reducing its environmental footprint but also represents an opportunity to innovate and create a more sustainable industrial model. As consumers demand more responsible products and companies adopt sustainable practices, the future of leather can be redefined toward a path that prioritizes both quality and sustainability (Singh et al. 2023; Díaz-Burgos et al. 2023).

The leather industry faces several significant environmental challenges that impact both the ecosystem and public health. One of the most critical issues is resource intensive. Leather production requires large amounts of water, with estimates suggesting that up to 15,000 L of water are needed to produce a single pair of leather shoes (Chapagain and Hoekstra 2004; CircumFauna 2024). This overuse not only depletes freshwater sources but also contributes to water pollution due to the chemicals used in the tanning process.

In addition, deforestation is an alarming challenge associated with the leather industry. The expansion of land for cattle ranching, which is the main source of leather, has led to the massive clearing of forests, exacerbating climate change by releasing large amounts of CO_2 stored in trees. This activity not only destroys natural habitats but also affects biodiversity and contributes to the loss of species.

Another area of concern is the generation of waste (FAO 2010). Although leather is largely obtained as a by-product of the meat industry, a significant portion is wasted during the production and finishing process. It is estimated that more than 90% of leather comes from livestock, but not all hides are used effectively, resulting in a considerable amount of waste. In addition, the waste generated by tanneries, which often contains toxic substances, can contaminate soils and water bodies if not properly managed (Kanagaraj et al. 2020).

Finally, a lack of awareness of sustainable practices and a resistance to change within the sector are additional barriers to addressing these challenges. Many companies still operate obsolete machinery and lack sustainable logistics, making the transition to greener methods difficult. In response to these problems, it is essential to encourage greater transparency in supply chains and to adopt innovative technologies that minimize the environmental impact of the leather sector (Takacs et al. 2022).

11.2 Fundamental Principles for Implementation

To implement a circular economy in the leather sector, it is essential to establish principles that focus on the reduction of resource consumption, as well as specific strategies to reduce the use of water, energy, and waste during the production processes (Karuppiah et al. 2021; Moktadir et al. 2020). One of the key principles is efficiency in the use of resources, which implies adopting technologies that optimize the use of water and energy; for example, the implementation of water recirculation systems in tanning allows the reuse of treated water, significantly reducing the volume required for production processes (Silva 2023; Cairone et al. 2024). Besides, it is crucial to minimize waste through processes that allow for proper waste separation, avoiding the mixing of hazardous waste with other wastes and thus reducing the hazardousness of the waste generated. Technological innovation also plays an important role, encouraging research and development of clean technologies that reduce the pollution load; for example, the use of alternative tanning methods, such as vegetable tanning, can reduce both waters use and the generation of toxic waste (Pučnik et al. 2024; Zhang et al. 2024). Another principle is to promote a closed cycle where all materials are kept in use for as long as possible, designing products for durability and repairability, which facilitates recycling at the end of their useful life. Finally, education and awareness of sustainable practices are vital to ensure effective adoption of these strategies. Specific strategies include optimizing water use through advanced technologies for water treatment and reuse in the tanning and finishing stages, which not only reduces overall consumption but also minimizes wastewater pollution. In addition, adopting renewable energy sources and improving energy efficiency in factories can significantly reduce greenhouse gas emissions associated with leather production (Illankoon and Vithanage 2023). The recovery and reuse of by-products generated during the production process, such as collagen or gelatin, can also maximize added value when used in other industries. Finally, fostering partnerships between companies, universities, and governments is essential to share knowledge and resources to facilitate the transition to more sustainable practices throughout the leather value chain. Effective implementation of these principles and strategies will not only contribute to reducing the environmental impact of the leather sector but will also improve its competitiveness by aligning with growing consumer demands for more sustainable products (Hassoun et al. 2023; Hamed et al. 2022).

On the other hand, in order to minimize the use of chemical products in the leather industry, it is essential to implement several strategies that promote more sustainable and environmentally friendly practices. One of the most effective strategies is to promote the use of less toxic dyes and chemicals. This involves selecting safer alternatives that reduce the toxicity of the waste generated during the production process. For example, natural dyes derived from vegetable sources can be used instead of synthetic dyes, which often contain heavy metals and other harmful compounds (Omoloso et al. 2021, 2024).

Another key strategy is the adoption of vegetable tanning processes, which use natural extracts such as tannins instead of synthetic chemicals. This method not only reduces water and soil pollution but also reduces workers' exposure to harmful chemicals. Studies have shown that the implementation of cleaner production processes (CPP) can lead to a significant reduction in the use of chemical inputs; for example, a 73% decrease in chemical inputs required for tanning has been reported when applying these strategies (Qiao et al. 2021; Hassan et al. 2023). It is essential to establish an environmental management system that includes regular audits to identify and eliminate unnecessary use of chemicals at all stages of the production process. This may include training personnel on safe and sustainable practices, as well as promoting technological innovation to develop alternative methods to reduce or eliminate the use of hazardous compounds (Mondejar et al. 2021; Ostad-Ali-Askari 2022).

Chemical recovery and reuse can also be an effective strategy. Implementing technologies that allow the recovery and recycling of chemicals used in the process can significantly reduce the amount needed for each production cycle (Jaffari et al. 2024; Chojnacka et al. 2021). Finally, fostering collaboration among companies to share best practices and innovative technologies can accelerate the transition to a more responsible and sustainable use of chemicals in the leather industry. These strategies will not only help mitigate the sector's environmental impact but can also improve its competitiveness by aligning with growing consumer demands for more sustainable products (Shahzad et al. 2022).

Finally, in order to establish effective recycling and reuse systems in the leather industry, it is essential to implement a series of strategies that make the most of the waste generated during production. One of the most important strategies is the reuse of by-products and waste as materials for new products. For example, the Portuguese company ERT has developed a hydrolysis technology that transforms leather waste into an insulating material used in the manufacture of components for the automotive sector. This process not only allows waste to be recycled but also contributes to reducing environmental impact by using a considerable percentage of recycled products in its production, which can receive sustainability certifications such as the Global Recycled Standard (Navarro et al. 2020).

Another effective strategy is the creation of recycled composite materials from leather scraps. A study developed in the leather industry has shown that it is possible to recycle 100% of the waste generated in the cutting phase by experimenting with organic and non-toxic binders. These new materials can be used in decorative and functional applications, thus increasing the useful life of the leather and reducing waste. In addition, other products such as glues, collagen, gelatin, and biofuels can be produced from the waste generated by tanneries, demonstrating the versatility of these wastes and their potential for reuse in different industries (Chojnacka et al. 2021).

The repairability of leather also plays a crucial role in its recycling. Leather products are easy to repair and restore, which helps to avoid premature disposal. This not only extends the useful life of the product but also encourages a culture of reuse among consumers. For example, sofas and car seats can be transformed into benches or tables, stimulating creativity and sustainable design (Pringle et al. 2016).

Finally, promoting collaboration among companies to share best practices and innovative technologies can facilitate the development of effective solutions for the recycling and reuse of waste throughout the leather value chain. These strategies will not only help to reduce the environmental impact of the sector but will also foster a more circular and sustainable economic model (Luthra et al. 2022).

11.3 Eco-design

Eco-design, or ecological design, is a comprehensive approach aimed at minimizing the environmental impact of products throughout their lifecycle (Bovea and Pérez-Belis 2012). In the context of the tanning industry, integrating eco-design principles is crucial for promoting sustainable practices and fostering a circular economy (Navarro et al. 2020). Eco-design offers several advantages for the leather industry. On one hand, it promotes resource efficiency, which can lead to significant cost reductions in production. By optimizing processes and using recyclable or low-impact materials, companies can lower their operating expenses. On the other hand, products designed with eco-design principles tend to be of higher quality and durability. This not only benefits the environment by generating less waste but also provides greater value to consumers who seek long-lasting products.

Integrating eco-design in the leather industry not only focuses on the sustainability of products but also fundamentally addresses environmental impact management and the handling of liquid and solid waste (Navarro et al. 2020; Jadhav and Jadhav 2020). From the inputs required for the industry, eco-design promotes the use of less polluting materials and processes. By selecting inputs that generate less waste and emissions, negative impacts are minimized. On the other hand, implementing eco-design practices allows companies to optimize the use of water and energy. This not only reduces pressure on natural resources but also lowers operational costs by decreasing the number of inputs required for production. Within eco-design practices in the leather industry, the following stand out: dry pre-scraping, water recirculation, collagen extraction, chromium-free tanning, the use of natural dyes, the search for new materials to replace leather such as natural fibers and other biotechnological materials, and the valorization of waste through thermal and biological technologies (Santos and Gutterres 2007; Aquim et al. 2019; Maistrenko et al. 2022; Yu et al. 2021).

The mentioned practices help significantly reduce the amount of waste generated during the production process. This is crucial for mitigating soil and water pollution, as well as reducing pressure on landfills. On the other hand, water recirculation and the efficient use of chemical inputs allow for lower consumption of natural resources, which is essential in a context where these resources are increasingly scarce. By eliminating and reducing the use of hazardous chemicals, toxic discharges into the environment are minimized. Finally, initiatives that promote eco-design help create a sustainable cycle where products are designed to be reused or recycled, contributing to a more circular economy.

11.4 Practical Implementation Examples

The adoption of circular practices in the leather industry has proven effective in enhancing sustainability and reducing environmental impact (Shen 2014; Waite 2014). Companies like Durli Leathers have implemented circularity practices aimed at reducing water and energy consumption. These technologies have enabled significant reductions in water and energy use. Thanks to its focus on sustainability and efficiency, Durli Leathers has optimized its production processes, resulting in lower greenhouse gas emissions and a more responsible use of natural resources. Another success story is that of Veja, a company that has adopted vegetable tanning as a fundamental part of its production process. This technique uses natural tannins extracted from plants, thereby eliminating and reducing the use of toxic chemicals. Additionally, Veja strives to use sustainable materials in its products, which not only reduces the environmental impact of the tanning process but also promotes a more responsible and ethical business model. Other companies, such as H&M, have launched programs that promote clothing recycling and the use of sustainable materials in their garments. These programs aim to reduce textile waste by encouraging reuse and recycling within the product's life cycle. Companies like Cueros Vélez have integrated sustainable fashion principles by combining leather with recycled materials and natural pigments. These practical examples already implemented in the industry demonstrate how circular practices can be effectively applied, generating both environmental and other benefits, such as reducing the use of natural resources, minimizing hazardous waste, and promoting recycling. Overall, adopting these strategies not only enhances the sustainability of the sector but also addresses the growing market demand for products that are environmentally and socially responsible (Barros et al. 2021).

11.5 Technological Innovations

The transition to a circular economy in the leather industry can be facilitated through the adoption of new technologies and digital solutions that optimize resource and waste management, especially in handling by-products such as hair, which poses a significant challenge due to its accumulation and the environmental problems it generates (Moktadir et al. 2020). One of the most promising strategies is the use of composting technologies to transform hair into a useful resource (Barrena et al. 2007a, b). Research has shown that hair can be co-composted with sludge from wastewater treatment plants, resulting in a stable and nutritious compostable product for agricultural use. This approach not only reduces the amount of waste generated but also turns a problematic by-product into a nitrogen-rich fertilizer, providing an effective solution for hair management in the industry.

Methods are being explored to extract keratin from hair, which can be used as an additive in tanning and retanning processes (Mengistu et al. 2024; Valeika et al. 2019). The extracted keratin has been shown to improve the absorption of chemicals used in tanning without significantly altering the physical properties of the leather. This approach not only helps reduce the use of synthetic chemicals but also allows for the reuse of a by-product that would otherwise be discarded.

The exploration of new technologies to facilitate the transition to a circular economy in the tanning industry includes the utilization of waste such as hair and sawdust through pyrolysis processes. This thermal method allows these waste materials to be transformed into biochar, a material that has proven useful in various applications, including its use as an adsorbent for the removal of emerging contaminants and as a substitute for cement in the construction industry. In the case of hair, which is generated in large quantities during the tanning process (about 100 kg/ton of raw hide), its utilization through pyrolysis not only helps reduce waste but also allows the recovery of valuable compounds. The biochar obtained can be used as an effective filtering medium for the removal of heavy metals and other contaminants from water, showing superior capacity compared to commercial activated carbon (Rodríguez et al. 2020; Herrera et al. 2023). A successful example can be found in research that has demonstrated that biochar produced from agricultural and agroindustrial waste has outstanding adsorbent properties, being capable of removing contaminants such as hexavalent chromium from water, or pharmaceuticals like amoxicillin and diclofenac. In a study, the efficiency of biochar was compared with other filtering media, and it was found that biochar had a higher affinity for contaminants due to its superior physicochemical characteristics. This highlights how waste generated in the tanning industry can be transformed into materials with practical and environmentally beneficial applications.

The use of biochar as a replacement material for cement in construction is another valuable application. Its incorporation not only reduces the carbon footprint associated with cement production but also improves the mechanical properties of concrete. Research has shown that the use of biochar derived from pyrolysis can contribute to carbon sequestration and sustainability in the construction sector (Rodríguez et al. 2020; Herrera et al. 2023; Morales et al. 2021).

In terms of digital solutions, the implementation of Internet of Things (IoT)-based platforms can facilitate real-time monitoring of resource use and waste generation during the production process. These platforms can help companies identify areas where significant improvements can be made, thus optimizing water and energy consumption. For example, IoT sensors can be used to track water flow and detect leaks or inefficiencies in real time (Soori et al. 2023; Ding et al. 2023).

Fostering collaborations between companies and academic institutions can accelerate the adoption of these innovative technologies and sustainable practices. The creation of networks that connect researchers with industry will enable the sharing of knowledge and the development of effective solutions to maximize the use of available resources and minimize waste. These strategies will not only contribute to reducing the environmental impact of the leather sector but will also improve its competitiveness by aligning with growing consumer demand for more sustainable products (Sánchez-García et al. 2024).

11.6 Value Chain Collaboration

The leather industry value chain refers to the set of activities and processes that transform raw materials, such as animal hides, into finished products delivered to the final consumer. This chain includes multiple links, each with its own role and contribution to the final product's value. In general, the links in the leather industry value chain are livestock production, collection and transportation, tanning, secondary transformation, commercialization, product use, and end of life (Omoloso et al. 2024).

Effective collaboration among all supply chain actors is essential for implementing sustainable practices in the leather industry. From suppliers to consumers, every link in the chain plays a crucial role in promoting a circular approach that steers the sector toward sustainability, enabling a comprehensive response to environmental and social challenges (Omoloso et al. 2024). For example, synergy among supply chain actors translates into collaboration between various stakeholders, enabling the creation of synergies that amplify the positive impact of sustainable initiatives. This type of collaboration facilitates the exchange of knowledge and resources, leading to continuous improvement in sustainable practices, positively affecting economic, environmental, and social aspects. A collaborative supply chain promotes greater transparency and traceability regarding the origin of raw materials. Collaboration also involves the continuous training of the actors involved in the value chain. También, la colaboración puede impulsar la innovación en el desarrollo de nuevos procesos y productos más sostenibles. Training programs on good manufacturing practices are essential to ensure that all links in the chain comply with environmental and social standards. This not only improves the quality of the final product but also fosters a sense of shared responsibility among all participants (Moktadir et al. 2018).

Another key actor is the final consumer. Involving consumers in the value chain is essential to encourage a shift toward more sustainable practices. Brands can educate their customers on the importance of choosing responsible products and how their decisions impact the environment (D'Adamo et al. 2024).

Circular economy practices in the leather industry value chain include, e.g., the reuse of leather scraps. These materials are used to create a variety of items, from wallets to keychains, reducing the need for new raw materials and preventing waste generation (Wrzesińska-Jędrusiak et al. 2023). Another practice is the recycling of hides discarded by the meat industry, integrating them as raw materials in the production process (Patel et al. 2022). Additionally, practices have been implemented to transform leather production scraps into edible toys for dogs. This approach not only reduces waste generated by the leather industry, but also creates innovative products for other industries (Flores Tapia and Brito Moina 2023). Collaboration within the value chain is essential for implementing sustainable practices in the leather industry. By working together, stakeholders can tackle complex challenges, improve transparency and traceability, promote education and training, drive innovation, and engage consumers in the process. This approach not only benefits the environment but also strengthens the sector's competitiveness by aligning with the growing demand for responsible and ethical products. Building a sustainable supply chain is a step toward a more responsible and efficient future for the entire leather industry.

11.7 Conclusions and Future Directions

The leather industry is at a turning point in its journey toward a circular economy, facing both significant opportunities and challenges. Among the opportunities is the growing demand for sustainable products, driven by increased environmental awareness among consumers. This has led companies to adopt more responsible practices, such as the use of biological tanning agents and technologies that reduce water and chemical consumption. Innovations such as the use of by-products, e.g., the pyrolysis of waste such as hair and sawdust to produce biochar, offer creative solutions that not only mitigate environmental impact but also generate new income by transforming waste into valuable resources.

However, the transition to sustainability also presents significant challenges. Implementing sustainable practices often requires significant investments in research and development, as well as changes in organizational culture and corporate policies. Small and medium-sized enterprises may face additional barriers due to lack of resources and training to adopt these new technologies. In addition, the need to establish clear standards and regulations that promote responsible practices is crucial to avoid greenwashing and ensure that initiatives are genuine.

To support these initiatives and foster an enabling environment for sustainable innovation, it is essential to develop public policies that facilitate the transition to a circular economy in the leather sector. These policies could include tax incentives for companies that adopt sustainable practices, as well as funding for research and development projects focused on clean technologies. The creation of training programs for workers and entrepreneurs on sustainable practices is also fundamental to building a culture of environmental responsibility within the sector.

In addition, it is important to promote collaboration between the different actors in the supply chain through platforms that facilitate the exchange of information and resources. Networking between companies, universities and government institutions can accelerate the development and implementation of innovative solutions that address the sector's environmental challenges.

While the path to a circular economy in the leather industry is fraught with challenges, it also offers numerous opportunities to innovate and improve the sustainability of the sector. With the right support through public policy and effective collaboration among all stakeholders, it is possible to transform the industry toward a more responsible and sustainable model that benefits both the environment and the economy.

References

Ahmed Z, Mahmud S, Acet DH (2022) Circular economy model for developing countries: evidence from Bangladesh. Heliyon 8(5):e09530. https://doi.org/10.1016/J.HELIYON.2022.E09530

Amobonye A, Lalung J, Awasthi MK, Pillai S (2023) Fungal mycelium as leather alternative: a sustainable biogenic material for the fashion industry. Sustain Mater Technol 38:e00724. https://doi.org/10.1016/J.SUSMAT.2023.E00724

Barrena R, Pagans E, Vázquez F, Artola A, Sánchez A (2007a) Full-scale cocomposting of hair wastes from the leather manufacturing industry and sewage sludge. Compost Sci Util 15(1):16–21. https://doi.org/10.1080/1065657X.2007.10702305

Barrena R, la Pagans E, Artola A, Vázquez F, Sánchez A (2007b) Co-composting of hair waste from the tanning industry with de-inking and municipal wastewater sludges. Biodegradation 18(3):257–268. https://doi.org/10.1007/S10532-006-9060-Z/METRICS

Barros MV, Salvador R, do Prado GF, de Francisco AC, Piekarski CM (2021) Circular economy as a driver to sustainable businesses. Clean Environ Syst 2:100006. https://doi.org/10.1016/J.CESYS.2020.100006

Bovea MD, Pérez-Belis V (2012) A taxonomy of ecodesign tools for integrating environmental requirements into the product design process. J Clean Prod 20(1):61–71. https://doi.org/10.1016/J.JCLEPRO.2011.07.012

Cairone S, Hasan SW, Choo KH, Lekkas DF, Fortunato L, Zorpas AA, Korshin G, Zarra T, Belgiorno V, Naddeo V (2024) Revolutionizing wastewater treatment toward circular economy and carbon neutrality goals: pioneering sustainable and efficient solutions for automation and advanced process control with smart and cutting-edge technologies. J Water Process Eng 63:105486. https://doi.org/10.1016/J.JWPE.2024.105486

Chapagain AK, Hoekstra AY (2004) Water footprints of nations value of water. Delft

Chojnacka K, Skrzypczak D, Mikula K, Witek-Krowiak A, Izydorczyk G, Kuligowski K, Bandrów P, Kułażyński M (2021) Progress in sustainable technologies of leather wastes valorization as solutions for the circular economy. J Clean Prod 313:127902. https://doi.org/10.1016/J.JCLEPRO.2021.127902

CircumFauna. Leather water footprint. CircumFauna. Accessed 19 Dec 2024. https://circumfauna.org/leather-water-footprint

D'Adamo I, Gagliarducci M, Iannilli M, Mangani V (2024) Fashion wears sustainable leather: a social and strategic analysis toward sustainable production and consumption goals. Sustainability 16(22):9971. https://doi.org/10.3390/SU16229971

de Aquim PM, Hansen É, Gutterres M (2019) Water reuse: an alternative to minimize the environmental impact on the leather industry. J Environ Manage 230:456–463. https://doi.org/10.1016/J.JENVMAN.2018.09.077

Díaz-Burgos Á, Castillo-Parra C, Fernández-Izquierdo P, Arturo-Perdomo D, Lozada-Castro JJ (2023) Obtaining biodiesel from fat extracted from solid waste produced in the fleshing stage of leather manufacturing. Ing Inv 43(1). https://doi.org/10.15446/ING.INVESTIG.97254

Ding S, Tukker A, Ward H (2023) Opportunities and risks of internet of things (IoT) technologies for circular business models: a literature review. J Environ Manage 336:117662. https://doi.org/10.1016/J.JENVMAN.2023.117662

Dragomir VD, Dumitru M (2022) Practical solutions for circular business models in the fashion industry. Clean Logist Supply Chain 4:100040. https://doi.org/10.1016/J.CLSCN.2022.100040

FAO (2010) Livestock information, sector analysis and policy branch animal production and health division

FAO (2017) Transforming the livestock sector through the sustainable development goals. Rome

Flores Tapia NE, Brito Moina H (2023) Exploring tannery solid wastes as a source of animal feed. Processes 11(10):2965. https://doi.org/10.3390/PR11102965

Gerber PJ, Mottet A, Opio CI, Falcucci A, Teillard F (2015) Environmental impacts of beef production: review of challenges and perspectives for durability. Meat Sci 109:2–12. https://doi.org/10.1016/J.MEATSCI.2015.05.013

Hamed I, Jakobsen AN, Lerfall J (2022) Sustainable edible packaging systems based on active compounds from food processing byproducts: a review. Compr Rev Food Sci Food Saf 21(1):198–226. https://doi.org/10.1111/1541-4337.12870

Hassan MM, Harris J, Busfield JJC, Bilotti E (2023) A review of the green chemistry approaches to leather tanning in imparting sustainable leather manufacturing. Green Chem 25(19):7441–7469. https://doi.org/10.1039/D3GC02948D

Hassoun A, Cropotova J, Trollman H, Jagtap S, Garcia-Garcia G, Parra-López C, Nirmal N, Özogul F, Bhat Z, Aït-Kaddour A et al (2023) Use of industry 4.0 technologies to reduce and valorize seafood waste and by-products: a narrative review on current knowledge. Curr Res Food Sci 6:100505. https://doi.org/10.1016/J.CRFS.2023.100505

Herrera K, Morales LF, López JE, Montoya-Ruiz C, Muñoz S, Zapata D, Saldarriaga JF (2023) Biochar production from tannery waste pyrolysis as a circular economy strategy for the removal of emerging compounds in polluted waters. Biomass Convers Biorefin 1:1–14. https://doi.org/10.1007/S13399-023-04261-2/FIGURES/5

Illankoon C, Vithanage SC (2023) Closing the loop in the construction industry: a systematic literature review on the development of circular economy. J Build Eng 76:107362. https://doi.org/10.1016/J.JOBE.2023.107362

Jadhav NC, Jadhav AC (2020) Waste and 3R's in footwear and leather sectors, pp 261–293. https://doi.org/10.1007/978-981-15-6296-9_10

Jaffari ZH, Hong J, Park KY (2024) A systematic review of innovations in tannery solid waste treatment: a viable solution for the circular economy. Sci Total Environ 948:174848. https://doi.org/10.1016/J.SCITOTENV.2024.174848

Kanagaraj JK, Panda RC, Vinodh Kumar MVK (2020) Trends and advancements in sustainable leather processing: future directions and challenges—a review. J Environ Chem Eng 8(5):104379. https://doi.org/10.1016/J.JECE.2020.104379

Karuppiah K, Sankaranarayanan B, Ali SM, Jabbour CJC, Bhalaji RKA (2021) Inhibitors to circular economy practices in the leather industry using an integrated approach: implications for sustainable development goals in emerging economies. Sustain Prod Consum 27:1554–1568. https://doi.org/10.1016/J.SPC.2021.03.015

Luthra S, Sharma M, Kumar A, Joshi S, Collins E, Mangla S (2022) Overcoming barriers to cross-sector collaboration in circular supply chain management: a multi-method approach. Transp Res E Logist Transp Rev 157:102582. https://doi.org/10.1016/J.TRE.2021.102582

LWG. Responsible leather. Leather Working Group. Accessed 19 Dec 2024. https://www.leatherworkinggroup.com/

Maistrenko L, Iungin O, Pikus P, Pokholenko I, Gorbatiuk O, Moshynets O, Okhmat O, Kolesnyk T, Potters G, Mokrousova O (2022) Collagen obtained from leather production waste provides suitable gels for biomedical applications. Polymers 14(21):4749. https://doi.org/10.3390/POLYM14214749

Mengistu A, Angassa K, Tessema I, Andualem G, Yiheyes B, Berhane D, Abewaa M, Kassie M, Telay B (2024) Keratin hydrolysate as a chrome exhaust aid and keratin filler in leather processing: a cleaner technology approach for tannery solid waste management and leather manufacturing. Heliyon 10(13). https://doi.org/10.1016/j.heliyon.2024.e34049

Moktadir MA, Rahman T, Rahman MH, Ali SM, Paul SK (2018) Drivers to sustainable manufacturing practices and circular economy: a perspective of leather industries in Bangladesh. J Clean Prod 174:1366–1380. https://doi.org/10.1016/J.JCLEPRO.2017.11.063

Moktadir MA, Ahmadi HB, Sultana R, Zohra FT, Liou JJH, Rezaei J (2020) Circular economy practices in the leather industry: a practical step towards sustainable development. J Clean Prod 251:119737. https://doi.org/10.1016/J.JCLEPRO.2019.119737

Mondejar ME, Avtar R, Diaz HLB, Dubey RK, Esteban J, Gómez-Morales A, Hallam B, Mbungu NT, Okolo CC, Prasad KA et al (2021) Digitalization to achieve sustainable development goals: steps towards a smart green planet. Sci Total Environ 794:148539. https://doi.org/10.1016/J.SCITOTENV.2021.148539

Morales LF, Herrera K, López JE, Saldarriaga JF (2021) Use of biochar from rice husk pyrolysis: assessment of reactivity in lime pastes. Heliyon 7(11):e08423. https://doi.org/10.1016/J.HELIYON.2021.E08423

Muralidharan V, Palanivel S, Balaraman M (2022) Turning problem into possibility: a comprehensive review on leather solid waste intra-valorization attempts for leather processing. J Clean Prod 367:133021. https://doi.org/10.1016/J.JCLEPRO.2022.133021

Navarro D, Wu J, Lin W, Fullana-i-Palmer P, Puig R (2020) Life cycle assessment and leather production. J Leather Sci Eng 2(1):1–13. https://doi.org/10.1186/S42825-020-00035-Y/FIGURES/8

Omoloso O, Mortimer K, Wise WR, Jraisat L (2021) Sustainability research in the leather industry: a critical review of progress and opportunities for future research. J Clean Prod 285:125441. https://doi.org/10.1016/J.JCLEPRO.2020.125441

Omoloso O, Mortimer K, Wise WR (2024) The influence of sustainability drivers on the implementation of social sustainability practices in the leather industry. Clean Prod Lett 6:100051. https://doi.org/10.1016/J.CLPL.2023.100051

Ostad-Ali-Askari K (2022) Management of risks substances and sustainable development. Appl Water Sci 12(4):1–23. https://doi.org/10.1007/S13201-021-01562-7/FIGURES/10

Patel K, Munir D, Santos RM (2022) Beneficial use of animal hides for abattoir and tannery waste management: a review of unconventional, innovative, and sustainable approaches. Environ Sci Pollut Res 29(2):1807–1823. https://doi.org/10.1007/S11356-021-17101-5/METRICS

Prev. The untold story of animal skin. Prev. Accessed 19 Dec 2024. https://prev.shop/blogs/plant-revolution/the-untold-story-of-animal-skin

Pringle T, Barwood M, Rahimifard S (2016) The challenges in achieving a circular economy within leather recycling. Procedia CIRP 48:544–549. https://doi.org/10.1016/J.PROCIR.2016.04.112

Pučnik R, Dokl M, Fan YV, Vujanović A, Novak Pintarič Z, Aviso KB, Tan RR, Pahor B, Kravanja Z, Čuček L (2024) A waste separation system based on sensor technology and deep learning: a simple approach applied to a case study of plastic packaging waste. J Clean Prod 450:141762. https://doi.org/10.1016/J.JCLEPRO.2024.141762

Qiao DW, Yao J, Song LJ, Yang JY (2021) Migration of leather tannins and chromium in soils under the effect of simulated rain. Chemosphere 284:131413. https://doi.org/10.1016/J.CHEMOSPHERE.2021.131413

Rodríguez F, Montoya-Ruiz C, Estiati I, Saldarriaga JF (2020) Removal of drugs in polluted waters with char obtained by pyrolysis of hair waste from the tannery process. ACS Omega 5(38):24389–24402. https://doi.org/10.1021/ACSOMEGA.0C02768/ASSET/IMAGES/LARGE/AO0C02768_0010.JPEG

Sánchez-García E, Martínez-Falcó J, Marco-Lajara B, Manresa-Marhuenda E (2024) Revolutionizing the circular economy through new technologies: a new era of sustainable progress. Environ Technol Innov 33:103509. https://doi.org/10.1016/J.ETI.2023.103509

Santos LM, Gutterres M (2007) Reusing of a hide waste for leather fatliquoring. J Clean Prod 15(1):12–16. https://doi.org/10.1016/J.JCLEPRO.2006.01.025

Shahzad M, Qu Y, Rehman SU, Zafar AU (2022) Adoption of green innovation technology to accelerate sustainable development among manufacturing industry. J Innov Knowl 7(4):100231. https://doi.org/10.1016/J.JIK.2022.100231

Shen B (2014) Sustainable fashion supply chain: lessons from H&M. Sustainability 6(9):6236–6249. https://doi.org/10.3390/SU6096236

Silva JA (2023) Wastewater treatment and reuse for sustainable water resources management: a systematic literature review. Sustainability (Switzerland) 15(14):10940. https://doi.org/10.3390/SU151410940/S1

Singh BJ, Chakraborty A, Sehgal R (2023) A systematic review of industrial wastewater management: evaluating challenges and enablers. J Environ Manage 348:119230. https://doi.org/10.1016/J.JENVMAN.2023.119230

Soori M, Arezoo B, Dastres R (2023) Internet of things for smart factories in industry 4.0, a review. Internet Things Cyber-Phys Syst 3:192–204. https://doi.org/10.1016/J.IOTCPS.2023.04.006

Takacs F, Brunner D, Frankenberger K (2022) Barriers to a circular economy in small- and medium-sized enterprises and their integration in a sustainable strategic management framework. J Clean Prod 362:132227. https://doi.org/10.1016/J.JCLEPRO.2022.132227

Valeika V, Širvaitytė J, Bridžiuvienė D, Švedienė J (2019) An application of advanced hair-save processes in leather industry as the reason of formation of keratinous waste: few peculiarities of its utilisation. Environ Sci Pollut Res 26(6):6223–6233. https://doi.org/10.1007/S11356-019-04142-0/METRICS

Waite M (2014) The SURF framework applied to the textile industry, pp 209–226. https://doi.org/10.1007/978-981-287-110-7_8

Wrzesińska-Jędrusiak E, Czarnecki M, Kazimierski P, Bandrów P, Szufa S (2023) The circular economy in the management of waste from leather processing. Energies 16(1):564. https://doi.org/10.3390/EN16010564

Yu Y, Lin Y, Zeng Y, Wang YN, Zhang W, Zhou J, Shi B (2021) Life cycle assessment for chrome tanning, chrome-free metal tanning, and metal-free tanning systems. ACS Sustain Chem Eng 9(19):6720–6731. https://doi.org/10.1021/ACSSUSCHEMENG.1C00753/SUPPL_FILE/SC1C00753_SI_001.PDF

Zhang Z, Chen Z, Zhang J, Liu Y, Chen L, Yang M, Osman AI, Farghali M, Liu E, Hassan D et al (2024) Municipal solid waste management challenges in developing regions: a comprehensive review and future perspectives for Asia and Africa. Sci Total Environ 930:172794. https://doi.org/10.1016/J.SCITOTENV.2024.172794

Chapter 12
Confronting Challenges in Technological Development Within Sustainable Leather Manufacturing: Barriers and Routes

Zainab Hassan, Nazia Jamil, and Rida Batool

Abstract The leather industry and its affiliated sectors encompassing numerous tanneries are contributing substantially to Pakistan's economy. These are predominantly found in three major urban centers of Pakistan: Kasur, Karachi, and Sialkot. In recent years, the leather manufacturing sector has come under scrutiny for issues related to animal welfare, environmental consequences, and labor conditions in certain regions. Nevertheless, initiatives are underway to tackle these challenges by enhancing sustainability measures and promoting greater transparency within the supply chain. In this chapter, leather supply chain is analyzed, revealing its complexity and the sustainability challenges it faces, including air and water pollution, waste generation, and reliance on harmful chemicals. This chapter emphasizes the importance of achieving a balance between social, economic, and environmental factors for a sustainable leather industry. Innovations such as bio-based tanning agents, microbial tanning, enzyme based tanning, ultrasound tanning, nanotechnology, masked silica tanning, compact glyoxal tanning and circular economy strategies, including zero liquid discharge systems, aim to minimize the leather industry impact on the environment. Noteworthy case studies show promising advancements in waste management and resource efficiency. However, barriers such as high initial costs, lack of skilled labor, and regulatory complexities hinder technology adoption. A collaborative effort involving governments, industries, and financial institutions is vital for overcoming these challenges, ensuring the leather industry progresses towards sustainability while meeting global demands for leather products.

Z. Hassan · N. Jamil · R. Batool (✉)
Institute of Microbiology and Molecular Genetics, University of the Punjab, Quaid-e-Azam Campus, Lahore, Pakistan
e-mail: rida.mmg@pu.edu.pk

Z. Hassan
e-mail: zainabhassan000111@gmail.com

N. Jamil
e-mail: nazia.mmg@pu.edu.pk

M. A. Aboulhassan and S. Souabi (eds.), *Sustainable Practices in the Tannery Industry*,
https://doi.org/10.1007/978-3-031-92037-0_12

Keywords Life Cycle Assessment (LCA) · Lethal Dose 50 (LD_{50}) · Gross Domestic Product (GDP) · Polyvinyl alcohol/polyethylene glycol (PVA/PEG) · International Council of Tanners (ICT) · Circular economy (CE) · Advanced oxidation processes (AOPs) · Zero liquid discharge (ZLD) · International Labour Organization (ILO)

12.1 Introduction

Leather and Leather merchandise are one of the most traded goods worldwide, with an annual international trade value of more than 80 billion US Dollars (Omoloso et al. 2021). China is thought to be the world's largest producer (18%), importer (63%) and exporter (43%) of leather. Other than China, the top producing countries include Italy (10%), Brazil (6%), Korea (7%), India (7%), Russia (6%), and Pakistan (2.5%). Footwear accounts for more than half of the world's market share and is the largest produced product made from leather (Memedovic and Mattila 2008), while leather constitutes the biggest single product produced from leather. Leather industry is the second largest contributor in terms of export earnings in Pakistan's economy. Besides employment of over 500,000 people (Siddiqui et al. 2016), it contributes to 5% of the economy in Gross Domestic Product (GDP) (Ghafoor et al. 2012). Shahab (2013) stated that it earned about 700 million US dollars for exports from Pakistan, while in 2016; this figure had increased to US dollars 724 million (Siddiqui et al. 2016). The leather industry operates extensively across Pakistan, with more than 2500 tanneries, both registered and unregistered. Key cities hosting these industries include Karachi, Sialkot, Kasur, Gujranwala, Korangi, Multan, Faisalabad, Lahore, Sahiwal, Peshawar, and Hyderabad. The country is home to 784 leather production facilities, 461 manufacturing units for leather garments, 348 glove production units, and over 524 footwear manufacturing establishments (Khan et al. 2024).

Hides of cattle, sheep, pigs, goats, and other slaughtered animals are used to produce leather, an intermediate "industrial" product (Jaegler 2016; Joseph and Nithya 2009). This is so, because leather is a renewable by-product that comes from slaughtering animals. The leather industry uses a waste (hides and skin) recycling from the meat and dairy industries, all within agriculture, rather than the popular misconception that says that animals are raised mainly for leather use (Dixit et al. 2015). Leather is therefore a material that is sustainable in and of itself. The leather supply chain is also globalized, with its members including breeders, slaughterhouses, tanneries, manufacturers, and suppliers of chemicals and machinery; it may also include retailers, service providers like logistics firms, marketing firms, information technology companies, and end users (Jaegler 2016). A basic model of the leather supply chain appears as shown in Fig. 12.1.

As per MVO Nederland (2013), the leather supply chain seemed to be delineated in four essential stages as shown in Fig. 12.1: upstream, downstream, market actors, and focal company. The first stage includes acquiring hides and skins; second stage: tanning and turning hides and skins into leather, the third stage includes making

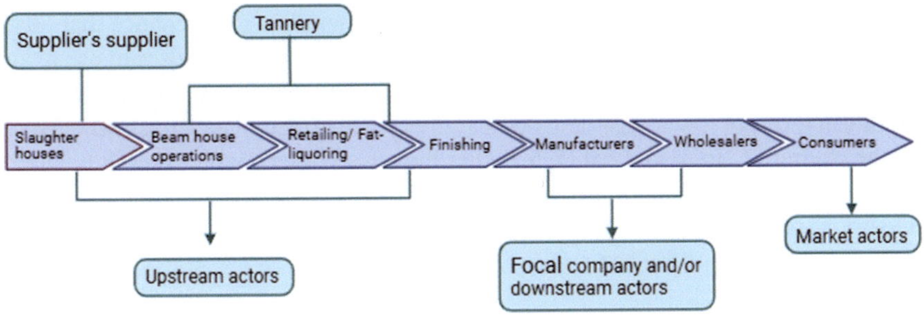

Fig. 12.1 Leather supply chain

different leather products and the last stage accounts for the consumers. The chain begins at the slaughterhouses, where after the animals are killed, hides and skins are gathered and salted. Reducing landfill dumping through the collection of wastes like hides and skin, which are the primary raw materials used in the production of leather, strengthens the industry's status as a "circular economy industry" that supports sustainable development. Once the salted or unsalted hides and skins arrive from slaughterhouses, they are directly sent to tanneries where they undergo several subsequent processing phases involving beam house operations and finishing operations to become leather (Daniels and Landmann 2010). Depending on how much is being allotted towards technology/innovation and what the use of the leather is, these processes can be either labor-intensive or capital-intensive along with requiring large amounts of chemicals, energy, and water. While the industry is based on a circular economy model, the process typically converts hides and skins into leather, and produces solid, liquid, and gaseous wastes likely to have harmful effects on the environment, society, and economy. Besides, the sustainability issue becomes very significant just because of the high complexity of the leather supply chain and the volumes of resources (both human and environmental) used in production processes (Moktadir et al. 2018).

12.2 Sustainability of the Leather Industry

The ability of the leather industry to function in a way that satisfies current demands without threatening the ability of future generations to fulfil their own needs is referred to as sustainability. Or, to put it another way, it refers to the economically, socially, and environmentally sound methods of producing leather goods. Most of the works on sustainability in the leather industry so far have focused on the green chemistry (Krishnamoorthy et al. 2012), pollution control, waste management (Kanagaraj et al. 2015), etc. All of these can be considered environmental sustainability practices (Glavic and Lukman 2007). The perception of sustainability as environmental sustainability may not always mean ignoring the other two dimensions of sustainability, which are social and economic, but there is a good chance that other relevant issues will go unaddressed. For instance, if a company chooses to use fewer chemicals in tannery operations, this could result in both economic, health and safety benefits in terms of lower costs and decreased exposure of employee to chemicals. As per the saying of Hu et al. (2011), a leather industry is sustainable when it balances triple aspects: social, economic, and environmental. Thus, understanding the leather industry's triple bottom line towards sustainability is very important.

12.2.1 Environmental Sustainability in the Leather Industry

Environmental sustainability has been known as the "planet" part of sustainability (Elkington and Rowlands 1999). This sustained environmental well-being is termed "maintenance of natural capital" in a sustaining environment that keeps a balance between source and sink services provided by environmental resources, Andersson and Mackenzie (2004) and Goodland (1995). The constraints of the biophysical environment must be learned by humanity. It implies that natural capital has to be preserved in its dual roles as a "sink" for waste and a source of inputs. Wolf et al. (2013) evaluated environmental sustainability in the leather industry, by utilizing a "sustainability footprint approach" to illustrate the potential environmental impacts that the leather supply chain may have, as seen in Fig. 12.2. These footprints indeed lead to a rich discussion among leather experts and leave well-recorded problems in the leather industry. Most of the earlier studies on sustainability in the leather industry reportedly refer to the green publications on the subject (Chen et al. 2014; Marconi et al. 2017). Indeed, established indicators are drawn to assess the environmental credentials for sustainability of leather-related businesses. This is also reflected in sustainability reports of various leather companies as well as in the environmental stewardship tool of the Leather Working Group. Major environmental indicators documented in sustainability reports from leather-related businesses include greenhouse gas emission reduction, energy efficiency, waste management, effective water management, and identification with Life Cycle Assessments (LCA) formulated by Omoloso et al. (2020). Since these dimensions are interrelated, it follows that environmental initiatives can produce a number of effects on social and economic dimensions and vice versa.

12.2.2 Social Sustainability in the Leather Industry

Consequently, without proper safety measures, hazardous materials that may be released during leather production may result in pollution in the immediate areas surrounding leather processing facilities and also adversely affect the health of employees and local residents (ICT 2018). Social sustainability is the point at which a community can meet the basic needs of its members while also adhering to ideals

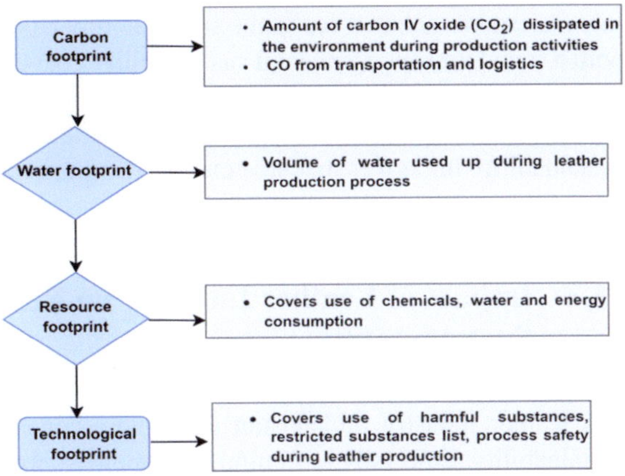

Fig. 12.2 Environmental impact of the leather industry (Wolf et al. 2013)

of equity, homogeneity, social justice, cohesion, diversity, sense of place, integration, social security, and social amenity for the current and future generations (Rasouli and Kumarasuriyar 2016).

As per literature of business management, social equity is about the fair and just treatment of staff members (Åhman 2013). Besides environmental sustainability, which really talks about managing natural resources, social sustainability is about managing social resources, such as those of people's competencies and abilities, as well as social values, institutions, and relationships (Sarkis et al. 2010). The leather industry is labor-intensive (ILO 2018) and thus requires focusing on the social dimension of its workforce. Key constituents of social sustainability will include product safety regulations; occupational health and safety standards and compliance with employment laws; together with resource use and practices considered animal welfare (ICT 2018).

12.2.3 Economic Sustainability in the Leather Industry

This aspect is recognized mostly under the dimension of sustainability as "profit". Economic sustainability is extended to future generations having access to a similar level and quality of capital that is available to current generations (Atkinson et al. 2007). In a business setting, economic sustainability is the process of examining how an organization attempts to augment the value it creates and delivers, without incurring costs that relate to the costs of its supply chain operations (Closs et al. 2011). In the leather industry, ICT (2018) pointed out a few characteristics of the economic sustainability; these include an adherence to fair trade; tracing hides and skins back to the slaughtering place; accessibility of the origin of leather production; and following proper labeling of leather and leather products whereby ICT definitions international standards and legal obligations will have been established. Therefore, organizations are encouraged to maximize profit while optimizing every possible resource and raw material to achieve economic sustainability (Singh and Gupta 2013). Practices with a social and environmental sustainability goal can be said to result in enhanced economic sustainability for the company or industry or supply chain, and vice versa. An organization that is financially successful and stable is more likely to develop sustainability initiatives that are concerned with social and environmental issues.

12.3 Impact of Leather Industry on Environmental Sustainability

The leather industry of Pakistan encounters considerable obstacles in achieving sustainability, especially in minimizing its environmental footprint, which includes issues related to water usage, chemical application, and waste production.

12.3.1 Air Pollution

The tanning process releases emissions which are mostly organic solvents employed during the final finishing stage of leather treatment and mainly while coating. Many of these solvents are hazardous and detrimental; therefore, reduction of these pollutants, is indispensable (Chiampo et al. 2023). During leather processing, the main air pollutants produced and released into the environment are chlorine (Cl_2), ammonia (NH_3), sulfur dioxide (SO_2), hydrogen sulfide (H_2S), carbon dioxide (CO_2), formic acid fumes, and volatile organic compounds. H_2S, NH_3, and Cl_2 are produced during liming, de-liming, and pickling stages of leather finishing.

12.3.2 Solid Wastes

In the leather processing industry, only 150 kg of finished leather is produced from every 1000 kg of raw hide, resulting in 850 kg being classified as solid waste in various forms. Among them are pre-fleshing wastes, lime-fleshing wastes, buffing wastes, shaving wastes and trimmings in leather (Ramamoorthy et al. 1989). Solid wastes contain various pollutants such as dichloromethane, sodium chloride, nitrogen, calcium, sulfides, ferric compounds and chromium and cadmium (Ozgunay et al. 2007).

12.3.3 Water Pollution

The leather industry is known for its significant water consumption, requiring as much as 4000 L of water for every ton of tanned leather produced. Additionally, the tanning process involves the use of numerous chemicals, which often enter local waterways such as streams, ponds, and rivers without proper treatment, leading to groundwater contamination. A single tannery has the potential to contaminate groundwater within a radius of approximately 7–8 km, resulting in severe consequences for aquatic ecosystems, avian species, wildlife, and human populations. Contaminated groundwater contains elements such as sodium, magnesium, calcium, sulfate ions, chloride, and chromium in its hexavalent form (Cr_6). Such components lead to hardness of water-causing health problems like kidney stones, gastrointestinal tract problems, and cardiovascular diseases (Hashmi et al. 2017).

12.3.4 Tanning Chemicals

Approximately 170 different chemical substances are utilized in the leather tanning process. These include sodium chloride, ammonium, sodium sulfate, fats, sulfuric acid, lime, non-ionic wetting agents, chromium sulfate, bactericides, calcium oxide, ammonium sulfate, soda ash, ammonium chloride, sodium bisulfate, sodium chlorite, formic acid, sodium bicarbonate, syntans, sodium hypochlorite, polyurethane, vegetable tannins, formaldehydes, fat emulsions, and various dyes (Chowdhury et al. 1996; Kankaria et al. 2011).

12.3.5 Energy Consumption

Production of leather consists of processes which are highly energy intensive and require a lot of energy resources. The non-renewable sources of energy such as fossil fuels utilized in the process contribute towards greenhouse gas emissions and increase the carbon footprint of the leather industry.

These challenges highlight the necessity for technological innovations in the leather manufacturing process (Thanikaivelan et al. 2005).

12.4 Technological Innovations for Sustainable Leather Production

A significant number of tanneries rely on chromium-based tanning agents, as they are highly effective in producing resilient leather that can be adapted for different applications. Unfortunately, chromium is extremely toxic to humans and can severely impact the environment if it is improperly disposed off in effluent. The chromium used in tanning is non-renewable, and Cr(III) atoms found in waste sludge have the potential to convert into Cr(IV), a substance that is both carcinogenic and mutagenic (Yu et al. 2021; China et al. 2020). Alternatives to chrome tanning include metal-free tanning, commonly referred to as vegetable tanning, as well as tanning processes that utilize different metals.

In preventing and addressing both social and environmental impacts, two important strategies are prevention through downstream processes-involvement in changing and improving tanning operations (prevention approach) and treatment through end-of-pipe processes dealing with the environmental challenge resulting from the tanning supply chain (treatment approach). Sustainable tanning practices are supported by advances in various technologies, including.

12.4.1 Bio-Based Tanning Agents

The technologies for bio-based tanning consist of vegetable tanning, enzyme-based tanning, and microbial tanning processes. Tannins derived from vegetables, plants, animals, or microorganisms are referred to as bio-tannins. These compounds are likely among the earliest reagents utilized in the tanning process of skin or hide, ultimately transforming them into the valuable end product known as leather (Mahdi et al. 2006).

12.4.1.1 Vegetable Tanning

Vegetable tanning, commonly known as bark tanning, is a traditional technique that utilizes plant-based materials to transform animal hides and skins into a durable, water-resistant, and non-decomposing material that is soft, flexible, and heat-resistant (Wemegah 2014). The process of vegetable tanning involves the use of tannin-rich leaves, roots, and barks to treat animal hides and skins. It is also considered "green tanning agent" due to its greener and more biodegradable nature compared with the usual conventional inorganic tanning process (Kuria et al. 2016). The attributes of vegetable tanned leather includes its high fullness, moldability, air permeability, and durability, highlights its role in the mitigation of chrome pollution in the leather industry (Wemegah 2014). In the vegetable tanning process, pickling preparation is no longer necessary and less sulfate salt pollution is generated. Furthermore, vegetable tannins provide a wide range of advantages, such as the production of bio-degradable-resistant leather which consequently causes a slower degradation of waste. By this method, a soft, lightweight leather with a natural and comfortable touch is obtained. With the passage of time, the durability of leather is maintained and the leather is sustainable and environmentally responsible, with recycling potential. All leather goods tanned with vegetable tanning are unique, with deep, rich colors that look absolutely natural, which can often be superior to chrome-tanned leather in quality. Numerous companies supply industrially produced tannin extracts from various plants, including *Schinopsis lorentzii* (Quebracho), *Uncaria gambir* (Gambier), *Acacia mearnsii* (Mimosa), and *Caesalpinia spinosa* (Tara). The tannin concentration of these extracts is 164.3 g/kg for Quebracho, 108.2 g/kg for Mimosa, 169.3 g/kg for Gambier and exciting 647.5 g/kg for Tara. These values are derived by photometric techniques and HPLC-ESI-MS analysis (Schweigmann et al. 2013).

12.4.1.2 Enzyme-Based Tanning

Leather processing using conventional chemical techniques contains several toxic chemicals, subsequently generate and release multiple pollutants. Such an activity results in significant rises in biological oxygen demand (BOD), chemical oxygen demand (COD), total dissolved solids (TDS), and total suspended solids (TSS) in the receiving water bodies (Kanagaraj et al. 2020). The technique of enzyme tanning is a novel approach that uses enzymes, which are naturally occurring biological catalysts, to hydrolyze protein content in animal skins and hides. This technique relies less on severe chemicals and thus, becomes an environmentally friendly choice compared to old-fashioned chrome tanning techniques. Extensive initiatives are being undertaken to integrate enzymes into numerous leather production processes. Leather technologists consider enzymes indispensable for the bating process (a process in which enzymes are used to soften or remove undesirable protein from the surface of leather), as enzymatic bating has become the standard practice. The major use of enzymes in the leather production industry is attributed to the two main sections: (i) that collagen—the protein which dominates the meat, as the raw material being processed—is or by nature readily attacked by enzymes. In spite of great modification in the process to make collagen resistant to thermal and bacteriological actions, it can still be recognized as a protein liable to some special enzymes under a controlled set of conditions. (ii) The exponentially growing industry of enzymes manufactures a varied range of products, thus pointing to its universality (Biškauskaitė et al. 2021). In the preliminary stage of skin treatment, enzymes are utilized for the soaking of skins and hides. Different types of enzymes, such as proteases, lipases, amylases, chondroitinases, amidases, and phospholipases, can be applied as soaking agents. The advantages of employing enzymes in this process are quicker rehydration and the efficient elimination of interfibrillary materials (Covington 2006).

12.4.1.3 Microbial Tanning

A highly varied collection of microorganisms has been identified on animal skins and hides, as well as in animal feces, preservation salt, dust, barns, water, air, soil, and feed associated with salted hides and skins. The rich organic and inorganic materials present in these salted hides and skins have facilitated the growth and metabolic activities of these microorganisms. *Bacillus brevis, Bacillus cereus, Bacillus licheniformis, Bacillus megaterium, Bacillus pumilus, Bacillus sphaericus, Bacillus subtilis, Kurthia variabilis, Micrococcus candidus, Micrococcus luteus, Micrococcus roseus, Micrococcus rubens, Staphylococcus aureus*, and *Staphylococcus epidermidis.* These microorganisms produce enzymes such as lipase, protease, caseinase, β-galactosidase, DNase, amylase, cellulase, lecithinase, and urease (Birbir and Caglayan 2018). Proteases are among the diverse range of enzymes utilized in bioprocessing, demonstrating considerable potential in multiple stages of leather processing. Their application offers notable advantages that enhance the quality of the final product throughout the biocatalytic process. Currently, investigations are underway regarding the use of proteases, both alone and in various combinations, which lead to diminished levels of usable chemicals in leather processing, particularly in degreasing and soaking operations. The predominant portion of proteases available on the market is mainly derived from microbial sources (de Castro Bizerra et al. 2024). The bio-catalytic leather processing industry shows significant interest in a blend of enzymes, including keratinases, lipases, and proteases, as they effectively remove globular proteins and hydrolyze various fats, oils, and greases, while also breaking down keratin in hair epidermis through the cleavage of disulfide bonds. Microbial keratinases are recognized as a valuable asset in modern eco-friendly cleaning technologies, particularly for softening hides to improve their malleability prior to the tanning process. Moreover, the application of keratinase to the leather industry is expected to eradicate about 40–50% of the pollution, primarily by facilitating fiber opening (reliming) and dehairing (liming) processes, while minimizing the reliance on harmful chemicals like sodium sulfide and lime (Srivastava et al. 2020).

12.4.2 Reduced Chemical Usage

Researchers have developed various technologies aimed at reducing chemical usage in conventional tanning processes. This approach is a proactive measure towards achieving environmental sustainability.

12.4.2.1 Ultrasound Tanning

Ultrasound-assisted tanning represents a novel and environmentally sustainable approach within the leather industry, focusing on minimizing ecological harm by significantly decreasing the reliance on chemicals and water throughout the tanning process. The application of ultrasound facilitates the absorption of tanning agents like chromium, natural tannins, and various alternative chemicals, resulting in a reduced requirement for these substances in the tanning process. The process of tanning hides and skins traditionally makes large use of chromium salts, hence leaving very high concentrations of chromium in waste water. Evidence suggests that ultrasound can decrease chromium consumption by up to 30%, addressing a critical environmental challenge linked to traditional tanning practices (Sivakumar et al. 2008).

Although chromium is frequently observed as an optimal tanning agent due to its leather performance, it carries certain negative characteristics that create significant challenges for its prolonged use. Therefore, one of the primary objectives in current leather research has been to decrease dependence on chromium as the main tanning agent. An effective substitute must be naturally abundant, easily accessible, cost-efficient, environmentally friendly, and capable of delivering competitive results in tanned leather. It should also be easy to handle and safe to use. Among the various metal salts known for their tanning efficacy, titanium is particularly advantageous as it is abundant in nature, readily accessible, and non-toxic. Titanium-tanned leather surpasses (when compared to) other metal tanning agents; thus, titanium tanning represents a promising alternative to chromium in the modern leather industry. The application of power ultrasound in titanium tanning seeks to improve the infiltration of titanium complexes into hides while also boosting the tanning effectiveness of titanium compounds (Peng et al. 2007).

12.4.2.2 Nanotechnology

Nanotechnology represents an emerging area within the leather industry, providing cutting-edge solutions aimed at minimizing the reliance on harmful chemicals while simultaneously enhancing the efficiency and quality of the tanning process. The process of tanning through nanotechnology incorporates the use of nanoparticles, nanocomposites, and nanoemulsions. The incorporation of nanoparticles and nanoscale materials enables tanneries to apply improvements in the process of leather making, in order to achieve a more sustainable and eco-compatible technique.

Because of the nanoscale size of nanoparticles, they are efficiently able to reach the leather surface. This allows them to occupy the interfibrillary cavities, thus enhancing the mechanical performance of the leather in a general way. Nanotechnology is used in the tanning process and it has a big impact on contributing to improvement of the physical and mechanical properties of leather. Nano-SiO_2 and its derivatives-treated leather has superior physiochemical characteristics in comparison with leather tanned in a conventional way. The new major advances include higher hydrothermal stability, shrinkage temperature, and better quality of smooth grains and fluffiness. On the other hand, the application of nanotechnology has been much enhanced in recent years with respect to biodegradability, wet heat resistance, antifungal properties and penetration of tanning agents. The incorporation of nano silica combined with the polymeric compounds further improved the leather properties, such as thickness increase to a whopping 190% when one uses PVA(PEG)-SiO_2-GO as a tanning agent. The high surface area of the nanomaterials also makes them good adsorbent materials, thus enabling effective removal of toxic metal ions, pathogenic microorganisms, and different organic and inorganic impurities from water (Sahu et al. 2024).

12.4.2.3 Masked Silica Tanning

In pursuit of enhancing occupational and environmental safety for workers, the leather industry is dedicated to developing eco-friendly tanning processes that minimize or eliminate toxicity, in accordance with Sustainable Development Goals. Several trials and investigations have been carried out for developing chromium-free leathers. In this exploration, silicates are emerging as significant candidates for skin and hide tanning agents, thanks to their precisely engineered surface functionalization. Silicates have been utilized for collagen stabilization since 1914. While the principles of silica tanning are well understood, uncertainties surrounding product preparation and processing conditions often lead to excessive in situ polymerization of silica. This incident affects the mechanical strength and performance of the leather, making it unviable as a pure tanning agent. Since silica is an alternative for chrome-free tanning, several combination tanning systems using different types of silicates have been developed, benefiting from their controlled diffusion, collagen stabilization, excellent biocompatibility, and enhanced mechanical performance of leather, among others (Srinivasan et al. 2024).

12.4.2.4 Compact Glyoxal Tanning

Compact glyoxal tanning is an emerging and eco-friendly technique in leather manufacturing, created as a substitute for traditional tanning methods, including chrome tanning. In this process, glyoxal, a small aldehyde, is the main tanning agent. By utilizing this method decreased consumption of chemicals and high environmental sustainability are achieved. Leather pretreated with glyoxal has excellent physical and chemical features e.g., toughness, long-term durability, and good fade resistance. With an LD_{50} value of 7070 mg/kg (oral-rat), glyoxal is considered to have low toxicity. According to the Hodge and Sterner scale, an LD_{50} of more than 5000 mg/kg indicates that a substance is practically non-toxic (Hodge and Sterner 1949). Glyoxal and glutaraldehyde are both extremely reactive but glyoxal, being the simplest dialdehyde with the two-carbon chain, lends itself easily to fit into the smaller pores of three-dimensional matrices for protein crosslinking. In addition, glyoxal may exist as dimer, trimer, and oligomer at high concentrations in aqueous solutions, which may also stabilize proteins through hydrogen-bonded networks. Its lower cost relative to glutaraldehyde also positions glyoxal as a promising alternative (Ramesh et al. 2022).

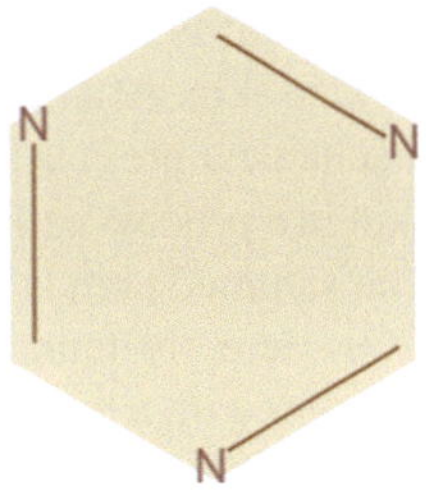

Fig. 12.3 s-Triazine compound

12.4.2.5 Triazine Derivative Tanning

Triazine is a heterocyclic compound notable for its six-membered ring structure, which contains three nitrogen atoms, contributing to its nitrogen-rich nature (Fig. 12.3). The tanning process utilizing triazine derivatives is a modern, chrome-free alternative that employs triazine-based compounds instead of conventional chromium salts. This innovative method provides an environmentally sustainable solution for leather processing by decreasing the reliance on toxic chemicals while still producing high-quality leather. Compounds like cyanuric chloride and cyanuric acid are particularly noteworthy for their effectiveness in cross-linking collagen fibers in hides and skins, resulting in leather that exhibits excellent strength and durability. The amino groups of collagen in hides interact with triazine derivatives to form covalent bonds, leading to a stable cross-linked network that significantly increases the leather's resistance to both thermal and mechanical stresses. The triazine ring features several reactive sites, enabling effective interaction with collagen fibers. The reactivity of triazine derivatives is affected by the pH level of the tanning solution. By optimizing the pH conditions, one can achieve maximum cross-linking and efficient tanning while minimizing chemical waste. Although triazine derivatives are generally less toxic compared to chromium salts, they tend to be more costly, especially for smaller tanneries. Nevertheless, the potential savings in effluent treatment and regulatory compliance could balance out these higher costs (Facchin et al. 2024).

12.4.2.6 Fast High Pressure Tanning

High-pressure tanning, characterized by its speed, is a developing approach in leather processing that effectively lowers the time, chemical inputs, and water usage necessary for transforming raw animal hides into finished leather goods. It is a specialized method that incorporates pressure in the tanning process, which aids in the better absorption of tanning agents such as synthetic tanning agents, vegetable tannins or chrome-free alternatives within the collagen structure of animal hides.

The CO_2-intensified tanning method integrates the beneficial features of traditional chromium tanning, such as superior performance and quality, while significantly reducing process time, tanning agents, and water usage, particularly in terms of wastewater. The process of CO_2-augmented tanning begins with a pretreatment phase that is identical to that used in traditional tanning, transitioning raw skin to pickled skin. After the pickling stage, the skin is extracted from pickle solution, and a mechanical press is used to clear weight to 15–30 wt% from the initial weight of the skin in a process called as samming, which means pressing out the aqueous solution. The sammed skin is then placed in a pressure-resistant drum where it interacts with CO_2. Here, the tanning component is prepared by dissolving basic chromium sulfate together with a basifying agent in the sammed pickle solution, accounting for almost 15–30 wt% of original skin weight. The basification (process of making a substance more basic) being made results into formation of oligo-nuclear complexes facilitating penetrability. The pump then pushes the basic solution from under pressure into the barrel so that the sammed skin can take the material into it. At the end of an hour, CO_2 starts to escape from the drum slowly. The wet blue leather is taken off. To get a comparable quality of leather with that manufactured using traditional tanning methods requires a minimum duration of 2.5 h. This method ensures that the tanning solution is completely absorbed, resulting in no chromium-contaminated wastewater and a reduction of approximately 40% in the use of tanning agents (Prokein et al. 2020).

12.4.3 Waste Reduction and Recycling

The leather manufacturing industry is responsible for generating large quantities of waste, which includes solid waste, wastewater, and emissions into the atmosphere. However, recent technological innovations and creative practices are helping tanneries to lessen their ecological impact through effective waste management and recycling efforts. Circular Economy, or CE, is one of several approaches for business. An approach for business, just like CE, promotes restorative or regenerative use of materials so that utility and value can be further maximized (Gusmerotti et al. 2019; Rajput and Singh 2019). This serves to decrease material consumption and waste generation through which value is derived from products. CE functions like a closed-loop value chain since wastes are consumed again for recycling into further processes. Value chains exist in two types: linear and closed-loop (Pomponi and Moncaster 2017).

In a traditional linear value chain, the process begins with the collection of raw materials, which are then transformed into products and ultimately distributed for sale. This manufacturing process generates significant waste, compounded by additional waste from packaging, logistics, and the disposal of products by consumers. Typically, this waste is overlooked and not subjected to reuse or recycling. In contrast, a CE focuses on waste reduction by promoting the recycling and reuse of materials. Instead of discarding waste, it is collected and returned to manufacturing facilities for re-manufacturing, aiming to extend the utility of resources for as long as possible. The closed loop supply chain embodies a circular framework for supply chain management. This approach is fundamentally aligned with the CE concept. Recycling and reusing products are critical for closing the loop. Strong coordination and governance are essential for strengthening the circular chain. It integrates both forward and backward supply chains to maximize an organization's profitability. Furthermore, the blue economy philosophy focuses on sustainable economic development by emphasizing local environmental and ecological attributes (Moktadir et al. 2020).

12.4.3.1 Zero Liquid Discharge System (ZLD)

The tanning industry is marked by its extensive water consumption, utilizing around 30–40 m^2 for every 1000 kg of hide or skin processed. This water subsequently becomes wastewater, which is laden with about 300 kg of diverse chemicals, primarily refractory organic compounds. This type of wastewater is identified by high chemical oxygen demand (COD), a total dissolved solids (TDS) level of more than 7000 mg/L, significant chromium concentrations, and the emission of toxic gases such as ammonia and sulfides (Ricky et al. 2022). Typically, end-of-pipe treatment alone does not adequately address environmental concerns. To promote cleaner production, it is essential to redesign current effluent treatment processes by incorporating alternative or supplementary techniques that facilitate resource recovery without the use of additional chemicals. This approach is anticipated to provide a sustainable solution for managing toxic effluents. The ZLD system is designed to achieve zero water emissions while also offering treatment facilities that recycle, recover, and reuse treated wastewater through advanced purification technologies. A pre-treatment arrangement of the conventional physicochemical processes constitutes the Zero Liquid Discharge system, appointing a tertiary treatment, effluent softening, reverse osmosis (RO) desalination, and thermal evaporation of the saline wastes generated from the RO treatment to achieve salt recovery. This approach yields extremely low water consumption. It is also one of the ways that ZLD could help in disaster prevention, especially in areas where the tannery industry significantly contributes to the economy (Ricky et al. 2022).

12.4.3.2 Solid Waste Recycling

The management of solid waste is a critical environmental challenge globally, with Pakistan facing significant issues in this area. The leather industry currently employs various waste disposal methods, which often present challenges such as high costs and lengthy processes. While researchers have explored alternative eco-friendly disposal methods, including phytoremediation and microbial remediation, these approaches also come with their own limitations. Vermitechnology is developed as a desirable green remediation technology for treating leather industry waste. Recent studies, particularly by Xiang et al., (2015), have highlighted a notable increase in earthworm research from 2000 to 2015, particularly in developed nations. All of this highlights the need for vermistabilization to control the organic solid waste resulting from the leather industry and the importance of earthworms in perceiving and reabsorbing or utilizing the nutrients in these wastes. Earthworms demonstrate a remarkable ability to thrive in soils contaminated with metals. Species of earthworms that exhibit resistance to heavy metals could be crucial in the bioremediation of waste generated by the leather industry. Many investigators found that there is significant value to be placed on small-scale vermicomposting, and that it plays an important role in solid waste management of the leather industry. However, so far, the realization of large-scale vermiremediation has eluded us. The process of vermiremediation applied to leather industry wastewater sludge leads to a significant decrease in heavy metal concentrations (Yuvaraj et al. 2020).

There is a growing recognition of organic waste as a valuable resource that should be redirected away from landfills. In the context of a Circular Economy focused on sustainable waste management, technology plays a crucial role in creating processes for the recovery of organic materials. Various strategies are being investigated and applied in different sectors, including: agricultural operations utilizing gasification, which combines the management of mixed agricultural waste with anaerobic digestion to optimize energy and material recovery, as well as the production of biogas from municipal solid waste. Existing research on the application of leather solid waste in building materials primarily concentrates on the development of panels or blocks that offer improved thermal or acoustic properties. The lightweight nature and irregular form of this waste make it particularly suitable for these uses. Among the most common adhesives that find application in producing these panels or blocks are mostly recycling and upcycling agents that have a direct link with the sector-from cement, gypsum to clay. Examples include utilizing waste gypsum and extruded polystyrene combined with animal feather-fiber reinforced cement mortars (Vidaurre-Arbizu et al. 2021).

12.4.3.3 Effluent Treatment

The treatment of tannery effluent typically consists of multiple stages, which can be classified into physical, chemical, and biological processes (Yusif et al. 2016).

Physical Treatment

Physical treatment processes for tannery effluents essentially do not involve a chemical agent or biological substances. This method involves the treatment of screened effluent to remove the larger particles effectively. Sedimentation can also be used, in which the solids settle to the bottom as a result of gravity, and then the water is drained. The main physical treatment methods using tannery effluent are mechanical treatment, post-treatment/purification, and sludge/waste treatment. Mechanical treatment is normally the first stage of treatment of raw effluent, including the preliminary screening to remove large constituents. Screens properly designed can remove 30–40% of the total suspended solids in the raw waste stream as solids are removed from it. This treatment stage might also include skimming to remove lipids, grease and oils, as well as gravimetric settling. After the mechanical process, sedimentation and sludge handling are the end steps in the treatment of effluent from the aqueous phase by gravity settling. Fractionalization of the solids content by filter presses (for dewatering) allows to obtain a sludge cake as high as 40% dry solids, whereas belt presses offer a sludge cake as high as 20–25% dry solids (Yusif et al. 2016).

Chemical Treatment

Chemical treatment of tannery effluent is a specialized treatment that is developed in order to remove waste products of toxic compounds in the wastewater that is generated in tanneries, enabling to recycle or discharge tannery effluent into natural water systems. This procedure uses a multitude of chemicals at different points of filtration to remove solids, destroy pathogenic microorganism, and remove toxic substance. Aluminous, ozone, peroxide, and lime are crucial chemicals in the treatment of tannery waste discharge. Neutralization of acidic water is one of the important treatments for chemical wastewater. Among the novel perspective in this area is chemical oxidation technology, employing a mixture of chemical oxidants, H_2O_2, O_3, ClO_2, $KMnO_4$, and K_2FeO_4 to degrade contaminants into less detrimental substances or to change them into an easier form to manage. Nevertheless, oxidative agents such as ozone and hydrogen peroxide have been widely employed in these technologies and generally achieves lower degradation rates. With the onset of advanced oxidation processes (AOPs), which can be harnessed by the high reactivity of hydroxyl radicals to promote oxidation, AOPs are positioned as a promising strategy to address the complex organic pollutants in tannery effluent. AOPs encompass a number of innovations, such as Fenton, wet oxidation, photo-Fenton, ozonation, and photocatalysis, with the key difference being the origin of the radicals. Notably, the solar Fenton process is recognized as one of the most effective techniques for managing tannery wastewater. Coagulation and flocculation technology involves the application of chemicals to aggregate suspended particles, facilitating their removal. An extensive advanced chemical treatment technology has been conceived for the management of tannery effluent, drawing from real-world experiences. In this situation, electrochemical treatment has emerged as a novel technology for tannery effluent treatment. Various researchers, such as Costa et al. (2008), Kongjao et al. (2008), and Sundarapandiyan et al. (2010) have studied the electrochemical treatment of tannery effluent to determine the optimal effectiveness of traditional coagulation and flocculation methods. Electrocoagulation (EC) is another modern technology, that effectively incorporates all the knowledge and advantages of conventional coagulation, flocculation, and electrochemistry in the effective treatment of tannery effluents.

Biological Treatment

The primary categories of living processes consists of aerobic, anaerobic, and combined biological methods, all of which can be effectively utilized during different stages of tannery effluent treatment. Aerobic treatment involves the utilization of microorganisms that thrive in oxygen-rich environments to break down organic materials. Bacteria and fungi have the potential to produce oxygenases and peroxidases, which can facilitate the oxidation of pollutants while also allowing for the capture of energy, nutrient elements and carbon released during this process. Anaerobic treatment of tannery effluent is a continuing process in which such microorganisms, metabolize biodegradable matter in the absence of oxygen. While anaerobic degradation is generally slower and less efficient than aerobic methods, it effectively reduces chemical oxygen demand (COD) and biological oxygen demand (BOD) in the effluent while generating renewable energy. Moreover, anaerobic processes can efficiently and economically handle effluents with high concentrations of easily degradable organic materials, such as those from the slaughter, food, sugar, paper, and tanning industries (Yusif et al. 2016).

12.4.3.4 Waste to Valuable Products

The composition of hair is predominantly keratin, a fibrous protein, supplemented by the cystine (7–12%) an amino acid, which enhances the structural integrity of keratin. It is vital to examine strategies for handling and reusing hair waste from the tanning sector to lessen its ecological impact and promote a circular economy. Hair collected from tanneries has been reused for several uses, which includes the production of fertilizers, facilitating the release of chrome byproducts in tanning processes, serving as biomaterials, and producing biochar. If hair waste is considered a meaningful resource, it is feasible to develop new products that have the potential to be used in the same or other industries, thus building a more sustainable system. Biochar is one of the options that is able to be exploited for more than one application, e.g., the treatment of polluted waters, soil and air. Due to its high surface area, this pyrolysis product is an outstanding adsorbent of a wide range of chemicals present in both water and soil grids (Herrera et al. 2023).

12.4.4 Digitalization and Automation

Technology is causing a major change in the leather industry. Technological advances, such as 3D printing, digital design, and automation, are changing the face of manufacturing processes and this has led to higher efficiency and reduced production costs for the manufacturers. In addition, the advent of highly advanced technologies like machine learning, Internet of Things (IoT), and artificial intelligence paved the way for the development of increasingly advanced and effective leather craftsmanship methodologies, allowing its entities to produce top-quality leather goods at an accelerated pace.

12.4.4.1 IoT

A strong case exists for migrating the current waste management infrastructure into an IoT-based system. Waste can now be collected with minimal human contact in real time using IoT technologies such as Radio-frequency identification (RFID) tags, sensors and Long Range (LoRa) technology. The data collected using these IoT technologies can be used for real-time monitoring of waste management processes and air quality by means of a shared system that combines RFID, General Packet Radio Service (GPRS), Geographic Information System (GIS), Global Positioning System (GPS), and web cameras. It is clear that the common systems are restricted to measuring and tracing of waste and air quality but do not have a single methodology in all of these areas, that gap exists in the current literature. Increasing the cleanliness of the environment depends on developing an effective mechanism to track and forecast toxic air pollutants that exist in waste (Hussain et al. 2020).

12.4.5 Alternative Leather Technologies

Against the backdrop of growing sensitivity with regard to sustainability, animal welfare and the ecological impact of traditional leather production, a number of alternative to leather technologies have been created. These advances are intended to provide eco-friendly, humane, and ecological materials that have the look, feel, and lifespan of conventional leathers. Some of the prominent alternative leather technologies are as follows.

12.4.5.1 Mycelium Based Leather

Mushroom leather, derived from mycelium, the vegetative hyphae of the fungi, is a sustainable and biodegradable alternative to conventionally obtained animal leather. This innovative material is cultivated in regulated settings using agricultural byproducts, which are subsequently processed to yield a product that impersonates the texture and look of animal leather. The production of mycelium-based leather is not only swift but also has a minimal environmental footprint and is fully biodegradable (Elsacker et al. 2023).

12.4.5.2 Pineapple Leather (Piñatex)

Piñatex® is a vegan leather alternative made from pineapple leaf fibers obtained from agricultural waste. The process involves taking the leaves, a byproduct of pineapple cultivation, and turning them into fibers. These fibers are processed into a nonwoven web that forms the basis of such leather products. The final product is strong, lightweight, and breathable (Duangsuwan et al. 2023).

12.4.5.3 Grape Leather (Vegea)

Vegea is derived from grape marc, the byproducts of the skins, seeds, and stems left after wine making. Grapevine residual biomass transformed into a leather-like product by embedding bio-based components and other non-toxic materials with the aid of integration, producing a durable and flexible product (Maskan et al. 2002).

12.4.5.4 Cactus Leather (Desserto)

Cactus leather, derived from the nopal cactus, is more durable and biodegradable than traditional leather. Production processes include the harvest of the mature cactus leaf, which is sun dried and processed so as to have a texture and appearance similar to that of usual leather (Rashid et al. 2024).

12.4.5.5 Lab-Grown Leather (Cell-Cultured Leather)

Lab cultured leather (also called bio-fabricated leather) is one produced by culturing animal dermis cells in vitro, giving the possibility to manufacture leather without necessarily having to slaughter animals. This is one of the novel technologies, which consists in the cultivation of animal cell, further transformed into the sheets of leather. Zoa™ biofabricated leather (Saxena et al. 2024) is at the forefront of this novel regime, produced by Modern Meadow, since it has the most experience in biofabricating leather.

12.4.5.6 Recycled Leather

Recycled leathers are manufactured from leather scraps being thrown away by consumers, which typically came from unused leathers and leftover materials of manufacturing. The manufacturing process consists in the collection of leather waste, grinding, and reforming it into new sheets with the use of a bonding agent or resin (Barbanera et al. 2020).

12.4.5.7 Soy Leather

Soy leather is produced from soybean oil or byproducts of the soybean, which is a biodegradable and plant-based alternative to conventional leathers (Anju et al. 2014).

12.5 Case Studies of Sustainable Tanneries

Many tanneries around the world have adopted new technologies and practices to reduce waste generation, save resources and comply with environmental regulations. Hermès (a renowned French luxury goods brand, was established in 1837) is dedicated to sustainable leather production by working with farms that take animal welfare and environmental importance into account. The company has also implemented energy-saving technologies in its tanneries and actively encourages measures to reduce the ecological impact of the leather supply chain. Tanneries de Saint-Jean-d'Angély has therefore developed a closed-loop wastewater recycling and reuse system, which greatly reduces this environmental footprint. Also, the company has capitalized on energy-saving equipment and practiced for sustainable sourcing of its raw materials. In Vietnam, PrimeAsia Leather Corporation has invested heavily in state-of-the-art technologies for wastewater treatment, relying on membrane filtration and reverse osmosis (RO) systems that can provide greater than 70% reuse of water within its production processes. The fish leather production at the Atlantic Leather tannery in Iceland is a work on fish skin discarded from the catches in the North Atlantic. This approach utilizes byproducts from the fishing sector, thereby minimizing waste and encouraging sustainable practices. The tannery operates with geothermal energy, a clean and renewable energy resource that is present in Iceland and, as a result, has low carbon emissions, and a low energy consumption. Not only water, also water is collected from the natural, geothermal spring, meaning that the complexity of the water treatment system is reduced.

12.6 Challenges and Barriers to Technological Adoption

The leather industry makes a significant contribution to the global economy, but it is overwhelmed with such challenges pertaining in the implementation of modern information and communication technologies. These challenges can hinder innovation, efficiency, and sustainability. Initial investment in new technology can be a lasting challenge for smaller leather businesses. Furthermore, a considerable proportion of industry workers may lack the technical skills to fully operate and service sophisticated machines. The management of personal information, such as customer information and category information, is not straightforward and thus it is an expensive pursuit for data privacy and security. In addition, it is not only slow, but also expensive to establish compatibility between different software and hardware units. Resistance to change by certain members of the professional body, who may favor traditional methods, can again hinder the revolutionization of new technologies. Terror of job displacement by automation, further resist this transition. Implementing new technologies at multiple points along the supply chain, particularly in overseas operations, poses further difficulties. There may be a lack of knowledge among many in the field of this technology as to its benefits and potential applications. In order to promote technology adoption, the industry needs to support educational programs and promote activities of awareness.

12.7 Conclusion

In the leather sector there are multiple financial, regulatory, and technical barriers in the implementation of sustainable technologies. It is a common issue, and therefore, often a roadblock, that substantial capital investments and uncertainties with regard to the return on capital invest keep tanneries, from the advanced solution mindset, away from adopting the best available technologies. Although regulatory guidelines are very important, they might lead to complexity, especially when policies are ambiguous or too demanding. Lack of infrastructure and also lack of finance makes these challenges even more complicated, especially in the developing countries i.e., Pakistan. Nevertheless, advantages of environmentally friendly and human friendly technologies are clear. For successful diffusion, an interdisciplinary team of governments, industry, and finance is needed. For instance, monetary incentives in the form of grants, tax incentives, and low-interest loans afford some relief from financial strain while training programs can help close the skills gap. In addition, clear regulatory routes that foster innovation will build confidence throughout the industry. Solving these challenges leather industry is advancing towards the sustainable production of goods and minimizing the ecological impact at a time when global demand for its products is high, with a focus on their authenticity and environmental responsibility.

References

Åhman H (2013) Social sustainability—society at the intersection of development and maintenance. Local Environ 18:1153–1166. https://doi.org/10.1080/13549839.2013.788480

Andersson AJ, Mackenzie FT (2004) Shallow-water oceans: a source or sink of atmospheric CO_2? Front Ecol Environ 2:348–353. https://doi.org/10.1890/1540-9295(2004)002[0348:SOASOS]2.0.CO;2

Anju B, Kumari KR, Anand V, Anjum MA (2014) Preparation, quality evaluation and storage stability of peach-soy fruit leather. SAARC J Agric 12(1):73–88. https://doi.org/10.3329/sja.v12i1.21114

Atkinson G, Dietz S, Neumayer E (2007) Handbook of sustainable development. Edward Elgar, Cheltenham, UK, Northampton, MA. https://books.google.com/books

Barbanera M, Belloni E, Buratti C, Calabrò G, Marconi M, Merli F, Armentano I (2020) Recycled leather cutting waste-based boards: thermal, acoustic, hygrothermal and ignitability properties. J Mater Cycles Waste Manage 22:1339–1351. https://doi.org/10.1007/s10163-020-01024-3

Birbir M, Caglayan P (2018) A review on catabolic activity of microorganisms in leather industry. Int Conf Adv Mater Syst 3:301–306. https://doi.org/10.24264/icams-2018.VI.3

Biškauskaitė R, Valeikienė V, Valeika V (2021) Enzymes for leather processing: effect on pickling and chroming. Materials 14(6):1480. https://doi.org/10.3390/ma14061480

Chen K-W, Lin L-C, Lee W-S (2014) Analyzing the carbon footprint of the finished bovine leather: a case study of aniline leather. Energy Procedia 61:1063–1066. https://doi.org/10.1016/J.EGYPRO.2014.11.1023

Chiampo F, Shanthakumar S, Ricky R, Ganapathy GP (2023) Tannery: environmental impacts and sustainable technologies. Mater Today Proc. https://doi.org/10.1016/j.matpr.2023.02.025

China CR, Maguta MM, Nyandoro SS, Hilonga A, Kanth SV, Njau KN (2020) Alternative tanning technologies and their suitability in curbing environmental pollution from the leather industry: a comprehensive review. Chemosphere 254. https://doi.org/10.1016/j.chemosphere.2020.126804

Chowdhury FJ, Inamul Haq SM, Aminul Islam M (1996) Accumulation of various pollutants by some aquatic macrophytes found in the Bariganga River. In: Proceedings of the 25th Bangladesh science conference

Closs DJ, Speier C, Meacham N (2011) Sustainability to support end-to-end value chains: the role of supply chain management. J Acad Mark Sci 39:101–116. https://doi.org/10.1007/s11747-010-0207-4

Costa CR, Botta CMR, Espindola ELG, Olivi P (2008) Electrochemical treatment of tannery wastewater using DSA® electrodes. J Hazard Mater 153:616–627. https://doi.org/10.1016/j.jhazmat.2007.09.005

Covington AD (2006) The chemistry of tanning materials. In: Conservation of leather and related materials. Routledge, pp 22–35. https://doi.org/10.4324/9780080454665

Daniels R, Landmann W (2010) Back to basics: a framework for leather manufacture

de Castro Bizerra V, Sales MB, Melo RLF, do Nascimento JGA, Junior JB, Silva MPF et al (2024) Opportunities for cleaner leather processing based on protease enzyme: current evidence from an advanced bibliometric analysis. Renew Sustain Energy Rev 191:114162

Dixit S, Yadav A, Dwivedi PD, Das M (2015) Toxic hazards of leather industry and technologies to combat threat: a review. J Clean Prod 87:39–49. https://doi.org/10.1016/J.JCLEPRO.2014.10.017

Duangsuwan S, Junkong P, Phinyocheep P, Thanawan S, Amornsakchai T (2023) Development of green leather alternative from natural rubber and pineapple leaf fiber. Sustainability 15(21):15400

Elkington J, Rowlands IH (1999) Cannibals with forks: the triple bottom line of 21st century business. Altern J 25(4):42

Elsacker E, Vandelook S, Peeters E (2023) Recent technological innovations in mycelium materials as leather substitutes: a patent review. Front Bioeng Biotechnol 11:1204861. https://doi.org/10.3389/fbioe.2023.1204861

Facchin M, Gatto V, Samiolo R, Conca S, Santandrea D, Beghetto V (2024) May 1,3,5-triazine derivatives be the future of leather tanning? A critical review. Environ Pollut 345:123472. https://doi.org/10.1016/j.envpol.2024.123472

Ghafoor A, Aslam M, Rasool S (2012) Determinants of leather goods exports: a case of Pakistan. J Bus Econ 4(2):256–269

Glavic P, Lukman R (2007) Review of sustainability terms and their definitions. J Clean Prod 15:1875–1885. https://doi.org/10.1016/j.jclepro.2006.12.006

Goodland R (1995) The concept of environmental sustainability. Annu Rev Ecol Syst 26:1–24. https://doi.org/10.1146/annurev.es.26.110195.000245

Gusmerotti NM, Testa F, Corsini F, Pretner G, Iraldo F (2019) Drivers and approaches to the circular economy in manufacturing firms. J Clean Prod 230:314–327. https://doi.org/10.1016/j.jclepro.2019.05.044

Hashmi GJ, Dastageer G, Sajid MS, Ali Z, Malik MF, Liaqat I (2017) Leather industry and environment: Pakistan scenario. Int J Appl Biol Forensics 1(2):20–25

Herrera K, Morales LF, López JE, Montoya-Ruiz C, Muñoz S, Zapata D, Saldarriaga JF (2024) Biochar production from tannery waste pyrolysis as a circular economy strategy for the removal of emerging compounds in polluted waters. Biomass Convers Biorefinery, 14(18):22867–22880. https://doi.org/10.1007/s13399-023-04261-2

Hodge HC, Sterner JH (1949) Tabulation of toxicity classes. Am Ind Hyg Assoc Q 10(4):93–96. https://doi.org/10.1080/00968204909344159

Hu J, Xiao Z, Zhou R, Deng W, Wang M, Ma S (2011) Ecological utilization of leather tannery waste with circular economy model. J Clean Prod 19:221–228. https://doi.org/10.1016/j.jclepro.2010.09.018

Hussain A, Draz U, Ali T, Tariq S, Irfan M, Glowacz A et al (2020) Waste management and prediction of air pollutants using IoT and machine learning approach. Energies 13(15):3930. https://doi.org/10.3390/en13153930

ICT (2018) Sustainability of the tanning industry | ICT leather. WWW document, 10.12.18. https://leathercouncil.org/introduction-to-leather/sustainability-ofthe-tanning-industry/

ILO (2018) Textiles, clothing, leather and footwear sector. WWW document, 5.8.18. http://www.ilo.org/global/industries-and-sectors/textiles-clothingleatherfootwear/lang-en/index.htm

Jaegler A (2016) A sustainable supply chain in the leather sector: dilemmas, challenges and learnings. Supply Chain Forum Int J 17:136–142. https://doi.org/10.1080/16258312.2016.1211833

Joseph K, Nithya N (2009) Material flows in the life cycle of leather. J Clean Prod 17:676–682. https://doi.org/10.1016/j.jclepro.2008.11.018

Kanagaraj J, Senthilvelan T, Panda RC, Kavitha S (2015) Eco-friendly waste management strategies for greener environment towards sustainable development in leather industry: a comprehensive review. J Clean Prod 89:1–17. https://doi.org/10.1016/J.JCLEPRO.2014.11.013

Kanagaraj J, Panda RC, Kumar V (2020) Trends and advancements in sustainable leather processing: future directions and challenges—a review. J Environ Chem Eng 8(5):104379. https://doi.org/10.1016/j.jece.2020.104379

Kankaria S, Andukur A, Hemamailin CG, Kisishnveni M (2011) Impact of tannery effluents on ground water and agriculture with a remedial measure. A case study. In: International conference on chemical biological and environmental science, Bangkok, pp 383–388

Khan Z, Shafique M, Tanoeyadi S, Solangi BA, Khan SA, Jabeen N et al (2024) Characterization and pilot-scale application of the ZMS-2 serine protease with novel properties for the eco-friendly leather processing. Appl Biochem Biotechnol 196(8):5045–5063. https://doi.org/10.1007/s12010-023-04790-6

Kongjao S, Damronglerd S, Hunsom M (2008) Simultaneous removal of organic and inorganic pollutants in tannery wastewater using electro coagulation technique. Korean J Chem Eng 25:703–709. https://doi.org/10.1007/s11814-008-0115-1

Krishnamoorthy G, Sadulla S, Sehgal PK, Mandal AB (2012) Green chemistry approaches to leather tanning process for making chrome-free leather by unnatural amino acids. J Hazard Mater 215–216:173–182. https://doi.org/10.1016/J.JHAZMAT.2012.02.046

Kuria A, Ombui J, Onyuka A, Sasia A, Kipyegon C, Kaimenyi P, Ngugi A (2016) Quality evaluation of leathers produced by selected vegetable tanning materials from Laikipia County, Kenya. IOSR J Agric Vet Sci 9(4):13–17. http://repository.dkut.ac.ke:8080/xmlui/handle/123456789/7854

Mahdi H, Palmina K, Glavtch I (2006) Characterization of Acacia nilotica as an indigenous tanning material of Sudan. J Trop Forest Sci 181–187. https://www.jstor.org/stable/43594670

Marconi M, Marilungo E, Papetti A, Germani M (2017) Traceability as a means to investigate supply chain sustainability: the real case of a leather shoe supply chain. Int J Prod Res 55:6638–6652. https://doi.org/10.1080/00207543.2017.1332437

Maskan A, Kaya S, Maskan M (2002) Hot air and sun drying of grape leather (pestil). J Food Eng 54(1):81–88. https://doi.org/10.1016/S0260-8774(01)00188-1

Moktadir MA, Rahman T, Rahman MH, Ali SM, Paul SK (2018) Drivers to sustainable manufacturing practices and circular economy: a perspective of leather industries in Bangladesh. J Clean Prod 174:1366–1380. https://doi.org/10.1016/j.jclepro.2017.11.063

Moktadir MA, Ahmadi HB, Sultana R, Liou JJ, Rezaei J (2020) Circular economy practices in the leather industry: a practical step towards sustainable development. J Clean Prod 251:119737. https://doi.org/10.1016/j.jclepro.2019.119737

Memedovic O, Mattila H (2008) The global leather value chain: the industries, the main actors and prospects for upgrading in LDCs. Int J Technol Learn Innov Dev 1(4):482–519. https://doi.org/10.1504/IJTLID.2008.021965

Omoloso O, Wise WR, Mortimer K, Jraisat L (2020) Corporate sustainability disclosure: a leather industry perspective. Emerg Sci J 4:44–51. https://doi.org/10.28991/esj-2020-01209

Omoloso O, Mortimer K, Wise WR, Jraisat L (2021) Sustainability research in the leather industry: a critical review of progress and opportunities for future research. J Clean Prod 285:125441. https://doi.org/10.1016/j.jclepro.2020.125441

Ozgunay H, Colak S, Mutlu MM, Akyuz F (2007) Characterization of leather industry wastes. Pol J Environ Stud 16(6)

Peng B, Shi B, Sun D, Chen Y, Shelly DC (2007) Ultrasonic effects on titanium tanning of leather. Ultrason Sonochem 14(3):305–313. https://doi.org/10.1016/j.ultsonch.2006.07.001

Pomponi F, Moncaster A (2017) Circular economy for the built environment: a research framework. J Clean Prod 143:710–718. https://doi.org/10.1016/j.jclepro.2016.12.055

Prokein M, Renner M, Weidner E (2020) Fast high-pressure tanning of animal skins by accelerated chromium sulphate complexation. Clean Technol Environ Policy 22:1133–1143. https://doi.org/10.1007/s10098-020-01855-6

Rajput S, Singh SP (2019) Connecting circular economy and industry 4.0. Int J Inf Manage 49:98–113. https://doi.org/10.1016/j.ijinfomgt.2019.03.002

Ramamoorthy G, Sehgal PK, Mahendra K (1989) Improved uptake of basic chromium salts in tanning operations using keratin hydrolysate. J Soc Leather Technol Chem 73:168–170

Ramesh RR, Ponnuvel M, Ramalingam S et al (2022) Compact glyoxal tanning system: a chrome-free sustainable and green approach towards tanning-cum-upgradation of low-grade raw materials in leather processing. Environ Sci Pollut Res 29:35382–35395. https://doi.org/10.1007/s11356-022-18660-x

Rashid A, Shah A, Khan S, Adamjee MA, Shahab A (2024) Potential of cactus leather in Pakistan's leather industry: a qualitative analysis. Int J Soc Sci Entrep 4(3):1–16

Rasouli AH, Kumarasuriyar A (2016) The social dimention of sustainability. Towards some definitions and analysis. J Soc Sci Policy Implicat 4:23–34. https://doi.org/10.15640/10.15640/jsspi.v4n2a3

Ricky R, Shanthakumar S, Ganapathy GP, Chiampo F (2022) Zero liquid discharge system for the tannery industry—an overview of sustainable approaches. Recycling 7(3):31. https://doi.org/10.3390/recycling703003

Sahu B, Ramesh A, Zameer F (2024) Impact of nanomaterials on leather: a nano-saga from processing to application. Clean Technol Environ Policy 1–28. https://doi.org/10.1007/s10098-024-02912-0

Sarkis J, Helms MM, Hervani AA (2010) Reverse logistics and social sustainability. Corp Soc Responsib Environ Manag 17:337–354. https://doi.org/10.1002/csr.220

Saxena D, Mehrotra R, Trivedi M (2024) Cellular milk and dairy products. In: Cellular agriculture for revolutionized food production. IGI Global, pp 128–148. https://doi.org/10.4018/979-8-3693-4115-5.ch007

Schweigmann M, Taube F, Schulz H, Schütze W, Gierus M (2013) Tannin content and structure of selected plant extracts, pp 261–263

Shahab S (2013) Comparative advantage of leather industry in Pakistan with selected Asian economics. Int J Econ Financ Issues 3:133–139

Siddiqui A, Uroos A, Nadeem F (2016) Sectoral competitiveness and value chain analysis: leather gloves value chain analysis in Pakistan. Trade Development Authority of Pakistan

Singh SC, Gupta D (2013) Sustainability: a challenge for Indian leather industry. J Supply Chain Manag Syst 2:37–43

Sivakumar V, Gopi K, Harikrishnan MV, Senthilkumar M, Swaminathan G (2008) Ultrasound assisted diffusion in vegetable tanning in leather processing. J Am Leather Chem Assoc 103(10):330–337

Srinivasan P, Rajagopal V, Shanmugam G, Madhan B (2024) Development of masked silica tanning system for sustainable leather production. Environ Sci Pollut Res 31(18):26567–26579. https://doi.org/10.1007/s11356-024-32567-9

Srivastava B, Khatri M, Singh G, Arya SK (2020) Microbial keratinases: an overview of biochemical characterization and its eco-friendly approach for industrial applications. J Clean Prod 252:119847. https://doi.org/10.1016/j.jclepro.2019.119847

Sundarapandiyan S, Chandrasekar R, Ramanaiah B, Krishnan S, Saravanan P (2010) Electrochemical oxidation and reuse of tannery saline wastewater. J Hazard Mater 180:197–203. https://doi.org/10.1016/j.jhazmat.2010.04.013

Thanikaivelan P, Rao JR, Nair BU, Ramasami T (2005) Recent trends in leather making: processes, problems, and pathways. Crit Rev Environ Sci Technol 35(1):37–79. https://doi.org/10.1080/10643380590521436

Vidaurre-Arbizu M, Pérez-Bou S, Zuazua-Ros A, Martín-Gómez C (2021) From the leather industry to building sector: exploration of potential applications of discarded solid wastes. J Clean Prod 291:125960. https://doi.org/10.1016/j.jclepro.2021.125960

Wemegah R (2014) Vegetable tanning in Bolgatanga: challenges and the way forward. Arts Des Stud 16:27–37

Wolf G, Meier S, Basf SE, Lin A (2013) Leather and sustainability: from contradiction to value creation. BASF China, Zuri

Xiang H, Zhang J, Zhu Q (2015) Worldwide earthworm research: a scientometric analysis, 2000–2015. Scientometrics 105:1195–1207

Yu Y, Zeng Y, Wang Y, Zhang W, Zhou J, Shi B (2021) Life cycle assessment for chrome tanning, chrome-free metal tanning, and metal-free tanning systems. ACS Sustain Chem Eng 9:6720–6731. https://doi.org/10.1021/acssuschemeng.1c00753

Yusif BB, Bichi KA, Oyekunle OA, Girei AI, Garba PY, Garba FH (2016) A review of tannery effluent treatment. Int J Appl Sci Math Theory 2(3):29–43

Yuvaraj A, Karmegam N, Ravindran B, Chang SW, Awasthi MK, Kannan S, Thangaraj R (2020) Recycling of leather industrial sludge through vermitechnology for a cleaner environment—a review. Ind Crops Prod 155:112791. https://doi.org/10.1016/j.indcrop.2020.112791

Chapter 13
Utilization of Tannery Solid Wastes for Eco-Friendly and Sustainable Leather Industry

Bedru Ahmed, Bashirunnisa Ananya, Wendwosen Ebabu, Zerihun Teshome, Zubaida Rukhsana Usha, and Md. Israil Hossain

Abstract Tannery solid waste disposal from the leather industry greatly threatens many developing nations due to the increase in output wastes. One of the major concerns of solid waste management is the proper utilization of industrial wastes, which imposed a financial burden with high expenses, creating a significant negative health and socioeconomic impact. Therefore, the conversion of tannery solid wastes to high value-added usage in the leather industry has sparked widespread interest in a great solution for solid waste management, which is still facing a proper lack of understanding of solid trash management. Herein, this study looked into details approaches to rationalize different hazardous chemicals and substances used during tanning processes and their suitable solution after discharging and the proper utilization of waste discharge. To do this review, different literature was sourced, a different source of waste was identified and then the information is analyzed in detail. Based on that, this article review gives a summary of tannery solid waste and their management strategies in the various leather manufacturing process as well as their negative

B. Ahmed · W. Ebabu · Z. Teshome · Md. Israil Hossain (✉)
Department of Leather Product Design and Engineering, Ethiopian Institute of Textile and Fashion Technology, Bahir Dar University, Bahir Dar, Ethiopia
e-mail: mohammadisrail.hossain@manchester.ac.uk

B. Ahmed · W. Ebabu
Department of Leather Engineering, Kombolcha Institute of Technology, Wollo University, Dessie, Ethiopia

B. Ananya
Department of Biomaterials, Faculty of Science and Engineering, University of Manchester, Manchester, UK

Z. R. Usha
National Synchrotron Radiation Laboratory, University of Science and Technology of China, Hefei, China

Md. Israil Hossain
Department of Materials, Faculty of Science and Engineering, The University of Manchester, Manchester, UK

M. A. Aboulhassan and S. Souabi (eds.), *Sustainable Practices in the Tannery Industry*,
https://doi.org/10.1007/978-3-031-92037-0_13

and potential beneficial implications of solid waste through the application of best practices.

Keywords Leather industry · Solid waste · Tannery · Sustainable management · Value-added products

13.1 Introduction

The leather industry plays a significant part in the global economy due to the high demand for leather goods (Masilamani et al. 2016). Raw hides and skins are by-products of the livestock and meat-processing industries that are used as raw materials in the leather industry. The hides and skins are trimmed during the leather-making process to make them suitable for productive activities. Although being the most well known in the international leather trade, is blamed for its harmful impact on the environment and ecosystem by releasing millions of pounds of leather waste to land places during the tanning process (John Sundar et al. 2011). The tanning industry produces both organic and inorganic solid, liquid, or gaseous pollutants, making it a highly polluting sector in the world (Hossain et al. 2019, 2020, 2021). Skin trimming, chrome shaving, buffing dust, leather scraps, and used leather waste (i.e., waste from leather products dumped once used up) are among the solid wastes generated by tanneries and leather goods industries (Ebabu et al. 2021; Chowdhury et al. 2019). Most of the leather processing procedures take place in an aqueous medium. To turn hides and skin into leather, more than twenty different chemicals are used. The waste from leather processing industry contains a variety of soluble and insoluble components as well as process chemicals to make up the effluent and wastewater (Babu 2004; Deb et al. 2017). According to research conducted by Teklay et al. (2017), the tannery generates a total of around 4049 tons of solid waste from beam house, tanning, re-tanning, and finishing processes, which is comparable to 55.8% of the rawhide and skin processed (Meşe et al. 2018). Furthermore, United Nations Industrial Development Organization (UNIDO) demonstrated that in the leather processing industry, 80% of rawhides or skins end up as solid wastes, which creates the leather sector a significant environmental burden on the environment and human health (Gökgöz and Güvercin 2018).

Conversely, disposing of these solid wastes without being treated poses a significant ecological hazard. Unfortunately, there is no naturally beneficial method of disposing of such solid trash. Widely and commonly used disposal methods for tanned leather wastes is burning and/or landfilling, which are not environment-friendly process at all and creates a more harmful effect on the environment. These direct disposal methods without any treatment can create a long-term negative effect on a country's growth potential (Lakdawala and Lakdawala 2013). Therefore, openly discharging these solid wastes such as shaving dust are highly consumed by poultry and fish, which converts the trivalent form of chromium (Cr^{3+}) of shaving dust to highly hazardous hexavalent chromium (Cr^{6+}) (Zhou et al. 2021). Lately, this carcinogenic hexavalent chromium entering the human body by the food chain causes severe diseases such as testicular cancer, respiratory problems, soft tissue sarcoma, tumor formation, etc. Additionally, trimming, fleshing, and splitting solid wastes discharged directly into the environment generate intense aromas and significant air pollution. Furthermore, the discharge of wastes, whether chemical, energy, or raw material, is a waste of scarce resources from an economic standpoint as well.

However, good waste management with proper compliance can reduce the harmful effects of trash and pollution (Aluyor and Badmus 2008; Islam et al. 2020). As a result, it can create a potential option to optimize the conversion of these wastes into valuable products, which is considered highly economical. Therefore, the demand for bioenergy is extensively marketed as a greener alternative to petroleum and other fossil fuels, especially considering the rising petroleum prices and growing concerns about fossil fuels' negative impact on global warming, which results to produce bioenergy from toxic leather solid wastes could assist to reduce pollution while also meeting the country's growing energy need. On the other hand, some of the solid wastes such as raw trimmings, fleshing, shaving, tanned trimmings, unusable splits, and buffing dust can be treated to produce glue and gelatin, animal feed, collagen, protein, etc. Additionally, brick production from tannery sludge and composite boards from chrome saving is highly studied. Whereas many researchers have given high concentrations to prepare adsorbent material from tannery waste to remove different heavy metals from industrial wastewater to protect soil and groundwater pollution. Unfortunately, there are still limitations in research on the management of leather solid wastes for an environmentally sustainable strategy. As a result, developing eco-friendly and sustainable technology to handle leather wastes sustainably is critical to conserve the environment, safeguard human health, and improve export revenues.

Herein, the main target of this current review article is to present and evaluate the different types of tannery solid waste and their management strategies. As well as their negative and potential beneficial implications of solid waste created to the ecosystem are also discussed. The review of this article is based on searching for documents according to the embodiment of terms in the online database. The objective of this review article was to compile, analyze, and filter the most relevant papers concerning tannery solid waste management. A systematic literature structure was followed during conducting this review study. The layout of the study mainly consists of three phases such as literature searching, identifying sources, data analyzing, accumulation, and reporting. The suitable sources of literature such as journal articles, review papers, master's and doctoral thesis, conference proceedings, and books, were put together from the digital libraries. The online search included Google Scholar, Springer, Research Gate, the Web of Science, Elsevier, Science Direct, and so on. A large number of up-to-date works of literature were collected and the most relevant sources were selected by filtering the less important articles. The emphasis of selection remained on tannery solid waste management and solid waste generated from the tannery was categorized and clustered in three ways like untanned solid waste, tanned solid waste, and tannery sludge. Their relevant form of title, abstract, picked the most closely connected articles and contents and irrelevant sources were

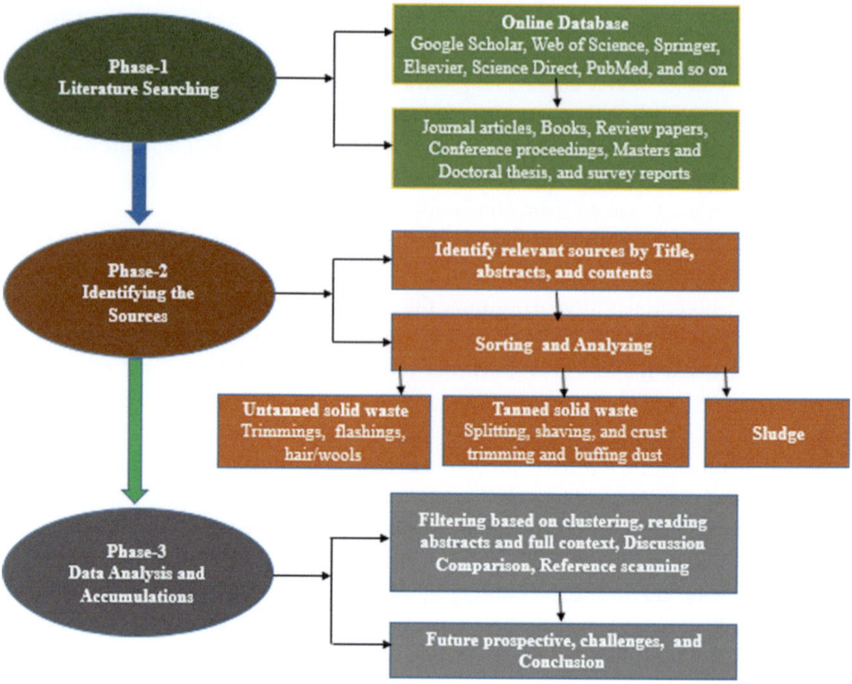

Fig. 13.1 Schematic representation of the methodology adopted for the literature review in this study

excluded. In the end future prospective, challenges and conclusions were manifested. Figure 13.1 shows a flowchart that summarizes the methodology of this study.

13.2 Generation of Leather Solid Wastes

Leather processing entails several steps. It is divided into four categories: pre-tanning, tanning, post-tanning, and finishing (Bitlisli et al. 2004). At various phases of the leather processing process, many types of solid waste are formed. Based on the tannery method, these solid wastes were categorized as untanned solid waste (before tanning) and tanned solid waste (after tanning). The untanned solid wastes are the followings trimmings waste, fleshing, and pelt splitting solid wastes. Chrome shaving waste, wet blue trimming, buffing dust, crust trimming, and final trimming are tanned leather wastes that include chromium sulfate chemicals (Nigam et al. 2015).

13.2.1 Untanned Solid Wastes

13.2.1.1 Trimming Waste (Raw Hides/Skins)

Trimming neck and tail parts from hides/skins to create a shape for the final leather is one of the first procedures. However, removing these wastes can either slow down the efficiency of machine operations. It is estimated that around 6% of trimming wastes are coming from around the head and belly rounds, and other trims (Buljan et al. 2000). The head/neck components can be turned into vegetable-tanned leathers, which are then turned into open-toe sandals. The smaller raw trimmings are gathered, stored, and transferred to units that make glue, gelatin, and animal feed. By substituting wet or dry fibrous collagen in meat-based products, the possibility of hiding collagen for human consumption as a nutritional supplement has been investigated.

13.2.1.2 Fleshing Waste

Raw hides are fleshed in several places, resulting in green fleshing and fatty tissues. This procedure is performed before washing and soaking. The yield of green fleshing is estimated to be 10% of the raw hide's weight (Buljan et al. 2000). These fleshing are typically rendered alongside the abattoir's other inedible offals, which include meat and bone because they are not tainted with chemicals. Meat-cum-bone meal is made from the finished result.

Limed flashings are obtained by scraping limed hides and skin using either hand or machine. The flashings are proteinaceous in origin, with cutaneous muscle layers and subcutaneous adherent tissues that are undesired in following leather-manufacturing activities. Limed flashings are available in 35% of the wet weight of raw hides (70% moisture). The limed flashings are gathered by glue producers as well as feed producers (Kumari et al. 2016). The limed fleshing's composition is following moisture content ranging from 78 to 80%, 8.3% ash content, and 15.2% total nitrogen (Fathima et al. 2005).

The hides and skins are trimmed after the fleshing operation to remove the ragged and damaged edges. These trimmings have been proven a good raw material for glue, technical gelatin, and animal feed because of their proteinaceous properties (Fathima et al. 2012). Limed trimmings range from 2 to 7% of the wet weight of the skins (Buljan et al. 2000). Pre-boiling the fleshing to separate the fats from the other components can also be used to make tallow. The procedure for obtaining tallow is then repeated in the same manner as for green fleshing.

13.2.1.3 Hair and Wool

To eliminate hair and wool, the hides and skins are treated with lime and sodium sulfide (Kamal Eldin Abu El Fatah). Because hair and wool are valuable commodities, the flesh side of goat and sheepskins is painted with lime and sulfide. Hair and wool are removed mechanically or manually after the treatment, which is known as the hair conserving process (Frendrup and Buljan 2000). Hair pulping occurs because of lime sulfide treatment of hides; as a result, if the pulping is not screened using appropriate mechanical techniques and prevented from entering the wastewater stream, it causes environmental concerns. Hair accounts for 10–12% of the animal's weight, which varies depending on the season of the year; for example, it is higher in the winter (Saravanabhavan et al. 2005). Depending on the breed and climatic conditions, wool accounts for 15–20% of the body weight. The coarse hair and wool are gathered, cleansed thoroughly, and used to make carpets (Lebedytė and Sun 2021). Hair could potentially be hydrolyzed for fertilizer production (Yılmaz et al. 2007).

13.2.2 Tanned Solid Wastes

Before splitting and shaving, excess water from chrome-tanned leather is removed. The hide forms pleats at the edges during this procedure, which are cut away for not to get in the way of future surgeries like splitting and shaving. During machine operation, they may harm the hides and skins, as well as the fleshing sides. These trimmings (estimated at 5.5–6% of the wet weight of hides) are collected from the tannery and used in the production of leather boards and bricks, utilizing a low-pressure and low-temperature technique to avoid Cr^{3+} to Cr^{6+} oxidation. Moreover, chromed trimmings are commonly used in fertilizer and leather meals for animal feed applications (Chaudhary et al. 2016).

13.2.2.1 Splitting Waste

The splits obtained during the splitting operation are extremely thin and cannot be used. Chrome splits account for 4–6% of total trash in the form of solid garbage (Fathima et al. 2012). In the leather board industry, they are gathered and used as raw material (Fathima et al. 2012). The chromium content of Cr_2O_3 is found to be 2–2.5% in the majority of chrome wastes (Kanagaraj et al. 2006). Currently, feed producers collect these wastes, which are steam cleaned and transformed into a leather meal (Chaudhary et al. 2016). Cattle splits with an area of 1–1.5 m^2 and a weight of 6–10 kg are retanned and colored to make chamois leather, which is also used to make casual and athletic shoes in various countries (Deb 2015).

13.2.2.2 Shaving Waste

Splits and trimmings from chrome-tanned leather have been used to make glue, gelatin, protein taste, and reconstituted collagen. The aforementioned items can be manufactured by saturating them with hydrogen peroxide, then grinding and extracting as by-products (95%) (Cot et al. 2003). When chrome-tanned leather is shaved, the shaving process produces chrome shavings as a waste product. Chrome shavings in the fibrous shredded form are available as a percentage of the raw hide's weight (1–2%). In chrome shavings, the chromium is in a trivalent form (Chaudhary et al. 2016).

13.2.2.3 Crust Leather Trimmings

Generally, crust leathers (leathers obtained after retanning, dyeing, and fatliquoring operations) are trimmed after staking, toggling, and nailing to remove the torn and ragged edges for aesthetic value and for processing the leather in the subsequent machine without any damage. The crust leather trimmings are also collected and used for making leather boards and the production of bricks (Kanagaraj et al. 2006).

13.2.2.4 Buffing Dust and Finished Leather Trimmings

The crust leathers are buffed on the flesh side and sometimes on the grain side. The buffing (done on the flesh side) and snuffing operation (done on the grain side to produce specialty leathers such as nubucks) generate fine lightweight leather fluffy mass. Developing low-cost composite materials for use in footwear, leather goods, household interiors, etc. (Senthil et al. 2015). After finishing, the leathers are trimmed at the edges uniformly. The purpose of trimming is aesthetic and for easy measurement of the leather.

13.2.3 Sludge

Sludge is produced during the wastewater treatment process. Because the chrome sulfate chemical used during tanning and retanning does not entirely bond to the leather, wastewater includes chromium. More than 40% of the chrome sulfate used in the leather manufacturing process ends up in the wastewater treatment system. The wastewater after treatment contains a significant amount of muck. The chromium in the sludge comes from the tanning and retanning processes (Jones and Bh 1979) (Fig. 13.2).

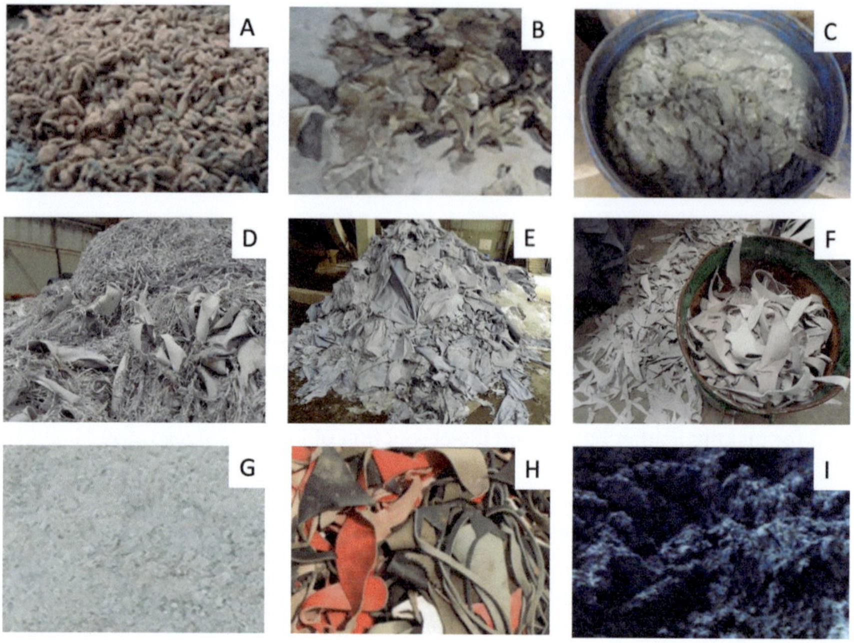

Fig. 13.2 Solid waste generated from tannery **a** raw trimmings, **b** hair waste, **c** fleshing, **d** chrome shavings, **e** splitting, **f** crust trimmings, **g** buffing dust, **h** finished leather trimmings, **i** sludge

13.3 Estimation of Solid Waste Generated in the Leather Manufacturing Process

Tannery originated a large volume of solid wastes during the leather manufacturing process and subsequent effluent treatment procedure. The reaction of collagen fibers in a hide with a chemical substance is referred to as tanning (tannin, alum, or other chemicals). About 850 kg of solid waste is generated for every ton of raw hides (Sawalha et al. 2019). However, the leather manufacturing process follows the following sequence of operations: The Beam-house (pre-tanning process) generates solid waste, which includes soaking, liming/unhairing, fleshing, splitting, deliming, and bating, reducing, and pickling. Conditioning, Staking, Milling, Spreading on toggle frame, Buffing, Impregnation, Coating, Ironing, and Polishing are all part of the tanning process; re-tanning includes shaving, neutralizing, re-tanning, fat liquoring, dyeing, setting out, and drying; and finishing includes conditioning, staking, milling, spreading on toggle frame, Buffing, Impregnation, Coating, Ironing, and polishing (Mohammed 2014) (Tables 13.1 and 13.2).

At least 300 kg of chemicals (lime, salt, etc.) are added to every ton of hides during the tanning process (De Haan et al. 1997). In wastewater, there are too many salts that have not been used (Verheijen et al. 1996). Different salts (Cl), fat, protein, preservatives (soaking), Lime and ammonium salts, ammonia, protein (hair), sulfides

Table 13.1 The number of different types of tannery solid wastes produced by processing 1 ton of raw hides (Thanikaivelan et al. 2005)

S. No.	Type (nature) of solid waste	Quantity (kg/ton)
1	Salt from handshaking	80
2	Salt from solar pans	220
3	Hair (pasting)	100
4	Raw trimming	40
5	Lime sludge	60
6	Fleshing	120
7	Wet blue trimming (for grain splits)	30
8	Chrome splitting (bovine)	65
9	Chrome shaving	95
10	Buffing dust	65
11	Dyed trimmings	35
12	Dry sludge from CETP	125

Table 13.2 The solid waste kg per ton of raw hides in beam house operations (Buljan et al. 2000)

Name of wastes	Quantity (kg/ton)
Raw trimmings	80–120
Wet limed fleshing's	250–300
Hair/wool	40–50

(fleshing, trimming, bating), Chromium (salts), polyphenolic compounds (tanning), color and solvent are all contained in the tannery wastewater.

13.3.1 Effects of Tannery Waste on the Environment

The pollution of the environment by tannery waste is one of the most serious environmental problems that the world faces today. Tannery waste pollutes the environment daily due to a lack of suitable management facilities. Approximately 95% of the tannery factories were established without proper systematic planning (Kaul 2005). These haphazard tanneries wreaked havoc on the ecosystem. These wastes have a significant impact on the environment's main elements, such as air, water, and soil, as well as the animals and plants that rely on these elements. The most damaging environmental effect was found to be a foul odor in the surrounding areas, which resulted in great pollution (Naidu 2000) (Table 13.3).

Table 13.3 The health risk of tannery wastes (Murad et al. 2018)

Inputs	Process	Emissions			Occupational hazards/ risks
		Effluent	Noise	Solid	
Water, biocides surfactants, enzymes, electricity	Soaking	Dirt, dung, salt, biocide insecticide	Drum/ paddle		Anthrax, skin allergies, hearing impairment, electrical shocks, accidents/injuries/falls
Water, lime, sodium sulfide, electricity	Liming, unhairing	Lime, sulfide	Drum/ paddle	Hair sludge	Anthrax, skin irritation, eye disease, accident/ injuries/fall
Water, electricity	Fleshing	Washing water, lime	Fleshing machine	Fleshing	Accident/injuries, skin irritation hearing impairment
Water, ammonium sulfate/ chloride, carbon dioxide, electricity	Deliming	Lime, nitrogen	Drum/ paddle	Scads	Accident/injuries, hearing impairment, respiratory problem
Water, electricity, enzyme	Bating	Water wash, liquor, enzyme	Drum	Hair roots	Skin, hearing irritation, impairment
Salt, acid, water, electricity, masking agent	Pickling	Acidic effluent	Drum		Skin irritation, acid burn, hearing impairment, respiratory problems
Solvents, salt water, surfactant, electricity	Degreasing	Effluent	Drum	Greasy residues	Skin irritation, respiratory problems
Water, chromium, sulfate, electricity	Tanning	Chrome floats	Drum	Chrome dust	Occupational asthma, injuries, skin irritation, chrome ulcer, chromosomal, aberration, hearing impairment

13.4 Main Applications Studied for Tannery Solid Wastes

The use of leather waste in the creation of adsorbent materials, biodiesel, biogas, biopolymers, and uses in agriculture are all discussed in this part. In addition, other broad uses, mostly include the extraction/recovery of industrial/commercially important chemicals, microorganism isolation, and manufacturing of enzymes and use in animal diets.

13.4.1 Adsorbent Materials

The option of manufacturing adsorbents from industrial or agricultural leftovers, referred to as bio sorbents, should also be explored, as it expands the application possibilities (Rigueto et al. 2020). In this context, companies have promoted the use of alternative precursors, such as sludge from tannery wastewater treatment, to produce adsorbents. Because this residue is carbonaceous, it can be used to make activated carbon, lowering sludge volume and producing a low-cost adsorbent that can be used by the industry to treat effluents (Kantarli and Yanik 2009). Adsorbent materials used a pyrolysis procedure to make activated carbon from tannery fleshing wastes, which had higher porosity (number of mesopores and macropores), a higher number of surface functional groups, and better thermal resistance than commercial activated carbon. The largest amount of chromium (VI) removed was roughly 97% (Palani et al. 2017). Activated carbon made from vegetable-tanned leather trimmings to remove three basic dyes, resulting in an adsorbent material that removed over 90% of the colors tested (Amalraj and Pius 2014). The removal of dyes used in leather dyeing was also researched using solid wastes from the shaving leather operation, and promising adsorbents were found in the decrease of color and the presence of organic and inorganic pollutants, allowing the reuse of leather industry effluents used a low-temperature, $ZnCl_2$-activated procedure to make activated carbons from chromium-tanned leather shavings. The produced materials had a stronger affinity for Acid Black 210 dye than a commercial activated carbon and nearly half the adsorption capacity, with a maximum experimental adsorption capacity of 44.4 mg/g (Arcibar-Orozco et al. 2019). When the adsorbent is contacted with the dye solution at pH 4.5, the free amino groups become protonated and form salt bridges with the dye molecules, which is shown in Fig. 13.3, and the dye uptake capacity is 23.32 mg/g a tan initial concentration of 400 mg/L and an adsorbent dosage of 6 g/L.

13.4.2 Biodiesel

Biodiesel is a liquid sustainable fuel made from vegetable oils and animal fats in general. Direct usage of these high viscosity oils and fats in vehicles can cause

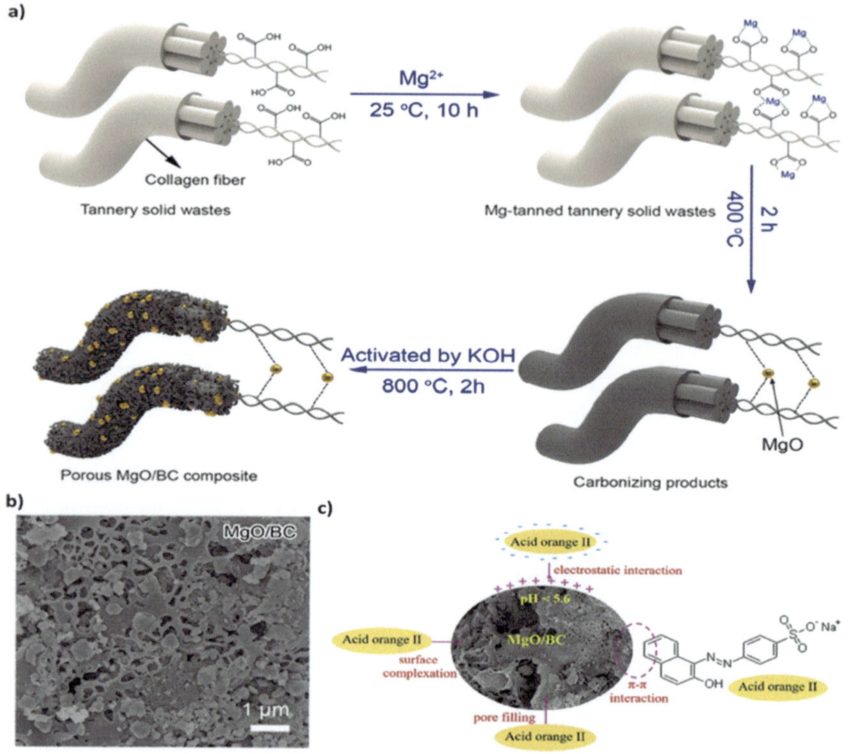

Fig. 13.3 **a** Synthesis process of porous MgO/BC, **b** SEM photographs of surface micromorphology of BC and MgO/BC, and **c** possible anionic dye removal mechanism of MgO/BC (Tang et al. 2021)

huge damage to the engine. However, the problem can be solved by turning them into biodiesel (fatty acid alkyl ester) via esterification or transesterification reactions (Booramurthy et al. 2022). Biodiesel has the advantages of being biodegradable, non-toxic, emitting less pollution, and being environmentally friendly (Boonyuen et al. 2018). Apart from all these advantages, the most significant disadvantage of using biodiesel is its high raw material and operation costs, which are more expensive than petro-diesel. However, the disadvantage of tannery industries to generating a substantial amount of solid waste during the pre-tanning process of hides can be turned into a favorable solution for producing biodiesel (Lazaroiu et al. 2018). The expense of treating the tannery wastes is putting a financial strain on the tanneries. So far, these wastes include a lot of fat materials, which can be removed and utilized to make biodiesel production. Fat removal not only minimizes pollution from these wastes but also can provide a huge economical profit. The high level of pollutants present in this feedstock makes it difficult to use as a biodiesel feedstock.

13.4.3 Biogas

Biogas is an essential part of the growing renewable energy market (Belostotskiy et al. 2015), which is produced by the biological breakdown of various organic wastes from landfills, tannery waste (Christy et al. 2014), sewage sludge, agricultural, animal manure, and municipal garbage (Bernat et al. 2015). It is estimated that the usage of biogas will double in the next years, with estimates ranging from 14.5 GW in 2012 to 29.5 GW in 2022 (Ullah Khan et al. 2017). However, this biogas can be utilized in various aspects such as to make fuel, electricity, chemicals, heat, and others. Due to regular human activity and the advancement of the industry, the amount of garbage produced is growing. Because of the enormous output of wastewater and solid waste, the tanning industry is one of the most polluting industries. Between 15 and 50 tons of liquid effluents and 450–730 kg of solid waste are predicted to be produced during the processing of 1 ton of rawhide, where only 250 kg of finished leather generates (Vico and Artemio 2017). The most important part of the tannery waste is animal fat, which has a high thermal value. Because these residues have a high protein concentration, they can be treated in an anaerobic digester (AD). AD is a biochemical process of organic matter breakdown carried out by a symbiotic community of microorganisms in the absence of oxygen.

13.4.4 Biopolymers

Transforming industrial waste into useful material boosts economic flow in this environment. The extraction of collagen and/or gelatin found in wastes from the leather processing industry for use as biodegradable materials is a sustainable alternative that can be used in the case of pollution caused by fossil polymers and is investigated the extraction of gelatin from raw bovine skin derived from tannery wastes using acid hydrolysis, which was then dried and crushed for use in film production (Vickers 2017; Sullivan et al. 2018). The scientists tweaked the film's formulation, using a polymer made up of polyvinyl alcohol (PVA), glycerol, and glutaraldehyde to get a performance comparable to that of the low-density polyethylene (LDPE) film used as a control.

13.4.5 Agricultural Fertilizer Production

Tannery solid wastes have been frequently employed as fertilizers to boost agricultural production due to the presence of nitrogen and organic matter (Zemleduch-Barylska et al. 2015). Studies suggest that chrome shaving, trimming, and fleshing waste can be used as a fertilizer by digesting it with the proteolytic enzyme in the presence of lime. The discharge of untreated tannery waste into soil and water

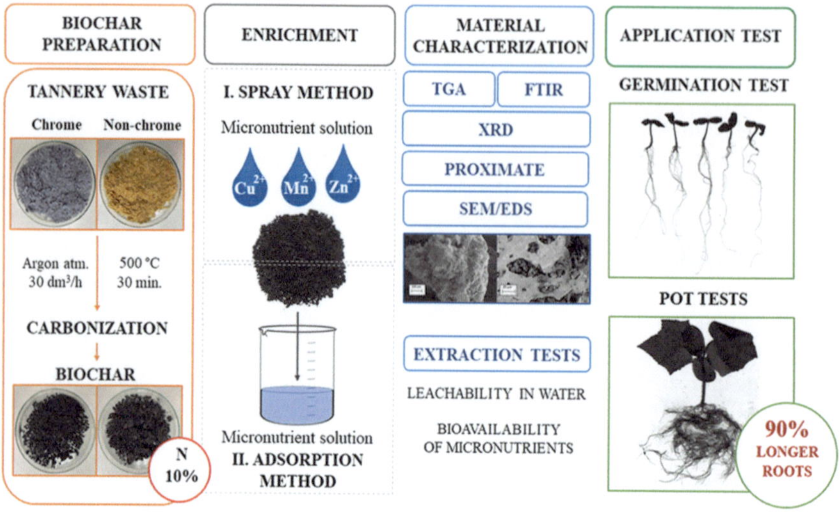

Fig. 13.4 Biochar fertilizer process by utilizing tannery solid wastes and cucumber plant growth (Skrzypczak et al. 2022)

bodies has become a significant environmental issue, whereas composting and vermicomposting can be environmentally acceptable solutions for managing these solid wastes that are extensively touted as effective and ecologically friendly solid waste management techniques (Ravindran et al. 2019). The fertilizer enhanced the nitrogen and phosphorus content in the soil, which increased the fresh and dry mass of the vegetable. Nunes et al. conducted a study to create organic sweet peppers using vermicompost tanned leather shavings mixed with cow manure and sawdust (Nunes et al. 2018). Furthermore, Skrzypczak et al. prepared micronutrient enriched biochar fertilizer for the production of cucumber, whereas the application of biochars stimulates plant growth showing about 20% longer stem and 70% longer root as compared with the control one, as shown in Fig. 13.4 (Skrzypczak et al. 2022).

13.4.6 Brick Production

High levels of heavy metals such as Cr, As, Ni, Co, Cu, Zn, Fe, and Cd contains in tannery sludge due to the usage of basic chromium salt, dyes, pigments, and retanning agents during the tanning process (Houshyar et al. 2012). Because of their non-biodegradability and ability to accumulate in biological systems, these heavy metals are extremely dangerous. However, these heavy metals contain tannery sludge that can be stabilized via valorization in construction materials such as brick or concrete (Wu et al. 2010; Cusidó and Cremades 2012). In certain, waste-water treatment sludge is mixed with clay to manufacture bricks, and in others, chrome shavings are

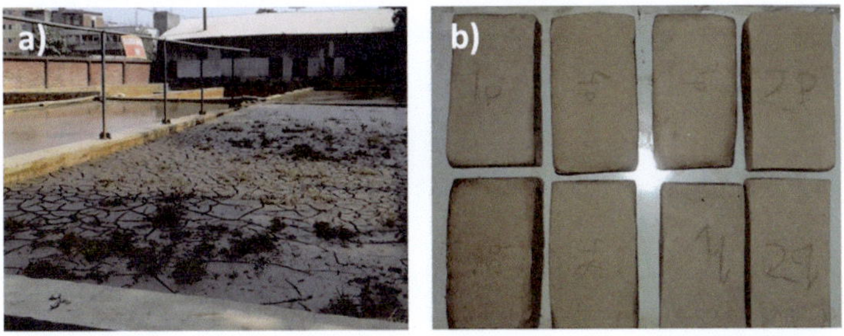

Fig. 13.5 **a** Raw tannery sludge in the sludge-drying bed, **b** raw TS bricks prepared in the laboratory (Juel et al. 2017)

mixed with clay to make bricks (Rajamani et al. 2008). However, to offer tannery sludge as a viable recycling option in the construction industry, the properties of the sludge-stabilized construction material must conform to local building material standards. Juel et al. (2017) constructed bricks by mixing tannery sludge and clay in different varying proportions from 0 to 40% by dry weight as shown in Fig. 13.5.

13.4.7 Collagen Protein Recovery

Chrome shaving waste has been used as a raw material for collagen and protein production (Sedliacik et al. 2011). By combining chrome shaving waste with magnesium oxide and raising the pH to 8, the chrome shaving waste is hydrolyzed. The collagen hydrolysis is separated from the chrome cake after mixing in a temperature-controlled system (72 °C) for 6 h and filtering with a hydraulic press. In this context, alkaline enzymatic hydrolysis of chrome-containing leather wastes (CCLW) such as chrome shaving, useless splits, and trims produces a collagen hydrolysate and an impure chromium hydroxide chrome cake. These items have the potential to be used in a variety of industries. Collagen has become widely employed in a variety of industries in recent years, including foods, cosmetics, medical materials, and cell culture technology (Bello et al. 2020). Collagen is a type of protein with a specified amino acid sequence, size, and structure, which is often made from animal pelts for industrial usage. Acidification can be used to extract tanning salts from these cakes. These can be reused in the tanning process (Collivignarelli and Barducci 1984) (Fig. 13.6).

13.4.8 Animal Diet

The impact of vegetable tanned wastes on the diets of broilers (Lee et al.) and cattle (Alam et al. 2010; Dettmer et al. 2012) has been explored by certain researchers. The researchers concluded that diets did not affect meat production characteristics

Fig. 13.6 Formation of collagen powder from chrome containing leather waste (Dwivedi and Saxena 2020)

in broilers, but that increased levels of tannery residues in the diet improved gizzard and ham weights. When compared to commercial protein concentrate, the high level of protein and availability of critical amino acids in leather waste resulted in body weight gain (Alam et al. 2010). In this regard, Jini et al. investigated the acute and subacute toxicity of delimed tannery fleshing (DTF) and hydrolysates generated by acid (AH) and fermentation (FH) hydrolysis in rats fed diets comprising DTF and hydrolysates (Jini et al. 2016). As indicated by assessments of body weight increase, feed efficiency, organ weight, hematological, histology, and serum biochemistry, both hydrolysates (AH or FH) and DTF can be supplemented in diets of up to 15% without generating toxicity-related alterations in birds and fish. The authors also stress the significance of continuing studies to verify the total nutritional safety of animal feed based on tannery waste and/or hydrolysates, thereby broadening the options for sustainable and environmentally responsible use of these residues.

13.4.9 Composite Board

Leather shavings have been proposed as a potential source of natural organic semi-conductors (Nashy et al. 2012). Asphalt combinations have also included leather dust, which was discovered that adding 0.3% leather particles to an asphalt micro-surface layer increased its technical qualities and reduced pavement layer cracking. Produced by the presence of leather fibers in the micro-surface of the asphalt (Krummenauer et al. 2009). Another interesting application of leather waste is the creation of pavement stones for pedestrian usage using leather fibers in combination with Portland cement (Zăinescu et al. 2018). Physical–mechanical examinations of the produced materials revealed values that were within the limitations set by standards for industrial use. Chrome shavings are used to make a leather board, where chrome tanned leather waste is consistently fibred and turned into a pulp-like paperboard pulp. The pulp charges can be adjusted as needed, after which they are mixed with rubber latex solution and the coagulated mass is pressed under different conditions. In the footwear business, the leather board is utilized as an insole. The leather board is often created by extracting fibers from shavings and other solid waste with hammer mills and Hollander beaters, then processing those fibers with chemicals and binders (Fig. 13.7).

13.4.10 Isolation of Enzymes and Microorganisms

For the handling of tannery residues, enzyme and fermentation-based approaches have recently been investigated. Enzymatic degradation aided by microbes is regarded as a long-term process since it breaks down complicated molecules into smaller intermediates with minimal environmental impact. Proteases account for 60% of global enzyme sales and are employed in a variety of industries, including food, leather, dairy, pharmaceuticals, and waste management (Ghasemi et al. 2011).

13.5 Tannery Solid Waste Management Methods

Solid waste management refers to the collection, transportation, storage, recycling, or disposal of solid waste, as well as the reuse of a decommissioned disposal site (Proclamation 2007). It is a complicated procedure that involves a variety of technologies and disciplines. These technologies include those related to solid waste creation (including source reduction), on-site handling and storage, collection, transfer and transportation, processing, and disposal. All these procedures must be carried out in accordance with existing legal, social, and environmental requirements that preserve public health and the environment while also being aesthetically and economically appealing.

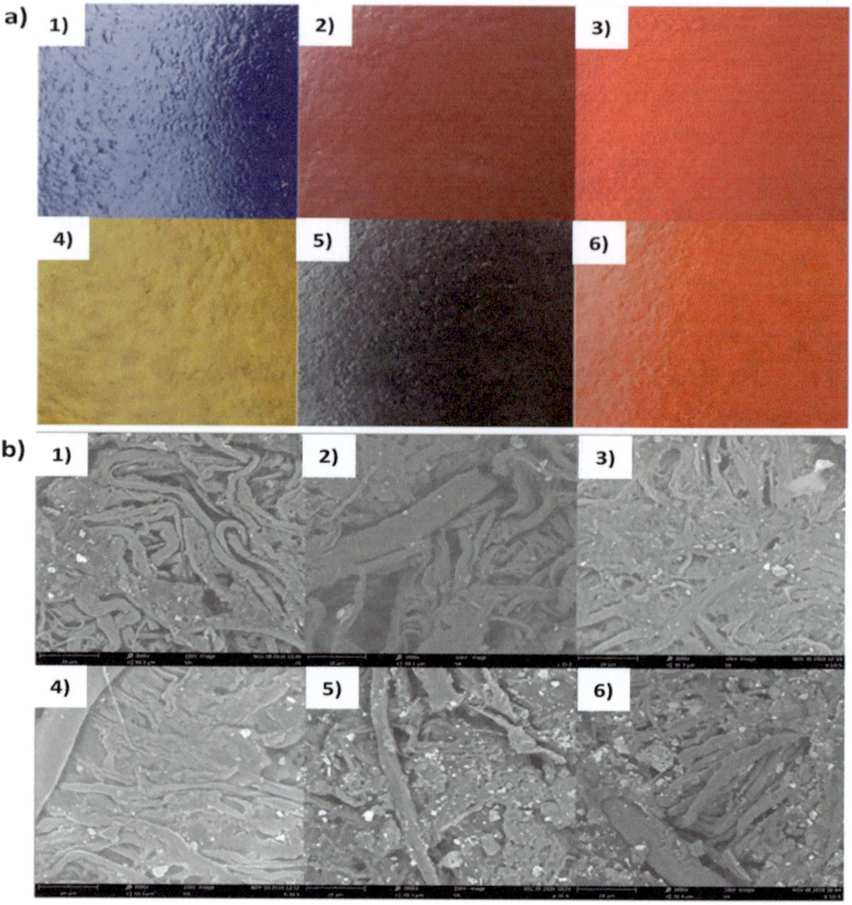

Fig. 13.7 **a** The digital images of finished leather composite boards and **b** scanned electron microscope images of (1) control RCL board; (2) RCL-jute; (3) RCL-hibiscus; (4) RCL-sisal; (5) RCL-palm; and (6) RCL-enset composite boards (Teklay et al. 2017)

13.5.1 Integrated Waste Management and Hierarchy of Tannery Waste Management

Integrated Waste Management (IWM) is described as the selection and deployment of appropriate techniques, technologies, and management programs to fulfil specified waste management objectives and goals (Bifulco et al. 2011). Administrative, financial, legal, architectural, planning, environmental, and engineering activities must all be considered in integrated solid waste management to be responsive to public opinions. All these disciplines must communicate and interact with one another in a constructive multidisciplinary relationship for an integrated solid waste management plan to be successful. (1) Source reduction, (2) recycling and composting, (3)

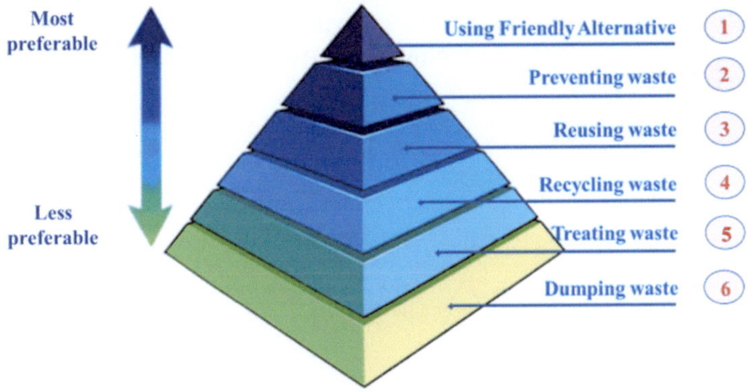

Fig. 13.8 Hierarchy of tanning waste management (Jayakumar et al. 2016)

combustion (waste-to-energy plants), and (4) landfills are the four primary waste management alternatives (strategies) for IWM. The order in almost all countries is similar, with the first entries having more importance than those below them. The hierarchies of tanning waste management are follows, using friendly alternatives (the most preferable), preventing waste, reusing waste, recycling waste, treating waste, and dumping waste (the least preferable). The tanning waste management hierarchy is depicted in Fig. 13.8.

13.5.2 Sludge Management and Disposal

The primary goal of sludge management is to lower the volume and weight of the material to be carried while also increasing the dry matter content. Sludge from primary clarification, secondary clarification, and tertiary clarification is pumped to the sludge thickener. Then it is dried to dewater it. After that, it is moved from the sludge thickener to the sludge drying beds (Spiegel et al. 2012). The biggest stumbling point, however, is the chromium content, which varies greatly depending on legislation and practice. Sludge utilization/application does not have a general solution. Each ETP generates sludge with distinct characteristics, and sludge utilization restrictions vary greatly between areas and countries. When building an ETP, a full review of choices should be prepared, and the most appropriate application should be presented. Zhou et al. conducted a study on tannery sludge disposal method by thermal decomposition technique. It was expected from their study that the energy in tannery sludge could be rebound, and the toxic chromium (Cr) could be settled in the solid residues, that might be reused as a tanning agent, and achieved the target of cleaner leather production. The result revealed that among the different thermal decomposition methods for the disposal of tannery sludge, incineration is risky, while pyrolysis is the secured. The adsorption of downdraft gasifier is the secured than that of updraft gasifier. The reaction type and sequence of raw materials in different regions of the thermal decomposition reactors are schematically illustrated in Fig. 13.9.

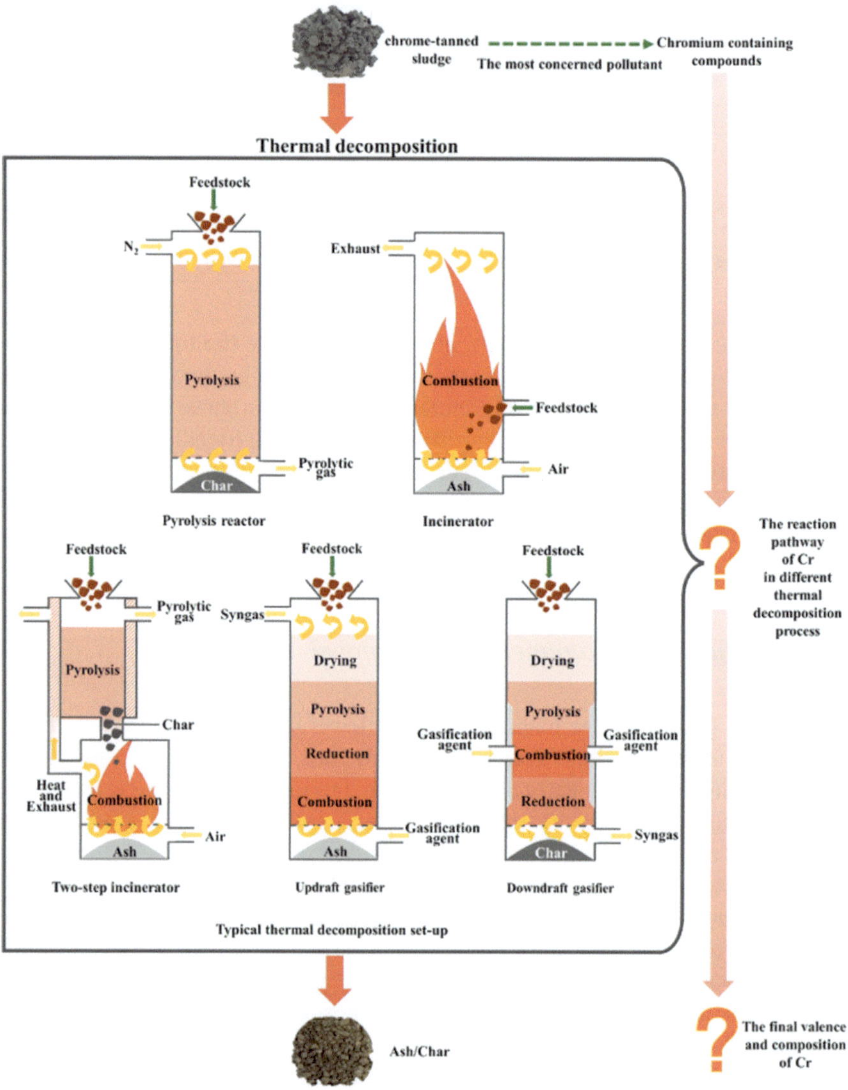

Fig. 13.9 Schematic illustration of thermal decomposition process of chrome tanned sludge (Zhou et al. 2021)

13.5.3 Waste-to-Energy (WTE) Technological Aspects

Different tannery wastes like raw skin/hide trimmings can be converted valuable resources. Many countries produce glue, collagen, peptides, feed, fertilizers, and industrial gelatin from various tannery wasted raw materials. Garbage to energy technology provides a means of managing solid waste while also generating the energy required to reduce reliance on foreign oil and other fossil fuels. Waste to energy conversion can be accomplished using a variety of methods and technologies. Energy can be recovered from organic waste (both biodegradable and non-biodegradable) using one of two processes (thermo-chemical conversion or bio-chemical conversion). A study conducted by Masilamani et al. (2016) on extraction of valuable collagen from tannery trimming waste (a waste to energy approach). The experimental study conducted by using acetic acid and propionic acid for the solubilization of skin matrix collagen. The result revealed that in acidic extraction technique, the acetic acid extraction showed less yield (85%) than that of propionic acid extraction (85%) method. It was anticipated from this study that the tannery trimming waste can be a feasible and economic source of collagen, that can be applied different fields such as biomaterials, cosmetics, tissue engineering, pharmaceutical industry, and so on. The raw hide/skins trimmings are possible substitute source to produce high worth collagen (as shown in Fig. 13.10) and the extraction method can be easily implemented on an industrial scale.

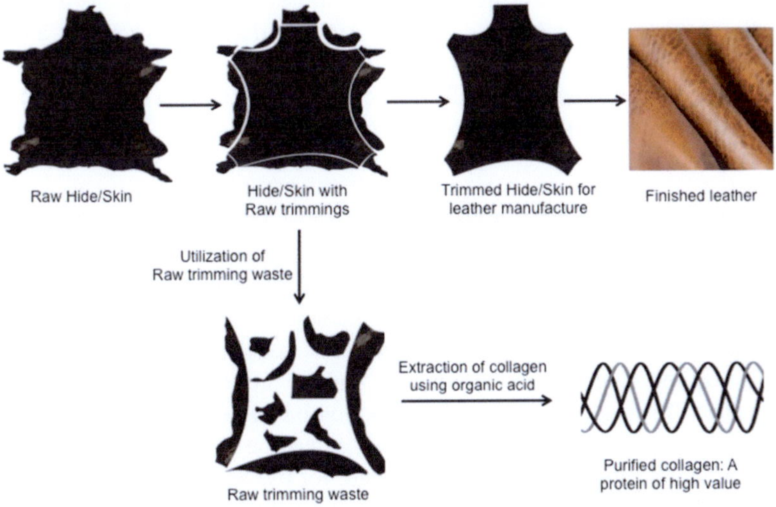

Fig. 13.10 Schematic representation on the utilization of raw trimmings for extracting high value collagen for a waste to wealth approach in sustaining leather manufacture (Masilamani et al. 2016)

13.5.4 *Thermo-Chemical Conversion*

Thermal decomposition of organic materials is used in the thermo-chemical conversion process to produce heat energy, fuel oil, or gas. Thermo-chemical conversion procedures are beneficial for wastes that contain a high percentage of organic non-biodegradable stuff and have a low moisture content. Incineration and Pyrolysis/Gasification are the two most common technological alternatives in this category (Edwin and Sekhar 2018).

13.5.4.1 Incineration

The tannery wastes were subjected to incineration at 800 °C in a Starved Air thermal incinerator under different flow rate of oxygen to optimize the flux of oxygen require preventing the conversion of chromium trivalent to hexavalent. The energy audit of incineration of wastes under the external supply of oxygen was carried out under different conditions. Using Portland cement and fine aggregate, the calcined wastes were effectively consolidated and stabilized. Unconfined compressive strength and heavy metal leaching were investigated on the solidified blocks. To measure the degree of leachate and metals, a TCLP leach ability research was performed on a solidified block. The TCLP leachate had a metal fixation percentage of 99.1–99.9% and a dissolved organic concentration of 55–66 mg/L (Sekaran et al. 2007). The ability to reduce the original number of combustibles by 80–95% is one of the most appealing qualities of the burning process. Air pollution control remains a critical issue in the deployment of solid waste incineration. The cost of the best available technology for the incineration facility in the United States might be as much as 35% of the total project cost in the United States.

13.5.4.2 Pyrolysis and Gasification

Gasification and pyrolysis are two processes that use high temperatures to break down organic waste. Both processes limit the amount of oxygen available during breakdown; gasification allows a tiny amount of oxygen, but pyrolysis does not. Gasification and pyrolysis, in other words, limit or prevent oxidation (Edward and Anbumalar). To make a gaseous fuel, a carbon-based substance is treated with a restricted amount of oxygen or steam. The gas can be purified and burnt in a gas engine or chemically converted into methanol for use as a synthesis product. This method is no longer employed for mass trash, but rather for waste that is reasonably homogeneous (Palani et al. 2017). Pyrolysis is the application of heat to a substance in a low-oxygen environment to produce chemical degradation rather than combustion. Organic material is heated in a confined chamber, releasing valuable gas while leaving residue particles, coke, and, in some circumstances, oil. The gas is then burned to generate energy. The remaining solids might be filled with soil or vitrified. Pyrolysis, like gasification, is a method that is only suitable for homogenous waste.

13.5.5 Bio-Chemical Conversion

The biochemical conversion process is based on the microbial action of enzymatic degradation of organic matter to produce methane gas or alcohol. Biochemical conversion procedures, on the other hand, are recommended for wastes with a high proportion of organic biodegradable matter and a high moisture/water content, which promotes microbial activity. Anaerobic Digestion (Bio-methanation) and fermentation processes, which can take biomass and convert it to ethanol utilizing cellulosic waste or organic material, are the most common technological alternatives in this area.

13.5.5.1 Anaerobic Digestion (Bio-Methanation)

The organic part of the waste is sorted and supplied to a closed container (biogas digester) where the organic wastes undergo biodegradation under anaerobic conditions, creating methane-rich biogas and effluent/sludge. Depending on the waste content, biogas output ranges from 50 to 150 m^3 per ton of garbage. Biogas can be used for cooking and heating, or it can be used to generate motor power or electricity using dual fuel or gas engines or gas/steam turbines. After stabilization, anaerobic digestion sludge can be utilized as a soil conditioner or even marketed as manure, depending on its composition, which is mostly determined by the input waste composition (Johannessen 1999).

13.5.5.2 Composting

Composting is a regulated biological process that accelerates the biological degradation of organic materials by utilizing natural aerobic processes. It is done by microbial populations that break down organic molecules into carbon dioxide, water, minerals, and stable organic matter. Minerals and organic waste are changed into a potentially reusable soil-like compound called compost, while carbon dioxide and water are discharged into the atmosphere. The amount of residual material is typically reduced by 25–60% because to the loss of water and carbon dioxide. Compost can be utilized as a soil amendment in a range of agricultural, horticultural, and landscaping applications as long as proper precautions are taken to ensure the finished product is free of contaminants and impurities.

13.5.6 Tannery By-Products Management Technique

By-products are products of a manufacturing process, usually industrial or biological, that is produced in addition to the main product. Solid wastes, as well as fleshing, trimming, and other tannery processes, are produced during the processing of hides/skins (Karthikeyan et al. 2014). Tannery by-products are a term for certain forms of solid waste. The best accessible waste management techniques and the summary of tannery solid waste and their applications are listed in Tables 13.4 and 13.5.

Table 13.4 A best available technique for waste management (Murad et al.)

Reuse/recycle/recovery and treatment	Type of waste
Leather fiberboard production	Splits, shavings, trimmings (tanned waste)
Leather production or leather making	Splits
Filling materials, wool	Hair and wool
Small leather goods	Tanned trimming and split
Gelatin and hide glue	Raw trimmings, limed fleshing, green and split
Sausage casings	Untanned splits
Protein hydrolysis	Hair raw and limed trimmings, green and limed fleshing, green limed and tanned split, and shavings
Fat recovery	Raw trimming, green and limed fleshing's
Collagen	Splits, limed trimmings
Agriculture and fertilizer	Hair for the nitrogen content, residues from composting and anaerobic digestion, sludges from wastewater treatment
Composition	Hair, green, limed fleshing, tanned splits and shavings, fats, grease and oil, sludge from wastewater treatment
Thermal treatment	Fats, grease, a mixture of no halogenated organic solvents, and oil
Anaerobic digestion	Limed fleshing and green, hair, tanned split and shavings, fats, grease, and oil, sludge from wastewater treatment
Recycling of organic solvents	Organic solvents (no mixture)
Regeneration of air abatement filters	Activated carbon filters
Reuse and recycle packaging materials by feeding them back to the supplier via an appropriate recycling system	Containers, pallets, plastics, cardboard

Table 13.5 The summary of tannery solid waste and their applications

S. No.	Solid waste from tannery		Applications	References
1	Untanned solid waste	Trimmings (raw hides/skins)	Glue	Buljan et al. (2000) Fathima et al. (2012)
			Gelatin	
			Animal feed	
			Cosmetics	
			Shampoo and lotions	
			Bleaching aids	
		Flashings	Cosmetics	Fathima et al. (2005)
			Cleaning agent	
			Cookies	
			Fuel	
		Limed trimming	Glue	Buljan et al. (2000) Fathima et al. (2012)
			Technical gelatin	
			Animal food Tallow making	
		Hair and wool	Carpet manufacturing Fertilizers	Yılmaz et al. (2007)
2	Tanned solid waste	Chrome trimming	Leather board	Taylor et al. (1999)
			Bricks production	Wu et al. (2010)
			Fertilizer and leather meal	Taylor et al. (1999)
		Chrome splitting	Leather meal	Wu et al. (2010)
			Chamois leather	
			Casual and sports shoes	
			Production of bricks	
		Chrome shaving	Leather meal	Wu et al. (2010)
			Protein source for livestock	Fathima et al. (2012)
		Crust leather trimmings	Leather boards and production of bricks	Fathima et al. (2005)
		Finished leather trimmings	Small leather articles	Fathima et al. (2012)
3	Sludge	Waste treatment sludge	Composting	Li et al. (2019)
			Landfill	Assefa et al. (2019)
			Brick manufacturing	Rajamani et al. (2008)
			Incineration (avoid fossil fuel burning)	Edwin and Sekhar (2018)

13.6 Conclusion

Tannery solid wastes are incorporating both untanned wastes (trimmings waste, fleshing, pelt splitting, etc.) and tanned wastes (chrome shaving waste, wet blue trimming, buffing dust, crust trimming, etc.). These solid wastes are constituted severe environmental and health concerns when it is disposed of untreated. However, based exclusively on waste, it is not possible to conclude that tannery leather manufacturing should be halted for the sake of the environment. Since it is possible to control produced waste at a marginal level, pollution can be mitigated, and the cost of living can be reduced by introducing reuse, recycling, minimization, and an integrated waste management system. Additionally, the trash generated from the tannery process can be used to create valuable products and can be utilized in subsequent processes. Nowadays, modern technology allows for the recycling of chemicals that are required in the facility but would be harmful to the environment if not recycled, and this is the way to take it, even though it is high-priced. Although the initial investment cost is significant, the long-term benefits will offset the expense and mitigate the negative environmental effects. That is the main reason tannery solid waste management is very essential because recycling and reuse of this huge waste of resources create additional profits as well as protect the environment from pollution.

Acknowledgements The authors wish to acknowledge the Ethiopian Institute of Textile and Fashion Technology, Bahir Dar University library for supporting and providing books, journals, and conference proceedings whenever needed and Wollo University Kombolcha Institute of Technology for giving the opportunity to take Master of Science in leather product design and engineering at Bahir Dar University.

Author Contributions Bedru Ahmed: Writing—original draft. Bashirunnisa Ananya: Writing—original draft, reviewing and editing. Wendwosen Ebabu: Writing—original draft. Zerihun Teshome: Writing—original draft, reviewing and editing. Zubaida Rukhsana Usha: Conceptualization, Formal analysis, Visualization, Investigation, Writing—original draft, reviewing and editing. Md. Israil Hossain: Supervision, Conceptualization, Formal analysis, Visualization, Investigation, Writing—original draft, reviewing and editing.

Funding Authors didn't receive financial support from any organization for this submitted manuscript.

Conflict of Interest The authors declare that they have no known conflict of interest associated with this publication.

References

Alam J, Hossain M, Beg AH, Nam K-C, Lee S-S (2010) Effects of tannery wastes on the fattening of growing cattle, carcass, and meat quality. Food Sci Anim Resour 30:190–197

Aluyor EO, Badmus OA (2008) COD removal from industrial wastewater using activated carbon prepared from animal horns. Afr J Biotechnol 7:3887–3891

Amalraj A, Pius A (2014) Removal of selected basic dyes using activated carbon from tannery wastes. Sep Sci Technol 49:90–100

Arcibar-Orozco JA, Barajas-Elias BS, Caballero-Briones F, Nielsen L, Rangel-Mendez JR (2019) Hybrid carbon nanochromium composites prepared from chrome-tanned leather shavings for dye adsorption. Water Air Soil Pollut 230:1–17

Assefa R, Bai R, Leta S, Kloos H (2019) Nitrogen removal in integrated anaerobic–aerobic sequencing batch reactors and constructed wetland system: a field experimental study. Appl Water Sci 9:1–11

Babu CJL (2004) Salinity reduction in tannery effluents: CLRI research

Bello AB, Kim D, Kim D, Park H, Lee S-H (2020) Engineering and functionalization of gelatin biomaterials: from cell culture to medical applications. Tissue Eng Part B Rev 26:164–180

Belostotskiy DE, Ziganshina EE, Siniagina M, Boulygina EA, Miluykov VA, Ziganshin AM (2015) Impact of the substrate loading regime and phosphoric acid supplementation on performance of biogas reactors and microbial community dynamics during anaerobic digestion of chicken wastes. Bioresour Technol 193:42–52

Bernat K, Zielińska M, Cydzik-Kwiatkowska A, Wojnowska-Baryła I (2015) Biogas production from different size fractions separated from solid waste and the accompanying changes in the community structure of methanogenic Archaea. Biochem Eng J 100:30–40

Bifulco I, Francese R, Lettieri M, Liscio L, Passero I, Tortora G (2011) The TIE project: agile development of a virtual world serious game on waste disposal. In: DMS. Citeseer, pp 204–209

Bitlisli BO, Başaran B, Sari Ö, Aslan A, Zengin G (2004) Some physical and chemical properties of ostrich skins and leathers

Boonyuen S, Smith SM, Malaithong M, Prokaew A, Cherdhirunkorn B, Luengnaruemitchai A (2018) Biodiesel production by a renewable catalyst from calcined *Turbo jourdani* (Gastropoda: Turbinidae) shells. J Clean Prod 177:925–929

Booramurthy VK, Kasimani R, Pandian S (2022) Biodiesel production from tannery waste using a nano catalyst (ferric-manganese doped sulphated zirconia). Energy Sources Part A Recov Util Environ Eff 44:1092–1104

Buljan J, Reich G, Ludvik J (2000) Mass balance in leather processing. In: United Nations industrial development organization. Regional programme for pollution control in the tanning industry in South-East Asia

Chaudhary R, Pati A (2016) Poultry feed based on protein hydrolysate derived from chrome-tanned leather solid waste: creating value from waste. Environ Sci Pollut Res 23:8120–8124

Chowdhury M, Hossain I, Deb AK, Biswas TK (2019) Removal of toxicants from leather industrial wastewater using sawdust filter media and ferric oxide (Fe_2O_3) coagulant. Orient J Chem 35:597

Christy PM, Gopinath L, Divya D (2014) A review on anaerobic decomposition and enhancement of biogas production through enzymes and microorganisms. Renew Sustain Energy Rev 34:167–173

Collivignarelli C, Barducci G (1984) Waste recovery from the tanning industry. Waste Manag Res 2:265–278

Cot J, Marsal A, Manich A, Celma P, Choque R, Cabeza L, Labastida L, Lopez J, Salmeron J (2003) Minimization of industrial wastes: adding value to collagenic materials procter memorial lecture. J Soc Leather Technol Chem 87:91–97

Cusidó JA, Cremades LV (2012) Environmental effects of using clay bricks produced with sewage sludge: leachability and toxicity studies. Waste Manage 32:1202–1208

De Haan C, Steinfeld H, Blackburn H (1997) Livestock & the environment: finding a balance. European Commission Directorate-General for Development, Development Policy

Deb AK (2015) Study on the treatment of tannery solid wastes of Bangladesh for their potential utilisation. University of Dhaka

Deb AK, Shaikh MAA, Sultan MZ, Rafi MIH (2017) Application of lac dye in shoe upper leather dyeing. Rev Pielarie Incaltaminte 17:97

Dettmer A, Cavalli É, Ayub MAZ, Gutterres M (2012) Optimization of the unhairing leather processing with enzymes and the evaluation of inter-fibrillary proteins removal: an environment-friendly alternative. Bioprocess Biosyst Eng 35:1317–1324

Dwivedi SP, Saxena A (2020) Extraction of collagen powder from chrome containing leather waste and its composites with alumina employing different casting techniques. Mater Chem Phys 253:123274

Ebabu W, Israil Hossain M, El-Naggar ME, Kechi A, Hailemariam SS, Ahmed FE (2021) Exploration of functional polymers for cleaner leather industry. J Inorg Organomet Polym Mater

Edward IA, Anbumalar V. Conceptual design of fluidized dual bed gasification system

Edwin M, Sekhar SJ (2018) Techno-economic evaluation of milk chilling unit retrofitted with hybrid renewable energy system in coastal province. Energy 151:66–78

Fathima NN, Aravindhan R, Rao JR, Nair BU (2005) Solid waste removes toxic liquid waste: adsorption of chromium (VI) by iron complexed protein waste. Environ Sci Technol 39:2804–2810

Fathima N, Rao R, Nair BU (2012) Tannery solid waste to treat toxic liquid wastes: a new holistic paradigm. Environ Eng Sci 29:363–372

Frendrup W, Buljan J (2000) Hair-save unhairing methods in leather processing. United Nations Industrial Development Organization

Ghasemi Y, Rsoul AS, Ebrahiminezhad A, Kazemi A, Shahbazi M, Talebnia N (2011) Screening and isolation of extracellular protease producing bacteria from the Maharloo Salt Lake

Gökgöz F, Güvercin MT (2018) Energy security and renewable energy efficiency in EU. Renew Sustain Energy Rev 96:226–239

Hossain MI, Deb AK, Chowdhury M, El-Naggar M, Sarker MR (2019) Removal of dye basic black 7 (BB7) from tannery wastewater using convenient modified graphene oxide (MGO). GSJ 7

Hossain MI, Deb AK, Sultan MZ, Chowdhury M, Shaikh AA, Sarker MR (2020) Synthesis and application of graphene oxide (GO) for removal of cationic dyes from tannery effluents. Text Leather Rev 3:146–157

Hossain MI, Soliman MM, El-Naggar ME, Sultan MZ, Kechi A, Abdelsalam NR, Abu-Saied MA, Chowdhury M (2021) Synthesis and characterization of graphene oxide-ammonium ferric sulfate composite for the removal of dyes from tannery wastewater. J Mater Res Technol 12:1715–1727

Houshyar Z, Khoshfetrat A, Fatehifar E (2012) Influence of ozonation process on characteristics of pre-alkalized tannery effluents. Chem Eng J 191:59–65

Islam MH, Sarker MR, Hossain MI, Ali K, Noor KMA (2020) Towards sustainable supply chain management (SSCM): a case of leather industry. J Oper Strateg Plan 3:81–98

Jayakumar GC, Kumar G, Tesema A, Dung T, Kobayashi T, Xu K (2016) Bioremediation for tanning industry: a future perspective for zero emission

Jini R, Bijinu B, Baskaran V, Bhaskar N (2016) Utilization of solid wastes from tanneries as possible protein source for feed applications: acute and sub-acute toxicological studies to assess safety of products prepared from delimed tannery fleshings. Waste Biomass Valorization 7:439–446

Johannessen LM (1999) Guidance note on leachate management for municipal solid waste landfills. World Bank

John Sundar V, Gnanamani A, Muralidharan C, Chandrababu NK, Mandal AB (2011) Recovery and utilization of proteinous wastes of leather making: a review. Rev Environ Sci Bio/Technol 10:151–163

Jones B, Bh J (1979) Chronium recovery through incineration of liquid and solid tannery wastes—"the ultimate solution"

Juel MAI, Mizan A, Ahmed T (2017) Sustainable use of tannery sludge in brick manufacturing in Bangladesh. Waste Manage 60:259–269

Kamal Eldin Abu El Fatah M. Recycling of unhairing–liming solution in leather industry. University of Khartoum

Kanagaraj J, Velappan K, Babu N, Sadulla S (2006) Solid wastes generation in the leather industry and its utilization for cleaner environment—a review

Kantarli IC, Yanik J (2009) Use of waste sludge from the tannery industry. Energy Fuels 23:3126–3133

Karthikeyan K, Aaron KP, Krishnaraj K, Chandrasekaran B, Krishna KJ (2014) Study and evaluation of work related musculoskeletal disorder risk in leather garments manufacturing industry. Int J Theor Appl Res Mech Eng 3:2319–3182

Kaul S (2005) Wastewater management: with special reference to tanneries. Discovery Publishing House

Krummenauer K, De Oliveira Andrade JJ (2009) Incorporation of chromium-tanned leather residue to asphalt micro-surface layer. Constr Build Mater 23:574–581

Kumari P, Sen S, Suneetha V (2016) Production of glue from tannery effluent by physical, chemical and biological methods. Res J Pharm Technol 9:1031–1035

Lakdawala M, Lakdawala J (2013) Comparative study of effect of PAC and GAC on removal of COD contributing component of sugar industry waste water. Res J Recent Sci 2:90–97

Lazaroiu G, Mihaescu L, Negreanu G, Pana C, Pisa I, Cernat A, Ciupageanu D-A (2018) Experimental investigations of innovative biomass energy harnessing solutions. Energies 11:3469

Lebedytė M, Sun D (2021) A review: can waste wool keratin be regenerated as a novel textile fibre via reduction method? J Text Inst

Lee M, Lin W, Yu B, Lee T. Animal breeding and genetics

Li Y, Guo R, Lu W, Zhu D (2019) Research progress on resource utilization of leather solid waste. J Leather Sci Eng 1:1–17

Masilamani D, Madhan B, Shanmugam G, Palanivel S, Narayan B (2016) Extraction of collagen from raw trimming wastes of tannery: a waste to wealth approach. J Clean Prod 113:338–344

Meşe P, Karaağaç B, Uyanik N (2018) Investigating effect of chrome tanned leather scraps in ethylene propylene diene monomer rubber. Prog Rubber Plast Recycl Technol 34:89–103

Mohammed FEF (2014) Development and application of cleaner production in rural tanneries. Sudan University of Science and Technology

Murad AW, Mia MAS, Rahman MA. Studies on thewaste management system of a tannery: an overview

Murad W, Mia MAS, Rahman MA (2018) Studies on the waste management system of a tannery: an overview. Int J Sci Eng Technol Res 7

Naidu R (2000) Towards better management of soils contaminated with tannery waste. Australian Centre for International Agricultural Research

Nashy E-SH, Al-Ashkar E, Moez AA (2012) Optical and spectroscopic studies on tannery wastes as a possible source of organic semiconductors. Spectrochim Acta Part A Mol Biomol Spectroscopy 86:33–38

Nigam H, Das M, Chauhan S, Pandey P, Swati P, Yadav M, Tiwari A (2015) Effect of chromium generated by solid waste of tannery and microbial degradation of chromium to reduce its toxicity: a review. Adv Appl Sci Res 6:129–136

Nunes RR, Pigatin LBF, Oliveira TS, Bontempi RM, Rezende MOO (2018) Vermicomposted tannery wastes in the organic cultivation of sweet pepper: growth, nutritive value and production. Int J Recycl Org Waste Agric 7: 313–324

Palani Y, Rao Jonnalagadda R, Fathima Nishter N (2017) Adsorption on activated carbon derived from tannery fleshing waste: adsorption isotherms, thermodynamics, and kinetics. Environ Prog Sustain Energy 36:1725–1733

Proclamation FCS (2007) Federal Negarit Gazeta of the Federal Democratic Republic of Ethiopia. Addis Ababa, Ethiopia

Rajamani S, Streit K, Casey P (2008) Environmental update on world leather sector from International Union of Environment (IUE) Commission of IULTCS. In: XVIII Congresso Latino-Americano dos Quımicos e Técnicos da Indústria do Couro

Ravindran B, Nguyen DD, Chaudhary DK, Chang SW, Kim J, Lee SR, Shin J, Jeon BH, Chung S, Lee J (2019) Influence of biochar on physico-chemical and microbial community during swine manure composting process. J Environ Manage 232:592–599

Rigueto CVT, Nazari MT, De Souza CF, Cadore JS, Brião VB, Piccin JS (2020) Alternative techniques for caffeine removal from wastewater: an overview of opportunities and challenges. J Water Process Eng 35:101231

Saravanabhavan S, Thanikaivelan P, Rao JR, Nair BU (2005) Silicate enhanced enzymatic dehairing: a new lime-sulfide-free process for cowhides. Environ Sci Technol 39:3776–3783

Sawalha H, Alsharabaty R, Sarsour S, Al-Jabari M (2019) Wastewater from leather tanning and processing in Palestine: characterization and management aspects. J Environ Manage 251:109596

Sedliacik J, Matyasovsky J, Smidriakova M, Sedliacikova M, Jurkovic P (2011) Application of collagen extracted from chrome shavings for innovative polycondensation adhesives. J Am Leather Chem Assoc 106:332–340

Sekaran G, Swarnalatha S, Srinivasulu T (2007) Solid waste management in leather sector. J Des Manuf Technol 1:47–52

Senthil R, Hemalatha T, Manikandan R, Das BN, Sastry TP (2015) Leather boards from buffing dust: a novel perspective. Clean Technol Environ Policy 17:571–576

Skrzypczak D, Szopa D, Mikula K, Izydorczyk G, Baśladyńska S, Hoppe V, Pstrowska K, Wzorek Z, Kominko H, Kułażyński M, Moustakas K, Chojnacka K, Witek-Krowiak A (2022) Tannery waste-derived biochar as a carrier of micronutrients essential to plants. Chemosphere 294:133720

Spiegel J, Alegret M, Clair V, Pagliccia N, Martinez B, Bonet M, Yassi A (2012) Intersectoral action for health at a municipal level in Cuba. Int J Public Health 57:15–23

Sullivan K, Thomas S, Rosano M (2018) Using industrial ecology and strategic management concepts to pursue the sustainable development goals. J Clean Prod 174:237–246

Tang Y, Zhao J, Zhang Y, Zhou J, Shi B (2021) Conversion of tannery solid waste to an adsorbent for high-efficiency dye removal from tannery wastewater: a road to circular utilization. Chemosphere 263:127987

Taylor MM, Cabeza LF, Brown EM, Marmer W (1999) Chemical modification of protein products isolated from chromium-containing solid tannery waste and resultant influence on physical and functional properties. J Am Leather Chem Assoc 94:171–181

Teklay A, Gebeyehu G, Getachew T, Yaynshet T, Sastry TP (2017) Preparation of value added composite boards using finished leather waste and plant fibers—a waste utilization effort in Ethiopia. Clean Technol Environ Policy 19:1285–1296

Thanikaivelan P, Rao JR, Nair BU, Ramasami T (2005) Recent trends in leather making: processes, problems, and pathways. Crit Rev Environ Sci Technol 35:37–79

Ullah Khan I, Hafiz Dzarfan Othman M, Hashim H, Matsuura T, Ismail AF, Rezaei-Dashtarzhandi M, Wan Azelee I (2017) Biogas as a renewable energy fuel—a review of biogas upgrading, utilisation and storage. Energy Convers Manage 150:277–294

Verheijen L, Wiersema D, Pol LH, Brandjes P, De Wit J, Van Der Meer H, Bos J, Westra P, Nell A, Jansen J (1996) Livestock and the environment: finding a balance

Vickers NJ (2017) Animal communication: when I'm calling you, will you answer too? Curr Biol 27:R713–R715

Vico A, Artemio N (2017) Biogas: production, applications and global developments

Wu TY, Mohammad AW, Jahim JM, Anuar N (2010) Pollution control technologies for the treatment of palm oil mill effluent (POME) through end-of-pipe processes. J Environ Manage 91:1467–1490

Yilmaz O, Kantarli IC, Yuksel M, Saglam M, Yanik J (2007) Conversion of leather wastes to useful products. Resour Conserv Recycl 49:436–448

Zăinescu G, Deselnicu V, Constantinescu R (2018) Composite structures containing leather fibers with applications in constructions industry. Proc ICAMS

Zemleduch-Barylska A, Lorenc-Plucińska G (2015) Populus × canescens grown on Cr-rich tannery waste: comparison of leaf and root biochemical and proteomic responses. Plant Physiol Biochem 90:1–13

Zhou Y, Chen Z, Gong H, Yang Z (2021) Chromium speciation in tannery sludge residues after different thermal decomposition processes. J Clean Prod 314:128071

Chapter 14
Fostering Circular Economy Principles in Tannery Sector: Accomplishment Strategies

Magesh Kumar Marichamy

Abstract Tannery industries play a vital role for economy booming but the activities related with leather processing pose severe threat to three fundamental elements of nature, namely land, water and air. Longevity and prosperity of living beings on earth depends on healthiness of these natural elements. Due to globalization and industrialization, the quality index of environmental factors is reduced. The complexity of contaminants produced during tannery operation challenges scientific community to develop a process that mitigates pollutants load in an economical way. Even though several methods like adsorption, coagulation, ultrafiltration, advanced oxidation process, photo degradation, etc., can liquidate primary pollutants but it ultimately leads to generation of secondary pollutants, exorbitant capital investment, operation and maintenance cost. In order to avoid piling up of "treatment plants" scale-up costs, the promotion of circular economy in industries is important. In present day, the natural resources are excavated deep beneath earth's lithosphere, converted them to useful products and throw them as waste once the material life is expired. This process is linear in contrast to circular economy (CE) where the used or worn-out materials are kept in utility loop through processes like overhauling, refurbishment, reuse and recycling. The natural resources are finite and if the consumption rate is not brought down to zero level, the human race and living beings will face the nature's fury like global warming, climate change, loss of biodiversity, health complications and ultimately extinction. It is the responsibility of policy makers and scientific community to work in-tandem with each other to disengage economic activities from natural resources consumption. Circular economy practices are important in tannery industries to meet the regulatory norms and to attain sustainability in product manufacture. The objective of CE is to valorize the by-products produced during tanning process in one stage by using them in subsequent stages of process. An implication of circular economy principles in tannery operation is a boon because it minimizes resource consumption, prevent environmental degradation, convince stakeholders and smoothen the environmentalists. It

M. K. Marichamy (✉)
Department of Chemical Engineering, SRM Institute of Science and Technology, Kattankulathur, Tamil Nadu, India
e-mail: mmagesh23@gmail.com

M. A. Aboulhassan and S. Souabi (eds.), *Sustainable Practices in the Tannery Industry*,
https://doi.org/10.1007/978-3-031-92037-0_14

is a tactics to attain equilibrium between rate of industrial growth, environmental health conservation and economic growth without compromising the grade of each. CE practices have gained momentum among industries and academicians owing to its resource and environmental benefits apart from showing the pathway to achieve millennium development goals (MDGs) put forwarded by United Nations for the wellbeing of global community. This particular chapter highlights the strategies to implement circular economy practices to optimize water and chemical use along with frameworks to attain sustainable development goals (SDGs).

Keywords Tannery wastes · Water reuse · Chemicals optimization · Circular economy · Cradle-to-cradle · Sustainable development goals

14.1 Introduction

The wastes generated from tannery industries are of grave-concern because of the hazardous nature of complex contaminants. The wastes encompass all phases, namely solid, liquid and gas which creates multidimensional problems for the waste processing units. Hence, a holistic approach is needed to tackle the detrimental pollutants either at upstream stage or intermediate stage or in downstream stage. Tannery industry is one of the oldest manufacturing units to convert fresh unprocessed hide into chemically modified leather products, namely footwear, bags, dress wear, etc. Though tannery industry is essential for economic growth of nations, it also affects the ecosystem to a significant extent. The raw hides, procured from slaughter houses, are physical treated and chemically altered to produce flexible good quality leather. Tannery wastes comprise mainly of organic biomass (skin, hides, hair, etc.) and toxic chemicals used in hair removal, skin softening and to avoid putrefaction process. Only 200 kg of usable leather is produced when 1000 kg of raw hide is process, which implies 80% of let-out from the process is considered as waste. The typical distribution of waste is: 120 kg untanned trimmings, 150 kg untanned flesh, 115 kg tanned wastes from splitter, 100 kg of trimmed wastes, 32 kg dyeing wastes, 150 kg suspended solids and 20 kg chemical wastes (Saira and Shanthakumar 2023). The aqueous wastes generated in tannery process contain chromium species, heavy metals, persistent dyes, sodium salts, pH modifier, suspended solids and biodegradable organics. Moreover, the process generates rotten-egg-like smelling hydrogen sulfide gas along with pungent smelling ammonia gas. Both the gases turn poisonous, if the threshold concentration in air exceeds the 100 ppm limits. If these wastes are expelled lethargically without any pre- or post-treatment, then the impacts are permanent and irreversible. Land disposal of solid wastes, permits the percolation of heavy metals chromium, lead, copper, barium, etc., to soil subsurface and contaminate underground water table. This causes bioaccumulation and biomagnification of heavy metals via the food chain from plants to human beings and animals. The solid sludges can be either minimized (by process up gradation) or reused or transformed into energy (by pyrolysis and incineration process). Liquid wastes should be dealt with stringent norms because the contaminants are easily transported by flowing water bodies. The solid wastes are location centric because the wastes are handled and treated in dump yards. The probability of pollutant spread and distribution is restricted in case of solid wastes because there is no carrier for these wastes to spread to various geospatial locations. Liquid effluents can be tackled with microbial actions and physico-chemical methods. Hence, the contaminants are present persistently in any one component of the ecosphere and it necessitates the scientific community to identify eco-friendly option to mitigate tannery pollutants for sustainability. The three major directions to support eco-friendly practices in tannery industry is levying of ecotax, analysis of product carbon footprint and replenishable feed materials. Ecotax, also called environmental tax, is an essential tool to improvise green practices and civilized ecological methods in industries. These activities accelerate the rate of promoting green manufacturing policies, optimizing energy consumption and switch over to circular economy principles (Shen and Zhang 2022).

Circular economy (CE) is a resource cum materials oriented management principle to keep the materials in loop in a process to preserve nature and maintain environmental requirements. CE has the ability to improve the environmental conditions, boost economic status and elevates social equity. The non-circular economy (linear economy) affects all aspects of environment, overburdens under privileged living beings and hamper the health progress (United States EPA 2024). CE is an ultimate attempt to balance the relationship between industrial growth and economic boom without affecting the environmental conditions. Economic development is a key driver for resource exploitation and environment decay. As of now, wealth recovery from waste, resource recycling for reuse and process modernization are the vital components to achieve tangible outputs for sustainable production practices, increase in green technologies and minimizing the dependency on earthy mineral sources. CE inculcates the knowledge of repairing, reusing, refurbishing and recycling of intermediates and final products. This approach extends the life cycle of materials to infinity or the material become unusable (European Parliament 2024). The waste generation is reduced to a minimum, if the circular economy principles are implanted in the minds of young generations, entrepreneurs, common public and corporates. The bottom-line is, when a product is reached life expiry then the base materials used to make the product are kept in user loop to create social and economic values of the refurbished product. "Planned obsolescence" is a strategy followed by product makers to deliberately limit the lifespan of product which mesmerizes the consumers to purchase new product. Due to globalization and open boundary trading, the significance of reduce, recycle and reuse has slowed down in recent decades but due to international pressure and environmental activist's voice, a reverse trend is observed. Successful implementation of CE would retard the exploitation of natural resources and also prevents further worsening of climatic conditions, biodiversity loss (Pringle et al. 2016). If the products and processes are more toward sustainability, then energy consumption and resource usage is reduced. Tannery process intensively uses natural resources and synthetic chemicals for wide range of products and generates voluminous wastes which needs to be addressed in accordance with circular economy principle. CE principles are applied in tannery sectors through biological processing of solid wastes to retrieve protease, chemical processing to segregate chromium and thermal treatment for power generation. Even though there are few challenges to implement leather product recycling like, use of obsolete technology and development of secondary markets for retrofitted products, the scorching pollution problems mandates to administer CE principles in process industries. The aim of this chapter is to highlight the importance of preserving water bodies, ecosystem and elements of nature for sustained future. The significance of water resources management and protection is highlighted in a classical Tamil language literature "Thirukkural" penned by Tamil philosopher, poet and saint Thiruvalluvar who lived 2000 years ago (National Informatics Centre, GOI). In chapter 2, verses 20, saint Thiruvalluvar emphasized the importance of water in his divine couplets saying "If there is no water, the functions of nature ceases and world cannot exist; if there is no rain, morality cannot exist" (Govt. of Tamil Nadu, India). This shows the vision of our ancestors who have coexisted and protected nature because they realized the fact that, human beings cannot progress in sustainable manner by disturbing the ecosystem balance. They appreciated the importance of water cycle (via rain), solar energy utilization and elements recycle (through agriculture). In this twenty-first century, even though the people have access to latest tools, mechanisms to protect and rejuvenate nature, a declining trend for environment health is observed. Hence, global efforts are needed to overcome the turbulences rising ahead of environment protection activities.

14.2 Tannery Industry Waste Types—Sources and Characteristics

Tanning industries are facing stringent legal issues and confronted by environmental activists even if the output from these industries are perturbed slightly from regulatory norms. Pollution prevention and cost optimization should be the motto of tanners to meet discharge standards and reduce cost of treatment. The interpretation of physico-chemical data's obtained in tannery industry should be done carefully because of non-simplicity of wastes generated. The types of pollutants generated in tanning operations are shown in Fig. 14.1.

To have a better insight of the problem, the tanners must understand the relationship between the leather processing principles and waste treatment methods followed. For producing fixed quantity of leather products, the amount of chemicals required is also fixed. If water rationing is followed in industries, the concentration of pollutants will spike in effluents and treatment become difficult and challenging (UNIDO 2011). Tanning wastes are generated from various unit operations and unit processes in leather industry utilizing different varieties of chemicals making the wastes more complex which need strenuous exercise to characterize the wastes. Integrated treatment strategies is necessary to tackle the deleterious contaminants but the process choice is decided by contaminant qualities and end-user requirements. The types of

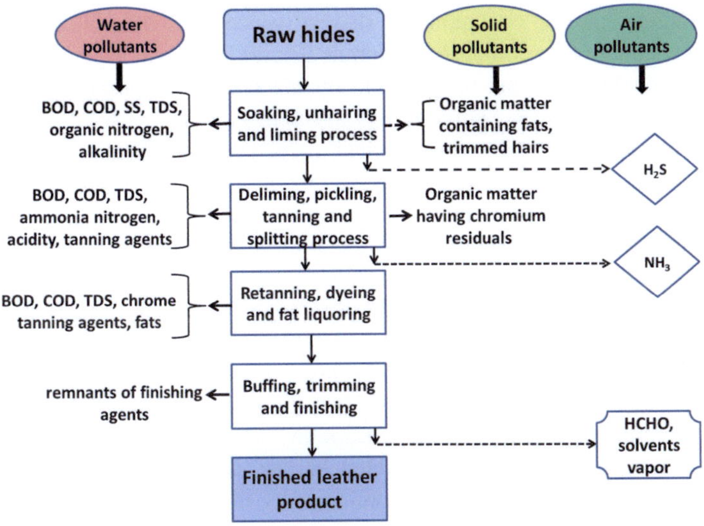

Fig. 14.1 Types of pollutants generated in tanning operations. BOD—biochemical oxygen demand, COD—chemical oxygen demand, SS—suspended solids, TDS—total dissolved solids

pollutants formed during leather processing include solid wastes, liquid effluents, air emissions, chemical wastes and sludges. It is estimated that 16% of solid waste is generated while processing fresh hides (FAO, United Nations). A statistics on typical solid wastes based on 1000 kg raw hides, from tanning operation is reported in Table 14.1.

The leather wastes after tanning operation contain splitted, shaved and trimmed let-out hides whose typical composition is 4.5 kg fat, 15 kg of minerals and 4 kg of chromium as Cr_2O_3 per 100 kg of solid wastes (Famielec and Wieczorek-Ciurowa 2011). Solid wastes are generated from trimming, fleshing, splitting and shaving operations. Remnants of trimmed hides and skins are rich in protein used in animal feeds and gelatin manufacture, while the fatty tissues removed in fleshing operation are used to extract animal oil. Low-quality leather products are made from splitting

Table 14.1 Amount of solid wastes in percentage generated in tanning operations

Process/operations	Pre-tanning	Tanning	Finishing
Trimmings (%)	12	11	3.2
Fleshings (%)	15		
Wet blue split (%)		11.5	
Buffing dust (%)			2
Operation wise total (%)	27	22.5	3.4
Grand total of solid wastes (%)	53		

Note All the quantities are based on mass % of wastes generated waste, whereas chrome contaminated shaving waste is used to fill the interstitial space between atoms in composite materials (Flores Tapia and Moina 2023). Water is used for soaking, liming, tanning and dyeing operations in tannery industries. The effluent from soaking operation is loaded with suspended solids (SS), Chemical Oxygen Demand (COD), Biochemical oxygen demand (BOD), total dissolved solids (TDS) and organic nitrogen. Liming operation uses lime and sodium sulfide thereby increased alkalinity, precipitate formation and salty taste is observed in effluent. Liquid effluent from tanning operation is saturated with chromium salts and tannins, whereas dyeing operation generates residual coloring impurities, resins and pigments. Chromium compounds are primarily found in exit stream from chrome tanning process. It is washed away from the leathers while carrying out re-tanning and dyeing operation. The chromium ions are present in soluble form when dispensed in effluent but the solubility decreases when the exit stream is mixing with other effluents containing proteins. Hence, chromium salts precipitates to form protein–chromium complex which become added burden as deposited sludge. Since the colloidal particles are stabilized by chromium salts, the probability of partial tanning of protein molecules become high. Chromium hydroxide precipitate is formed if the chromium–protein complex is degraded but the precipitate remain in ecosystem for long period of time. The possible oxidation states of chromium are Cr(II), Cr(III) and Cr(VI), in which trivalent chromium is 500 times less toxic than the hexavalent form. Cr(VI) is carcinogenic, damages vital organs of all living beings, whereas trivalent form has more stability in aqueous phase leading to prolonged stay in water bodies (Nur-E-Alam et al. 2020). A full fledge research and continuous preaching is needed to understand and combat the negative impacts of tannery wastes. Liquid effluent is generally characterized by extreme BOD, COD, total dissolved solids, sulfides, chlorides, metallic and non-metallic components. Organic-rich untreated effluent from tanneries cause increased uptake of oxygen for aerobic decomposition and growth of flora. The turbulence generated by randomized inflow of wastewater enhance solubility and diffusion coefficient of oxygen, thereby leading to a process called eutrophication. This abnormal growth of plants and exotic weeds due to excess nutrients form a non-penetrative layer for sunlight thus inhibiting the progress of aquatic life cycle (Bhardwaj and Kumar 2023). The physico-chemical qualities of effluent depends on geological place of source and the characteristics varies dynamically with respect to time. The liquid effluent from soaking, liming and deliming processes has elevated alkaline value whereas the wastewater from pickling, chrome tanning process is loaded with acidic ions and coloring impurities. The COD of wastewater produced in liming operation (15,000–30,000 mg/L) is more than that of COD pollutant present in effluent from soaking and deliming operation. The complexity and types of contaminants in tannery wastewater is primarily because of alteration in process conditions, variation in types, quality and quantities of chemicals used. Hence an understanding about the sources and characteristics of effluents is crucial to design optimized plant for water recycle and reuse.

14.3 Scope for Waste Minimization and Volarization in Tanneries

Leather production is a valorized output of hide and skin processing but at the cost of negatively impacting the environment. There are scopes to minimize the detrimental impacts to environment due to leather production. The quantum and quality of waste generated can be minimized by cleaner technologies, process modifications and use of advanced processes. The major pollutants produced from soaking process are dissolved salts, dirt, globular proteins, etc. It is imperative to reduce chloride concentration because it is not removed easily by any one of the standard treatment strategies. The salt load in effluent can be minimized by shaking-off the extra salts attached physically to the surface of raw skin. The methods used for desalting operation include salt shaker machine, brush-type desalter and desalting drum. The major focus of these methods is to remove physically adsorbed salt from the hide surface. The salt shaker machine is made up of inclined stainless steel drum rotates at 6 rpm. The time required for salt removal from hide depends on inclination angle made by steel drum with base surface and rotation speed. In brush-type desalter, nylon bristles are embedded on interior lateral surface of cylindrical drum. The adhesive force between salt molecule and hide surface is weakened due to pull force exerted by bristles on adhered particles. The efficiency of process primarily depends on drum rotating speed, angle of inclination and gravitational force. The removal efficiency of salt shaker machine is more than brush-type and rotating drum-type desalter (Department of Industrial Policy and Promotion, Government of India). By these methods, the salt load is decreased to the tune of 8% which chloride ion is removed on the basis 91 kg Cl^- ions per ton of raw hide being processed. The other way to reduce salt load is the usage of eco-friendly antiseptics like boron compounds, sodium sulfite or by means of commercial bactericides. This method reduces the salt pollutant to the tune of 50 kg Cl^- ions per ton of raw hide but it suffers from poor preservation efficiency. Instead of using salts to prevent decay of hides, green methods like preserving the hides by normal cooling and chilling operations can be followed. Removal of excess salts may offer environmental benefits like drastic reduction in TDS value and hence the biological treatment process efficiency increases because of its reduced salinity. An added advantage due to desalting is the increase of recovery rate and membrane lifetime in reverse osmosis plant. Moreover less concentration of salt eases the disposal of salt-laden solid concentrates from the evaporator. The major pollutants discharged along with liming effluents are suspended solids, chemical oxygen demand (COD), biological oxygen demand (BOD) and total Kjeldahl nitrogen. It is observed that, recycling of spent floats can minimally reduce the BOD and COD concentration in soaking process but quality of leather is negatively impacted by recycling operation. To overcome this quality degradation, unhairing and opening-up operations are finished in two steps in a discrete manner. Almost 73% of suspended solids and 40% COD pollutant load is decreased when unhairing and liming process is based on sulfide-lime-based process (UNIDO Programme 2000). Pollution abatement, being the top priority in any industries, minimizes pollutants load and reduce

Table 14.2 Typical pollution parameters in tannery effluents and their environmental impacts

Pollution parameter	Environmental impacts
Biochemical oxygen demand (BOD_5)	Elevated BOD decreases the concentration of dissolved oxygen in water bodies and affects aquatic ecosystem
Chemical oxygen demand (COD)	Oxygen for aquatic life is depleted when COD is increased. The responsible chemical compounds entering recipient water bodies are not biodegradable
Total suspended solids (TSS)	Magnitude of BOD and COD increases due to particles contributing for TSS. In effluent treatment plant, sludge generation is accelerated due to TSS
Total dissolved solids (TDS)	High TDS causes elevation of osmotic pressure and makes the water even unfit for irrigation and livestock watering
Oil and grease	It forms floating layer thereby preventing light and oxygen to enter water bodies
Chromium, Cr^{3+} (III)	Responsible for chromium hydroxide formation which persistently present in ecosystem in the form of precipitates
Chlorides (Cl^-)	Overgrown concentrations prevent growth of plants, microbes and other water living beings

treatment costs. In tannery industry, the quantity of pollutants is minimized using chilled-cooled raw hides, skins instead of salt preservatives, ammonia-free deliming, bating operation and novel techniques for chemicals management (Buljan and Rajamani 2023). In spite of all precautions taken to curtail the pollutants nuisance, a meager amount of environmental disruptors has to be tackled by end-of-pipe (EOP) methods. Typical pollution parameters in tannery effluents and their environmental impacts are summarized in Table 14.2.

Solid wastes generated in tannery process can be used as secondary raw materials in adsorption, biopolymer making, and agriculture land manure and biopolymer manufacture. Tannery sludge subjected to anaerobic digestion process can degrade organic constituents to biogas and compost. Solidified tannery sludge is a suitable alternate for cement concrete in construction industry. Solid wastes vented from tannery processing plant are tested as viable alternate for sustainable bio-cementation process (CPCB India 2019).

14.4 Circular Economy Principles for Industries

A circular economy is an ethical approach to conserve natural resources from being exploited by humans. The current system of natural resources usage obeys "linear economy" principle in which non-replenishable natural reserves are exploited, converted them into products and discard the products as waste once they become old. The concept of recycle, reuse and refurbish is considered as obsolete in linear economy principle, whereas these are the key ingredients for circular economy. The over-utilization of natural resources in linear economy is sole responsible for climate change, global warming, biodiversity collapse, environmental crisis. In contrary to this, circular economy promotes sustainable growth which aims at minimization of waste, reduced use of fresh natural resources, promoting recycling, reuse concepts, etc. It is easy to follow circular economy practices by individuals as well as industrial manufacturers, corporates, etc., and even it can be followed in day-to-day activities also. In agriculture sector animal wastes, biomass wastes can be used instead of chemically synthesized fertilizers like urea, super phosphate, etc. For super phosphate manufacture, phosphate rocks are mined from source place using explosives which creates dust and noise pollution. In textile industries, efforts are taken to use organic cotton, natural fibers and natural dyes to safeguard end-users health and maintain environmental standards. The use of plant and animal-based materials for fiber and cloth making is a sustainable practice to reduce environmental quality degradation. In construction industries, the use of novel materials, organic wastes, and carbon-based materials and reuse of broken concrete wastes is considered as outcome of circular economy model (United Nations Development Programme). Sustainability in production and environment is a burning issue among scientist, industrialists, nature lovers and policy makers. There is a need to achieve reasonable environmental and economic goals by following greener economy (circular economy) principles which in turn mitigates rate of greenhouse gases emission, hazards pollutants release and by using renewable sources for energy need instead of fossil fuels. The production process should use the resources meticulously and advanced technologies, keeping in mind about the havoc caused by inefficient process (Shahzabeen et al. 2023).

14.4.1 Methods to Achieve Circular Economy

It is estimated that, for each year in-excess of 100 billion tons of natural resources enter into economic activities to make products but on an average 60% of under-used products is sent to trash as wastes or emitted as greenhouse gases. Neither the supplier nor the consumers worry much about "take – make – waste" model in which the natural resources are mined, converted to under-used products and disposed as wastes, well before they lose shine and usage potential. The linear model depletes natural resources at an accelerated rate, acts as driving force for climate crisis and pushes our mother earth to the verge of extinction. There is a direct correspondence between linear economy and "earth overshoot day" because erratic and unethical exploitation of natural resources will lead to extinction of living beings and ultimately converts earth as a waste barren land. Human being greediness and crazy consumer attitude is cited as prime factor responsible for overshoot. It occurs when human's exploitation of natural resources exceed bio-capacity of earth which literally means the rate of consumption of natural resources exceeds the rate at which it is formed. If necessary steps are not taken at this instant then it is impossible to develop equitable and prosperous communities in the future. Circular economy is built in such a way that products stay in loop for long time thus minimizing wastes and pollution problems. If circular economy principle is followed by every individual meticulously then ample time will be available for nature to rejuvenate and refresh. Migrating toward circular economy would make a significant contribution for environment preservation and slow down of climate change. It is estimated that CO_2 gas emission is reduced by 3.7 billion tons by 2025 when circular economy principles are followed in cement, aluminum, steel production, plastics and food industries. Circular economy not only addresses environmental concerns but also takes care of sensible use of scarce, non-replenishable resources by designing a production process and usage pattern to keep the materials in end-user phase till it long lasts for lengthier duration. As a matter of fact, one ton of thrown-away cellular phones have 100 times more gold than that would be present in one ton of raw gold ore. These statistics open a new dimension for wealthiest job opportunity that converts waste to wealth. The prime objective of recycle, reuse, retrofit, refurbish, etc., is to break the bond between economic growth and natural resources usage pattern. Precautionary measures to cut-off the unnecessary activities should be highlighted along with untying the relationship between economic growth and environmental destruction (WRI 2024). The products need to be used very intensively in circular economy, to minimize the raw material consumption for new product making. The efficacy of circular economy principles depends on policy makers, law enforcers and mainly on consumer behavior. Since, the end products are used by consumers they have a major role to prevent environmental degradation and successful implementation of circular economy principles. Industries and companies are facing stringent norms and guidelines from law makers to follow zero discharge policy, severe penalties for violating the norms and even taken into custody if they industrial personnel are found guilty. Moreover, the proportion between the number of process industries and population growth is incomparable and hence the quantum of waste generated by common public overwhelms the amount of waste produced due to industrial activities. Hence for a successful waste mitigation process, the tail ender's (common public) role is vital and indispensible. Consumers can contribute to the efforts by choosing nature's friendly products, prolonged use of currently using products, repair the faulty product and distribute to second-hand vendor. These attitudes can be inculcated even from the childhood days by incorporating the importance of life-value of nature in both educational institutions and in awareness campaigns (Government of the Netherlands 2024).

The concept of circularity should begin from individuals then expands to family, village, small town, cities, province (state), country and then to global level. These steps are considered as "bottom-up" approach and may attain the target successfully if the line and delivery of action is perfect. If the products are used within a small region, then it accounts for reduced emissions, less dependency on material import and lower usage costs because the materials are kept in loop in usage cycle. Waste volarizing industries collect the superficially used products, refurbish and then resell to improve local community job opportunities and financial status.

14.4.2 Importance and Benefits of Circular Economy

The meticulous use of natural resources is influenced by circular economy principles because the management and utilization of non-renewable resources is highly chaotic and dynamic in nature. Circular economy principle consciously focus on reduction of fresh raw material use, recycle of partially worn-out products and recover value added products from industrial wastes. The conventional methods of industrial process produce primary, secondary and tertiary contaminants inevitably which burdens environmental policy implications. Although industrial activities boost economy, it violates nature's principle by means of ecological degradation and natural chain destruction. The concept of circular economy when applied to industrial equipment's and processes meliorate the influence on ecosystem (Hu et al. 2011). A successful transition from linear economy to circular economy is in the hand of consumers, government agencies and stake holders. The end-users, primarily consumers, should be responsible while procuring the products and disposing to the environment safely. Consumers have a mindset that, products made using recycled materials and second-hand products are inferior qualities. Also the attitudes to change the products like cell phones, cars, etc., every year should be considered. This lethargic mental attitude and spillover in utilizing the eco-friendly products acts as a barrier to attain sustainability in material flow in economic growth (Stangherlin et al. 2023). For implementing CE in industries and organizations, necessary actions should be initiated since the inception of an industry from nascent stage.

Circular economy offers both direct and indirect benefits to all forms of living beings and to ecosystem. The three R's: reduce, reuse and recycle is the guiding principle for framework formation and implementation of circular economy practices. These three factors when integrated into production and consumption stages, reduces wastes, minimize greenhouse gas concentration and preserve natural resources. It forms a closed loop system that promotes efficiency in resource usage pattern and takes care of environmental aspects. A shift to circular economy from linear economy offers multitude benefits encompassing environmental, economic and social domains. It promises a bright future by taking care of environmental well-being and sustainable growth is achieved harmonically. CE has synergistic effects on environment because following three R's principle reduces the amount of waste generated due to industrial and human activities. As a consequence of this, pollutants load decreases, natural resources are preserved and biodiversity is conserved. Durable and easily repairable new products are designed and developed to minimize the frequency of product replacement and discardment. An implication of CE principles contributes equally to economic benefits for business doers and common public. It fosters new skills and talents to repair and refurbish life-expired products. Moreover following circular economy practices will decouple economic booming from resource utilization rate. The other benefit is the development and strengthening of local communities and promotes the shared responsibility of concerning for the environment. Linear economy causes human health deterioration and quality of life is spoiled because each and every time fresh resources are mined, processed and utilized. All these activities cause air, land and water pollution. Strengthening the tools of circular economy gives major boost to combat climate change, global warming, abnormal pollution, resource depletion rate, etc. Proper implementations of circular economy principles have pragmatic influence on environment, preserve non-replenishable raw materials, reduce waste generation and pollution, improve health and life quality, create new job opportunities in waste handling sectors, attain breakthrough point in sustainable production practices and enhance start-up ideas for eco-friendly process design. The benefits of following the circular economy depends on how much faster the awareness is created among the students, common public and manufacturing sectors.

14.4.3 Relationship Between Circular Economy and Resource Recycle

Government agencies, authorities, bureaucrats and legal policies have a major role in developing and stringent implementations of circular economy framework. A government has to encourage better product design, facilitate better choice of consumption, increase business competition, cluster policies formation and industrial estate development to create an ambience of industrial symbiosis for sharing and exchange of raw materials and products within an industrial complex to reduce logistics and operating costs (Brears 2018). In the past decades, due to industrialization and urbanization, there is an unprecedented erratic consumption of natural raw materials. Utilizing the resources efficiently and shifting toward circular economy not only guarantees material security but also mitigates the challenges to environment and economy. Economic growth is directly proportional to raw material consumption and environmental degradation. The need of the hour is to decouple the interrelationship between material usage and economic activities all at industrial level, residence level and individual human level. The wastes generated from a man-made process are highly dreadful and multi-complex nature which requires advanced resource recovery system. The current perspective on circular economy is projected by Ellen MacArthur Foundation's butterfly diagram, which separates 'biological' and 'technical' materials flow (Velenturf et al. 2019). Technical materials are referred as non-renewable sources which are kept in human usage loop by sharing, recycling, reusing and refabricating the products. On the contrary, biological materials are replenishable and utilized in an open loop system by sequentially arranging extraction, bio-material production, energy recovery stage and giving back the nutrients to the substrate for the next cycle of primary bioprocess. It is estimated that United States citizens generate more than 15 million metric tons of "used textile waste" annually which is equivalent to 47 kg of textile waste per person per year. The Environmental Protection Agency (EPA) indicates that out of total textile wastes generated; only 15% is recycled and the rest 85% is thrown directly as landfills or pyrolyzed (NIST 2022). Since circular economy looks for eco-friendly materials in all production processes, it is impeccable to look for a substitute for concrete because it is the second most used material on earth (Schumacher et al. 2023).

Because of easy availability, durability, fire resistance, resilience, other thermal and physical properties, concrete is the most widely used construction material. In 2020, an estimated 14 billion cubic meters (m^3) of concrete was utilized in new construction globally, with more than 40% of that used in residential infrastructure. Even though concrete occupies privileged position in construction sector, the composition elements follow take, make, and use and discard path. Moreover the raw materials like limestone, clay and other siliceous materials extraction have an impact on environment and biological systems. Moreover, the demolition and end stage destruction of concrete materials is highly energy intensive and generates massive amounts of solid wastes. As per the CE policy, the concrete debris can be used for paving basement for roadways and recycled concrete making. The Environmental Protection Agency (EPA) estimates that Americans generate more than 15 million metric tons of textile waste annually, which is about 47 kg (104 pounds) per person per year. Only 15% of that is collected for resale or recycling. The remaining 85% is directly discarded in landfills or incinerated. This suggests that most people throw away unwanted textiles, regardless of condition. That said, only about 20% of textiles donated to charities and thrift stores are resold domestically, as these entities receive far too many donations to sell. The remainder is sent to sorters and graders who evaluate the textiles based on quality and sell to various end markets. This material is then exported for resale overseas, cut and packaged as wiping materials, turned into shoddy (shredded fibers used for stuffing and insulation), or recycled back into fibers. An unknown quantity of textiles is exported without being sorted or graded. The major importing countries are in Africa, Asia and Central America, and there is a lot of discussion and differing opinion around the benefits and detriments of exporting used textiles.

14.4.4 Circular Economy Applied to Tanneries

Ecological utilization of tannery wastes using circular economy principle focuses on comprehensive way of resource handling and reuse. Implementation of circular practices during leather processing and production can minimize the damage to environment and ecosystem. One of the suggested ways is to increase the usage of eco-friendly and bio-based materials for leather softening, dehairing, dyeing process, etc. A familiar proverb "old is gold" is most appropriate for tannery industries which empathize to follow traditional vegetable tanning process to reduce pollutants, CO_2 emissions and conserve water instead of chemical based chrome tanning process. In post-tanning process, final color and texture is given to leather products using harmful dyes and softeners. But the threat to environment can be mitigated using enzymatic reactions; plant-based natural dyes and proteins for leather softening. Banana wastes can be transformed into leather like products and support as sustainable alternative to animal hides. In the circular economy, the intention is to keep the materials in process loop as long as possible. The solid wastes (mostly skins) are used as substrate for methane gas production by fermentation process. This exhibits the chance of tapping the energy from solid wastes without discarding. Leather processing industries embrace circular economy principle to focus on sustainable production, reduce and recycle wastes. The key drivers of CE in leather industries include recycling of leather wastes, chemical recycling, upcycling of solid wastes, take-back policies, modular design of products, etc. Recycled leather is made by augmenting leather scraps and offcuts from various stages of production process. This recycling step minimizes landfills and produce second quality products with minimal environmental impacts. The breaking down of leather wastes into its chemical constituents is achieved in chemical recycling stages which assist to recover wealth from wastes thereby reducing the need for virgin raw materials. "Take-back" policy encourage the customers to return their used leather products to waste management agencies for making recycle products and second-hand products. "Modular design" of products ensures that, only the damaged portion on leather material is replaced or repaired successfully to prolong the life time of entire product rather than discarding it waste. In modular design, the products can be dismantled into various segments and repair performed only at the damaged area. This modular design concept promotes sustainability and mitigates wastes. The biodegradability of finished leather products can be accelerated by using metal free tanning process and biomass based polymers. A leather product is susceptible to biodegradable action when the microbes break the complex structure into simpler molecules without any negative impact on environment and society. The factors that decide the biodegradability index of leather products are natural factors, physico-chemical properties of leather and degree of decomposition. The natural factors include soil type, texture and composition, water salinity, oxygen level in decomposition, humidity and microbe's biological features. The properties of finished products depend on raw material composition, chemicals used during manufacture and post-tanning process conditions. As per the standards, a product is considered biodegradable if 90% of obsolete product is degraded in 6 months. Gradually the petrochemical-based binders and finishing agents are replaced by plant-based or biomass alternatives.

To achieve sustainability in leather manufacture, the main leather product is made from recycled leather and producing a durable, long lasting product. Leather products can be made biodegradable, recyclable and brought under the ambit of sustainability when attention paid to raw materials type and chemicals used. Nowadays, the negative image of leather products is due to elevated presence of Cr(VI) molecules. Almost 85% of cattle hides are tanned using chromium salts because the process is faster, save raw materials and the finished product is immune to user ambience. Even though chrome tanned products enjoy inevitable position, soon or later the chromium based process may become obsolete due to global sustainability efforts. The raw materials needed for vegetable tanning process is natural tannin, obtained from various parts of plant including woods, barks, leaves, dried fruit pods, etc. Some of the bio-based tannins are chestnut wood (*Castanea sativa*), oak wood (*Quercus* sp.), quebracho wood (*Schinopsis lorentzii*), Tara pods (*Caesalpinia spinosa*), etc. In spite of environmental advantages, not enough research output and supporting evidences available for large scale implementation of vegetable tanning process. But of course, there are wide range of scope and opportunities available in tannery sector for sustainability. The hazards of chemical agents used in tanning process are lime lighted in social media's, information messenger tools and thereof eco-friendly technology is warming up to replace dangerous processes in the upcoming decades.

14.5 Challenges Associated with Implementation

For achieving a true circular economy, the global community should overcome the barriers namely, convenient lifestyle of consumers, diluted government regulations, poor waste management practices, non-compliance of recycle schemes and wrong business model. To overcome all these blockades, partnership between industries, government agencies and common public is crucial. At present, industries are blamed for pollution problems but the myth is that even common public is responsible for global level pollution. Government norms are made flexible to foster innovative and start-up ideas to move forward toward circular solutions.

14.5.1 Strategies to Improve and Promote Circular Economy

The simple strategies to follow, for the transition to circular economy is minimize the per-capita usage of natural resources and choosing best product for daily activities, feasible government rules and drastic change in mental status and system. Circular economy principles are applied to global food system because it contributes to 30% of global emissions and almost one-third of food is wasted. When circularity is followed in food industries, more population can be fed with food with limited agricultural lands. CE principle is also used to produce bio-based fertilizers which directly improve soil health for better, quality yield. In textile industries, the implication of circularity reduces carbon footprint by increasing the use of recycled and durable materials. There is a need to synchronize agriculture and textile industry for promoting cultivation of natural fibers like cotton, jute rather than using synthetic fibers. The biodegradability of textile fibers is increased by preparing varying proportion of composite fibers between natural and synthetic materials. With respect to plastic industry, applying circularity promotes plastic making from recycled materials and thermoplastic materials. Development of industrial estates and symbiotic relationship among producers utilizes raw materials effectively without escalating logistics cost, raw materials cost and waste management tools. The concept of "co-generation" is widely used in energy intensive industries and power plants where desired product is manufactured along with electricity generation.

Recent generation of software's, modeling algorithms, Artificial Intelligence (AI) and Machine Learning (ML) tools can help industrial processes at each and every stage to attain circularity and provide holistic solutions to resource and environmental problems. AI and ML can support circularity by using pattern analysis to produce either recyclable or eco-friendly base materials besides identifying innovative ways to reuse post-consumer wastes and undesired by-products. AI technologies and online exchange Apps help for raw materials, assets sharing in a wholesale and retail manner. Creating network between supplier and user increases the sharing of resources and recycling of wastes. Excess production of leather products more than consumer demand jeopardizes the efforts to achieve production and economic circularity. But AI-enabled production process using predictive analysis algorithm can make the products based on data analysis from consumer demand forecasts. This reduces wastes and reduces the risk of overstocked products that go as waste ultimately (SAP Germany 2024). AI facilitated supply chain management system for raw materials and products close the circularity loop by optimizing the logistics routes. Streamlining of supply chains effectively manages transportation paths, chemical storage conditions, equipment maintenance and quality control measures. Core industries can slash the quantities of product loss and minimize carbon emissions when all the processes and operations are fine-tuned with the aid of AI tools. Integrating AI in waste collection and management system can identify valuable materials in mixed waste streams using artificial sensors, analytical instruments and pattern recognition method. Due to inefficient waste collection system, tons of recyclable wastes are exited without proper monitoring and control. AI integrated smart waste management technology can monitor the composition and quantities of wastes and robots segregate wastes with accuracy and increasing yield of desired compound from wastes thus reducing loss via exiting contaminated stream.

14.5.2 Action Plan and Monitoring Framework for Circular Economy

As the industries are forced to follow circularity principle in all stages of production, to cut down the costs linked with production, raw materials, inventory, etc., the data collection and analysis tools are needed to monitor the progress after implementing circularity. For analyzing the betterment due to CE, full-fledged monitoring framework is necessary to keep tabs on the progress toward predefined targets and goals set in road map. The monitoring process not only sets goals for future prospects but also imitates the progress trend in current situation. The key indicators defined in frameworks are material footprint, waste generation per capita, recycling rate of wastes; circular materials use rate and greenhouse gas emissions. Two general approaches, namely sensu stricto and sensu latu are used to classify the CE indicators. Sensu stricto approach has confined focus for resource management and characterized by slowing and closing of materials usage loop. Slowing refers to development of durable materials to extend the product life, whereas closing refers to minimizing the gap between post-use end and production starting point in a circular path. This results in circular flow of resources without eliminating or throwing away any materials from the loop. On the contrary, sensu latu has deeper vision which focuses on designing each and every step in a process starting from material allocation, process development and product distribution. It follows cradle-to-grave approach in material production starting from raw materials to end-user products keenly focusing on ecosystem effectiveness and human wellbeing (Moraga et al. 2019).

Modernization of recycling facilities with the motivation of developing high-quality, high-performing recycle system should be carried out on a war-footing phase. Government has to develop comprehensive waste management strategies along with industry-wise plans for pollution abatement and control. To achieve the targets set in goal-12, of sustainable development goals (SDG), the production and consumption pattern of materials should be in constant rate manner which lays the foundation for sustained life of present and future generations.

14.6 Conclusion

The adoption of circular economy requires a paradigm shift to a "complete multi-faceted life cycle" of a raw material from conventional "use and throw model". For solid wastes circularity, the equipment design and material properties related to process must be highlighted. Recycling is an important step in circular economy but the efficacy of "true recycling" depends on the quality and integrity of renewed product. The recycled material should have the qualities on par with that of natural raw materials. Sludge's and wet wastes from industries can be processed applying CE principles to produce compost for soil enrichment and "energy rich gas" for heating purposes. A framework for best manufacturing practices in industries is provided by integrating Industry 4.0 technologies into the existing units. Implementation of artificial intelligence (AI), machine learning (ML), Internet of Things (IoT), cloud computing, big data analytics (BDA) in industrial processes improves circularity in raw materials procurement, supply chain management, maintenance of equipment, monitoring of process conditions and generation of usable material from discarded wastes. The integration of advanced technologies with traditional manufacturing practices can contribute significantly to attain Sustainable Development Goals (SDGs) declared by United Nations. Interlinking CE principles and modern software tools not only solves environmental crisis but also gives major thrust to economic and societal growth. Even though chromium is considered as a major pollutant from tannery industries, augmented treatment methods like thermal-pressure hydrolysis process followed by membrane separation process can separate chromium species selectively from processed wastes. The rejuvenated chromium once again the process line and used for raw hide curing and tanning. Compendium of modern tools, digitalized processes, electronic gadgets and plant-based raw materials in leather industry successfully applies theoretical principles of circularity to real-world industrial challenges.

References

Bhardwaj A, Kumar S (2023) Tannery effluent treatment and its environmental impact: a review of current practices and emerging technologies. Water Qual Res J 58(2):128–152. https://doi.org/10.2166/wqrj.2023.002

Brears RC (2018) Natural resource management and the circular economy, 15 pp. https://doi.org/10.1007/978-3-319-71888-0

Buljan J, Rajamani SG (2023) Some challenges in the treatment of tannery effluents. United Nations Industrial Development Organization, Sustainable Development Goals (SDG). https://leatherpanel.org/sites/default/files/publications-attachments/challenges_in_the_treatment_of_tannery_effluents_p2.pdf

Central Pollution Control Board (CPCB) (2019) Guidelines for environmental improvement in leather tannery sector. Central Pollution Control Board (CPCB), New Delhi, India, pp 16–17. https://cpcb.nic.in/openpdffile.php?id=TmV3c0ZpbGVzLzcyXzE1NTQ0NTY2NDhfbWVkaWFwaG90bzIyMTYwLnBkZg

Circular Economy (2024) Government of the Netherlands. https://www.government.nl/topics/circular-economy/from-a-linear-to-a-circular-economy

European Parliament (2024) https://www.europarl.europa.eu/topics/en/article/20151201STO05603/circular-economy-definition-importance-and-benefits

Famielec S, Wieczorek-Ciurowa K (2011) Waste from leather industry. Threats to the environment. Czasopismo Techn Chem 43–48. ISSN 1897-6298. https://cris.pk.edu.pl/info/article/CUT82cb4b8627a84a44afc9e8cc332cee68/

Flores Tapia NE, Moina HB (2023) Exploring tannery solid wastes as a source of animal feed. Processes 11(10):2965. https://doi.org/10.3390/pr11102965

Food and Agriculture Organization (FAO) of the United Nations. Viale delle Terme di Caracalla. Rome, Italy. https://www.fao.org/4/x6114e/x6114e05.htm

Hu J, Xiao Z, Zhou R, Deng W, Wang M, Ma S (2011) Ecological utilization of leather tannery waste with circular economy model. J Clean Prod 19(2–3):221–228. https://doi.org/10.1016/j.jclepro.2010.09.018

Moraga G, Huysveld S, Mathieux F, Blengini GA, Alaerts L, Van Acker K, de Meester S, Dewulf J (2019) Circular economy indicators: what do they measure? Resour Conserv Recycl 146:452–461. https://doi.org/10.1016/j.resconrec.2019.03.045

Nur-E-Alam M, Abu Sayid Mia M, Ahmad F, Mafizur Rahman M (2020) An overview of chromium removal techniques from tannery effluent. Appl Water Sci 10:205. https://doi.org/10.1007/s13201-020-01286-0

Pringle T, Barwood M, Rahimifard S (2016) The challenges in achieving a circular economy within leather recycling. Procedia CIRP 48:544–549. https://doi.org/10.1016/j.procir.2016.04.112

Saira GC, Shanthakumar S (2023) Zero waste discharge in tannery industries—an achievable reality? A recent review. J Environ Manage 335:117508. https://doi.org/10.1016/j.jenvman.2023.117508

Schumacher GK (2022) National Institute of Standards and Technology (NIST), U.S. Department of Commerce. https://www.nist.gov/blogs/taking-measure/your-old-clothes-can-come-full-circle-textile-recycling

Schumacher K, Platt S, Newman A, Garboczi E, Beers K, Chu P (2023) Fostering a circular economy and carbon sequestration for construction materials workshop report: a focus on concrete.

National Institute of Standards and Technology (NIST). https://doi.org/10.6028/NIST.SP.1500-21
Shahzabeen A, Ghosh A, Pandey B, Shekhar S (2023) Green circular economy. A new paradigm for sustainable development, pp 51–52. https://doi.org/10.1007/978-3-031-40304-0
Shen Y, Zhang X (2022) Study on the impact of environmental tax on industrial green transformation. Int J Environ Res Public Health 19(24):16749. https://doi.org/10.3390/ijerph192416749
Stangherlin IC, Thøgersen J, de Barcellos MD (2023) Behavioral spillover in the circular economy: the importance of consumer goals. J Environ Psychol 91:102123. https://doi.org/10.1016/j.jenvp.2023.102123
Systems Applications and Products in Data Processing (SAP) (2024) Walldorf, Germany. https://www.sap.com/products/sustainability/what-is-circular-economy.html
Thiruvalluvar Statue, Department of Tourism, Government of Tamil Nadu, Chennai, Tamil Nadu, India. https://www.tamilnadutourism.tn.gov.in/destinations/thiruvalluvar-statue
UNIDO (2011) Introduction to treatment of tannery effluents: what every tanner should know about effluent treatment. United Nations Industrial Development Organization (UNIDO), Vienna. https://www.unido.org/sites/default/files/2011-11/Introduction_to_treatment_of_tannery_effluents_0.pdf
United Nations Development Programme (UNDP). Global climate promise. https://climatepromise.undp.org/news-and-stories/what-is-circular-economy-and-how-it-helps-fight-climate-change
United Nations Industrial Development Organization (UNIDO). Kanpur leather development project. Department of Industrial Policy and Promotion, Government of India. https://leatherpanel.org/sites/default/files/publications-attachments/factsheet_6_desalting-raw_hides_p.pdf
United Nations Industrial Development Organization (2000) Regional programme for pollution control in the tanning industry in South-East Asia. The scope for decreasing pollution load in leather processing. https://www.unido.org/sites/default/files/2009-05/The_scope_for_decreasing_pollution_load_in_leather_processing_0.pdf
United States Environmental Protection Agency (2024). https://www.epa.gov/circulareconomy/what-circular-economy
Valluvar Kottam, National Informatics Centre, Ministry of Electronics & Information Technology, Government of India. https://chennai.nic.in/tourist-place/valluvarkottam/
Velenturf APM, Archer SA, Gomes HI, Christgen B, Lag-Brotons AJ, Purnell P (2019) Circular economy and the matter of integrated resources. Sci Total Environ 689(1):963–969. https://doi.org/10.1016/j.scitotenv.2019.06.449
World Resources Institute (WRI) (2024) Washington, DC. https://www.wri.org/insights/how-build-circular-economy

MIX
Papier aus verantwortungsvollen Quellen
Paper from responsible sources
FSC® C105338

If you have any concerns about our products,
you can contact us on
ProductSafety@springernature.com

In case Publisher is established outside the EU,
the EU authorized representative is:
Springer Nature Customer Service Center GmbH
Europaplatz 3, 69115 Heidelberg, Germany

Printed by Libri Plureos GmbH
in Hamburg, Germany